BAYESIAN ANALYSIS for POPULATION ECOLOGY

CHAPMAN & HALL/CRC
Interdisciplinary Statistics Series

Series editors: N. Keiding, B.J.T. Morgan, C.K. Wikle, P. van der Heijden

Published titles

AN INVARIANT APPROACH TO STATISTICAL ANALYSIS OF SHAPES	S. Lele and J. Richtsmeier
ASTROSTATISTICS	G. Babu and E. Feigelson
BAYESIAN ANALYSIS FOR POPULATION ECOLOGY	Ruth King, Byron J.T. Morgan, Olivier Gimenez, and Stephen P. Brooks
BAYESIAN DISEASE MAPPING: HIERARCHICAL MODELING IN SPATIAL EPIDEMIOLOGY	Andrew B. Lawson
BIOEQUIVALENCE AND STATISTICS IN CLINICAL PHARMACOLOGY	S. Patterson and B. Jones
CLINICAL TRIALS IN ONCOLOGY SECOND EDITION	J. Crowley, S. Green, and J. Benedetti
CLUSTER RANDOMISED TRIALS	R.J. Hayes and L.H. Moulton
CORRESPONDENCE ANALYSIS IN PRACTICE, SECOND EDITION	M. Greenacre
DESIGN AND ANALYSIS OF QUALITY OF LIFE STUDIES IN CLINICAL TRIALS	D.L. Fairclough
DYNAMICAL SEARCH	L. Pronzato, H. Wynn, and A. Zhigljavsky
GENERALIZED LATENT VARIABLE MODELING: MULTILEVEL, LONGITUDINAL, AND STRUCTURAL EQUATION MODELS	A. Skrondal and S. Rabe-Hesketh
GRAPHICAL ANALYSIS OF MULTI-RESPONSE DATA	K. Basford and J. Tukey
INTRODUCTION TO COMPUTATIONAL BIOLOGY: MAPS, SEQUENCES, AND GENOMES	M. Waterman

Published titles

MARKOV CHAIN MONTE CARLO IN PRACTICE	W. Gilks, S. Richardson, and D. Spiegelhalter
MEASUREMENT ERROR AND MISCLASSIFICATION IN STATISTICS AND EPIDEMIOLOGY: IMPACTS AND BAYESIAN ADJUSTMENTS	P. Gustafson
META-ANALYSIS OF BINARY DATA USING PROFILE LIKELIHOOD	D. Böhning, R. Kuhnert, and S. Rattanasiri
STATISTICAL ANALYSIS OF GENE EXPRESSION MICROARRAY DATA	T. Speed
STATISTICAL AND COMPUTATIONAL PHARMACOGENOMICS	R. Wu and M. Lin
STATISTICS IN MUSICOLOGY	J. Beran
STATISTICAL CONCEPTS AND APPLICATIONS IN CLINICAL MEDICINE	J. Aitchison, J.W. Kay, and I.J. Lauder
STATISTICAL AND PROBABILISTIC METHODS IN ACTUARIAL SCIENCE	P.J. Boland
STATISTICAL DETECTION AND SURVEILLANCE OF GEOGRAPHIC CLUSTERS	P. Rogerson and I. Yamada
STATISTICS FOR ENVIRONMENTAL BIOLOGY AND TOXICOLOGY	A. Bailer and W. Piegorsch
STATISTICS FOR FISSION TRACK ANALYSIS	R.F. Galbraith

Chapman & Hall/CRC
Interdisciplinary Statistics Series

BAYESIAN ANALYSIS for POPULATION ECOLOGY

Ruth King
Byron J. T. Morgan
Olivier Gimenez
Stephen P. Brooks

CRC Press
Taylor & Francis Group
Boca Raton London New York

CRC Press is an imprint of the
Taylor & Francis Group, an **informa** business

A CHAPMAN & HALL BOOK

Chapman & Hall/CRC
Taylor & Francis Group
6000 Broken Sound Parkway NW, Suite 300
Boca Raton, FL 33487-2742

© 2010 by Taylor and Francis Group, LLC
Chapman & Hall/CRC is an imprint of Taylor & Francis Group, an Informa business

No claim to original U.S. Government works

Printed in the United States of America on acid-free paper
10 9 8 7 6 5 4 3 2 1

International Standard Book Number: 978-1-4398-1187-0 (Hardback)

This book contains information obtained from authentic and highly regarded sources. Reasonable efforts have been made to publish reliable data and information, but the author and publisher cannot assume responsibility for the validity of all materials or the consequences of their use. The authors and publishers have attempted to trace the copyright holders of all material reproduced in this publication and apologize to copyright holders if permission to publish in this form has not been obtained. If any copyright material has not been acknowledged please write and let us know so we may rectify in any future reprint.

Except as permitted under U.S. Copyright Law, no part of this book may be reprinted, reproduced, transmitted, or utilized in any form by any electronic, mechanical, or other means, now known or hereafter invented, including photocopying, microfilming, and recording, or in any information storage or retrieval system, without written permission from the publishers.

For permission to photocopy or use material electronically from this work, please access www.copyright.com (http://www.copyright.com/) or contact the Copyright Clearance Center, Inc. (CCC), 222 Rosewood Drive, Danvers, MA 01923, 978-750-8400. CCC is a not-for-profit organization that provides licenses and registration for a variety of users. For organizations that have been granted a photocopy license by the CCC, a separate system of payment has been arranged.

Trademark Notice: Product or corporate names may be trademarks or registered trademarks, and are used only for identification and explanation without intent to infringe.

Library of Congress Cataloging-in-Publication Data

Bayesian analysis for population ecology / authors, Ruth King ... [et al.].
  p. cm.
"A CRC title."
Includes bibliographical references and index.
ISBN 978-1-4398-1187-0 (hardcover : alk. paper)
 1. Population biology--Mathematics. 2. Bayesian field theory. I. King, Ruth, 1977-

QH352.B38 2010
577.8'801519542--dc22   2009026637

Visit the Taylor & Francis Web site at
http://www.taylorandfrancis.com

and the CRC Press Web site at
http://www.crcpress.com

Library
University of Texas
at San Antonio

Contents

Acknowledgments		xi
Preface		xiii
I	**Introduction to Statistical Analysis of Ecological Data**	**1**
1	**Introduction**	**3**
	1.1 Population Ecology	3
	1.2 Conservation and Management	3
	1.3 Data and Models	4
	1.4 Bayesian and Classical Statistical Inference	6
	1.5 Senescence	9
	1.6 Summary	11
	1.7 Further Reading	11
	1.8 Exercises	12
2	**Data, Models and Likelihoods**	**15**
	2.1 Introduction	15
	2.2 Population Data	15
	2.3 Modelling Survival	21
	2.4 Multi-Site, Multi-State and Movement Data	28
	2.5 Covariates and Large Data Sets; Senescence	30
	2.6 Combining Information	33
	2.7 Modelling Productivity	35
	2.8 Parameter Redundancy	36
	2.9 Summary	40
	2.10 Further Reading	40
	2.11 Exercises	42
3	**Classical Inference Based on Likelihood**	**49**
	3.1 Introduction	49
	3.2 Simple Likelihoods	49
	3.3 Model Selection	51
	3.4 Maximising Log-Likelihoods	56
	3.5 Confidence Regions	57

3.6	Computer Packages	58
3.7	Summary	61
3.8	Further Reading	62
3.9	Exercises	62

II Bayesian Techniques and Tools   67

4 Bayesian Inference   69

4.1	Introduction	69
4.2	Prior Selection and Elicitation	75
4.3	Prior Sensitivity Analyses	80
4.4	Summarising Posterior Distributions	85
4.5	Directed Acyclic Graphs	89
4.6	Summary	93
4.7	Further Reading	95
4.8	Exercises	96

5 Markov Chain Monte Carlo   99

5.1	Monte Carlo Integration	99
5.2	Markov Chains	101
5.3	Markov Chain Monte Carlo	101
5.4	Implementing MCMC	124
5.5	Summary	141
5.6	Further Reading	141
5.7	Exercises	142

6 Model Discrimination   147

6.1	Introduction	147
6.2	Bayesian Model Discrimination	148
6.3	Estimating Posterior Model Probabilities	152
6.4	Prior Sensitivity	170
6.5	Model Averaging	172
6.6	Marginal Posterior Distributions	176
6.7	Assessing Temporal/Age Dependence	178
6.8	Improving and Checking Performance	189
6.9	Additional Computational Techniques	192
6.10	Summary	193
6.11	Further Reading	194
6.12	Exercises	195

7 MCMC and RJMCMC Computer Programs   199

7.1	R Code (MCMC) for Dipper Data	199
7.2	WinBUGS Code (MCMC) for Dipper Data	205
7.3	MCMC within the Computer Package MARK	209

7.4	R code (RJMCMC) for Model Uncertainty	213
7.5	WinBUGS Code (RJMCMC) for Model Uncertainty	226
7.6	Summary	231
7.7	Further Reading	232
7.8	Exercises	233

III Ecological Applications — 239

8 Covariates, Missing Values and Random Effects — 241
- 8.1 Introduction — 241
- 8.2 Covariates — 242
- 8.3 Missing Values — 251
- 8.4 Assessing Covariate Dependence — 259
- 8.5 Random Effects — 264
- 8.6 Prediction — 270
- 8.7 Splines — 270
- 8.8 Summary — 273
- 8.9 Further Reading — 274

9 Multi-State Models — 277
- 9.1 Introduction — 277
- 9.2 Missing Covariate/Auxiliary Variable Approach — 277
- 9.3 Model Discrimination and Averaging — 286
- 9.4 Summary — 303
- 9.5 Further Reading — 304

10 State-Space Modelling — 307
- 10.1 Introduction — 307
- 10.2 Leslie Matrix-Based Models — 314
- 10.3 Non-Leslie-Based Models — 332
- 10.4 Capture-Recapture Data — 339
- 10.5 Summary — 342
- 10.6 Further Reading — 342

11 Closed Populations — 345
- 11.1 Introduction — 345
- 11.2 Models and Notation — 346
- 11.3 Model Fitting — 347
- 11.4 Model Discrimination and Averaging — 354
- 11.5 Line Transects — 356
- 11.6 Summary — 359
- 11.7 Further Reading — 360

Appendices — 363

A	**Common Distributions**	**365**
	A.1 Discrete Distributions	365
	A.2 Continuous Distributions	368
B	**Programming in R**	**375**
	B.1 Getting Started in R	375
	B.2 Useful R Commands	376
	B.3 Writing (RJ)MCMC Functions	378
	B.4 R Code for Model C/C	379
	B.5 R Code for White Stork Covariate Analysis	387
	B.6 Summary	398
C	**Programming in WinBUGS**	**401**
	C.1 WinBUGS	401
	C.2 Calling WinBUGS from R	410
	C.3 Summary	415
References		**417**
Index		**435**

Acknowledgments

We would like to thank all those people who have provided both moral support to each of the authors of the book and feedback on early drafts of the manuscript. The feedback provided has been invaluable to us and helped to improve the material contained within the book significantly. We would particularly like to thank Steve Buckland, Rachel McCrae, Charles Paxton, Chiara Mazzetta, Jeremy Greenwood, Gary White, Leslie New and Stuart King. We also thank here many ecologists with whom we have worked, and continue to work, who have patiently explained to us the problems of the real world, generously provided us with data and introduced us to the rigours of field work. An incomplete list of our ecological colleagues includes Steve Albon, Mike Harris, Tim Clutton-Brock, Tim Coulson, Steve Dawson, Marco Festa-Bianchet, Jean-Michel Gaillard, Jean-Dominique Lebreton, Alison Morris, Kelly Moyes, Daniel Oro, Josephine Pemberton, Jill Pilkington, Giacomo Tavecchia, David Thomson, Sarah Wanless, members of the British Trust for Ornithology, the Royal Society for the Protection of Birds, the EURING community, the field workers on the Kilmory deer project, the National Trust for Scotland and the Scottish Natural Heritage, and the volunteers who work on the Soay sheep project.

Preface

This book is written at a crucial time of climate change, when many animal and plant populations are under threat. It is clearly therefore a very important time for the study of population ecology, for the precise estimation of demographic rates, and application of the best methods for statistical inference and forecasting. Modern Bayesian methods have an important role to play. The aim of this book is to provide a comprehensive and up-to-date description of Bayesian methods for population ecology. There is particular emphasis on model choice and model averaging. Throughout the theory is applied to a wide range of real case studies, and methods are illustrated using computer programs written in WinBUGS and R. Full details of the data sets analysed and the computer programs used are provided on the book's Web site, which can be found at http://www.ncse.org.uk/books/bayesian. Asterisked sections denote that these sections may be omitted on the initial reading of the book.

The target audience for the book comprises ecologists who want to learn about Bayesian methods and appreciate how they can be used, quantitative ecologists, who already analyse their own data using classical methods of inference, and who would also like to be able to access the latest Bayesian methods, and statisticians who would like to gain an overview of the challenging problems in the field of population ecology. Readers who complete the book and who run the associated computer programs should be able to use the methods with confidence on their own problems. We have used the material of the book in workshops held in Radolfzel (Germany), Dunedin (New Zealand), and Cambridge (England); the last such workshop had the financial support of the Natural Environmental Research Council. We are grateful for the valuable input of workshop participants. The book may be used for a university final year or master's-level course in statistical ecology.

PART I

Introduction to Statistical Analysis of Ecological Data

This first part of the book provides an introduction to a range of different types of ecological data. This includes a description of the data collection processes, the form of the data collected and corresponding presentation of the data. We focus on the particular forms of ecological data that we study throughout the remainder of the book. We also discuss the associated models that are typically fitted to the data and define the corresponding likelihood functions. These models make particular assumptions regarding the underlying data collection process and the system under study. The models simplify reality, but can be very useful in focussing on the primary factors of the system under study and in providing an understanding of the system itself.

The models and corresponding likelihood functions lay the foundations for the remainder of the book. Within Part I of the book we consider the traditional (classical) methods of inference that are used to estimate the model parameters. These typically include the calculation of the maximum-likelihood estimates of parameters, using numerical optimisation algorithms. We build on these foundations in the remainder of the book, when we develop the Bayesian approach for fitting the same, as well as more complex, models to the data. The likelihood will be seen to be integral to both methods of inference. In later parts of the book we shall also undertake comparisons between the Bayesian and traditional approaches to model-fitting and inference.

CHAPTER 1

Introduction

1.1 Population Ecology

Population ecology is devoted to the study of individuals of the same species, how they make up the populations in which they exist and how these populations change over time. In population ecology we gather information on key demographic parameters, typically relating to mortality, movement and productivity, and use these to predict how the populations will develop in the future. We are also interested in trying to relate demographic parameters to external influences, such as global warming, changes in habitat and farming practice, and changes in hunting regulations.

The emphasis in this book is on wild animals of various kinds, but the ideas apply more widely, for example, to the study of plant populations. Furthermore, the definition of population ecology given above includes human demography, which shares the same attributes of determining vital rates, and then using these to predict how populations will behave. At the start of the 21st century many Western populations are shrinking, as birth rates decline. For example, the British Census of 2001 revealed that for the first time since records began in 1841, the number of people older than 60 years of age exceeds the number aged less than 16. Combined with increased longevity, this results in the demographic time-bomb in which more and more individuals in retirement are effectively supported by fewer and fewer individuals of working age. As we shall see, data on human populations are gathered in different ways from data on wild animal populations, resulting in different statistical methods being used. However, there are common features in the two areas; for example, there is now currently much concern with the extinction of wild animal species.

1.2 Conservation and Management

The Great Auk *Pinguinus impennis* was the last flightless seabird of the Northern hemisphere, and became extinct in 1844. This is just one example of the plant and animal species that have declined in recent years, with many either extinct or facing extinction. There is no shortage of further examples. In America, the extinction of the Passenger Pigeon, *Ectopistes migratorius*, has been well documented. Just over 100 years ago it was the most numerous species of bird in the world, with there being more of this single species than of all other species of North American birds in total. The last Passenger Pigeon died on the 1st September 1914, at the age of 29. In the last 100 years, Holland has lost half of its butterfly species. According to Kennedy (1990), Australia has had more mammal extinctions over the past 200 years than any other continent, and that book describes over 4000 individual plant and animal species of conservation concern. In Britain, the Royal Society for the Protection of Birds maintains a *red list* of bird species that are regarded as under threat, including having declined by more than 50% in numbers in the previous 25 years. Thus there is at present an appreciable international awareness of the dangers facing many different species and of the need to conserve as much as possible of the current biological diversity in both the plant and animal kingdoms. Indeed, the World Summit on Sustainable Development held in 2002 regarded the loss of species as one of the main issues to be tackled for the future.

In contrast, some wild animal species are seen as growing too fast, and this is currently true of red deer, *Cervus elaphus*, in Scotland, for example. In such cases it may be possible to manage populations, perhaps by a program of culling, or by inhibiting reproduction. Culling and conservation have in common the same aim, of managing and maintaining a sustainable population size. This is especially important in the area of fisheries, where it has long been realised that without proper management, such as setting of quotas or moratoria on fishing, then it is possible to fish stocks to extinction.

1.3 Data and Models

Almost by definition, population ecology is quantitative. Long-term studies of wild animal populations result in data which need to be summarised and suitably analysed. One approach is to construct probability models for the population processes being studied, and then use particular methods of statistical inference in order to estimate the model parameters. Predictions can then be based upon the models that have been fitted to the data in this way. We now outline two examples to illustrate these ideas.

Example 1.1. European Dippers
The data of Table 1.1 are taken from Lebreton et al. (1992) and describe the recaptures of European dippers, *Cinclus cinclus*, which are birds that are typically found in fast-flowing mountain streams. Table 1.1 is called an *m-array*. These data result from a small-scale study of adult birds of unknown age which have been marked with rings which identify individuals uniquely. In the first year of the study, 22 birds are marked and released. Of these, 11 are seen again in their first year of life, while 2 were not encountered again until a year later. Birds that are recaptured have their rings recorded and are then re-released. For instance, of the 60 birds released in 1982, 11 will have been captured from the number released in 1981, and the remaining 49 birds of the cohort will be newly marked and released.

The data are somewhat sparse, but they may be described by suitable mod-

DATA AND MODELS

Table 1.1 Capture-Recapture Data for European Dippers in 1981–1986

Year of Release	Number Released	Year of Recapture (1981+)					
		1	2	3	4	5	6
1981	22	11	2	0	0	0	0
1982	60		24	1	0	0	0
1983	78			34	2	0	0
1984	80				45	1	2
1985	88					51	0
1986	98						52

Source: Reproduced with permission from Brooks et al. (2000a, p. 366) published by the Institute of Mathematical Statistics.

els which involve parameters which are probabilities of annual survival, ϕ, and recapture, p.

By making different assumptions regarding these parameters, we produce different models for the data. For example, in 1983 there was a flood in the area in which these data were gathered, and one might suppose that the flood temporarily resulted in a reduced probability of annual survival. We can see from Table 1.1 that years of release 1982 and 1983 result in the smallest proportions of recaptures. Additionally, survival and recapture probabilities might vary over time.

□

Example 1.2. Soay Sheep on Hirta

The population of Soay sheep, *Ovis aries*, studied on the island of Hirta in the St. Kilda archipelago in Scotland, is one in which density of animals plays a role in producing the fluctuations in population size, seen in Figure 1.1.

In this case the population is constrained to a relatively small area, and the values shown in Figure 1.1 are deduced from a number of censuses taken by ecologists walking around the island. Models for the behaviour of this population are necessarily complicated; for example, one has to model separately males and females, as they have quite different life styles. Also, in contrast to the dippers, it is important to allow the model probabilities to vary with age, as well as time. Because of the intensity with which this population is studied, it is also possible for probabilities of annual survival to be functions of environmental covariates, which can for instance measure the weather and the population density, as well as individual covariates, which vary from animal to animal, and might record the colour of coat, shape of horns, weight, and so forth.

□

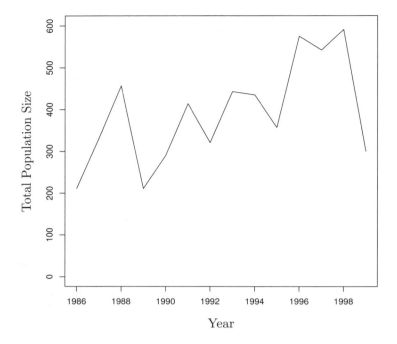

Figure 1.1 Size of the total population of Soay sheep on Hirta, 1986–1999.

We shall now outline in quite general terms the different ways in which statisticians fit probability models to data.

1.4 Bayesian and Classical Statistical Inference

We shall describe the methods of inference used in this book in detail in later chapters. Our starting point is a probability model, which is designed to take account of important aspects of the system being described. Models typically contain parameters, and in the classical approach, parameters are regarded as fixed quantities which need to be estimated. For any particular model and data set, the classical approach is usually to form a likelihood, which is to be regarded as a function of the model parameters. The likelihood is then maximised with respect to the parameters, resulting in maximum-likelihood estimates (MLEs) of those parameters. This is typically done for a range of different models, and models are compared in terms of their maximised likelihood values, or in terms of various information criteria, to be described later. Classical inference usually depends upon the application of powerful results which allow statisticians to test hypotheses and construct confidence intervals and regions. However, these procedures are usually dependent upon asymptotic assumptions being satisfied, which means that the results of the procedures are only valid if sample sizes are sufficiently large. The end-product of a classical analysis is a single, best, model to describe the data, chosen from the range of models considered. Usually little attention, if any, is paid to the fact that the best model is only one of many, and that while it may be best for one particular data set, an alternative model might be best for a different, replicate data set. Additionally, the different models may well differ in the estimates that they produce and in the predictions that result.

In the Bayesian approach, the model parameters are treated on the same basis as the data. Thus the parameters are regarded as having distributions, which provide information about the parameters. Before data are collected the distributions are described as prior distributions, while after data collection the distributions are described as posterior distributions. The aim of a Bayesian analysis is to estimate the joint posterior distribution of all of the model parameters. This is done by means of a very simple relationship, known as Bayes' Theorem, which is given in Chapter 4. When certain of the parameters have particular importance, then it is their marginal posterior distribution that is the end-point of the analysis. For example, in the case of the dippers, we might in some circumstances focus on the probability of annual survival, while in other cases, we might concentrate on the probability of recapture of a live animal in any year. In the case of the Soay sheep, we might be interested in how adult survival depends on the weights of the animals, or on the weather, and possibly an interaction between these two features. Explicit analytic forms for the posterior distributions in any analysis are usually not available, so instead we employ modern devices for simulating from those distributions, and then use the simulations to provide appropriate summaries of the posterior distributions. For example, we may estimate parameters by obtaining sample means from the simulations, and we may estimate variation by forming sample standard deviations from the simulations.

The Bayesian framework extends to encompass more than one model at a time. Just as a prior distribution can be formed for model parameters, we can also have prior distributions for alternative models, and furthermore obtain posterior model probabilities. The model probabilities then have a direct interpretation. Thus model comparisons can be made in terms of the posterior model probabilities, in analogy with the classical approach, when comparisons are made in terms of maximised likelihoods or information criteria. However, a great attraction of the Bayesian approach is that once one has model probabilities, then one can draw conclusions by averaging over models, rather than focussing on a single model, which is typically the case in classical inference. Consider, for example, two alternative models for the dipper data: in one model there is no allowance for the flood in 1983, whereas in the other model there is a separate survival probability for the survival in years affected by the flood. Both models contain probabilities of survival in non-flood years, and if

8   INTRODUCTION

one was interested in estimating the probability of surviving a non-flood year, then using the Bayesian approach one could obtain an estimate averaged over the two models. Averaging over models has the advantage that it takes account of the fact that several models might describe the data, with varying degrees of accuracy. Averaged estimators can therefore be more representative of the variation in the data, and we would expect them to be less precise than estimators obtained from just a single model. Of course, if different models provide interestingly different perspectives on the data, then it is important to realise that, and not blindly average out the differences. In the example of Soay sheep, models are far more complicated than in the case of dippers. Because we would like our models for sheep survival to include variation with regard to sex, age and time, there are potentially too many parameters to be estimated. One way around this difficulty is to include age classes, within which survival is constant over age. Different definitions of the age classes can then readily give rise to many different alternative models, and it is useful to be able to average over the component models, which can be numerous.

The advantages of the Bayesian approach are in fact many, and not at all limited to being able to average over models. For instance, one does not have to assume particular distributional forms, in order to ensure that classical procedures work; one does not have to appeal to often dubious asymptotic properties of maximum-likelihood estimators; one does not have to struggle with sometimes intransigent optimisation procedures in order to maximise likelihoods, or in some cases minimise sums of squares; one does not have to rely on piecemeal tests of goodness-of-fit. Instead, it is possible to include prior information in the analysis through the prior distribution, when that is felt to be appropriate, one can gauge the goodness-of-fit of models in routine ways, without having recourse to asymptotic considerations, and one can construct and fit complex, realistic models to data. A fundamentally important aspect of Bayesian analysis is the way in which it readily deals with different types of random variation. An illustration of this comes from modelling the survival of northern lapwings, *Vanellus vanellus*, which are birds of farmland which have seen a drastic decline in numbers in Britain in recent years.

Example 1.3. Lapwings
What is plotted in Figure 1.2 is a national index, which is the result of fitting a generalised linear model to count data from a range of sites distributed over Britain. The counts are obtained by volunteers visiting the sites during the breeding season, and estimating the numbers of birds present by observations on features such as nests and bird song. The smoothing produced by the generalised linear model is necessary because the raw site data contain large numbers of missing values. The index is called the Common Birds Census (CBC). It is not a proper census, but it may be regarded as an estimate of the number of birds present on the set of sites in the CBC. As can be seen from Figure 1.2, since the mid-1980s, the lapwing has been diminishing in numbers. It is an *indicator* species for farmland birds, which means that if we

can understand the reasons for the decline of the lapwing, then that might help us to understand the reasons for the decline of other farmland birds.

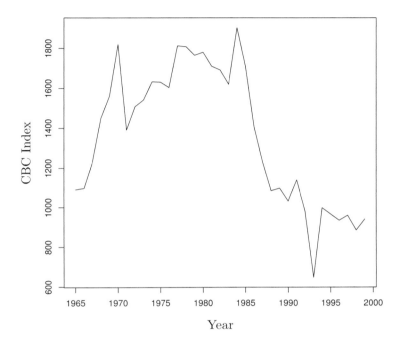

Figure 1.2 Common Birds Census index for lapwings over the period 1965–1999.

A symbolic representation of a model for the survival of lapwings is written as:
$$\text{survival} \sim A + Y + E,$$
where A is a fixed factor denoting age, Y is a factor denoting a random year effect, and E is a further random term, which might correspond to individual variation. To fit this kind of model involving a mixture of fixed and random effects is certainly possible using classical methods. However, as we shall see in detail later, in Section 8.5, the Bayesian analysis is conceptually much simpler. □

1.5 Senescence

In many human societies, we see steady increases in life expectancy over time. In order to try and understand better this important demographic change, it is of interest to study the effect of ageing on both the survival probability and the reproductive success of wild animals. Age dependence in demographic rates is of general importance and something that we shall incorporate into models of later chapters. We therefore present one illustration here.

Example 1.4. The Survival Probabilities of Soay Sheep

Soay sheep were introduced in Example 1.2. As well as census data, the observations on Soay sheep also include life history data which allow us to estimate probabilities of annual survival. We shall consider a Bayesian approach to modelling the sheep life history data in Chapter 8. A first stage in the analysis of Catchpole et al. (2000), which used methods of classical inference, was to examine age dependence in annual survival. Shown in Figure 1.3 are the classical estimates of female annual survival probabilities as a function of age. There appears to be an indication of senescence, with survival probabilities being smaller for the oldest animals in the study.

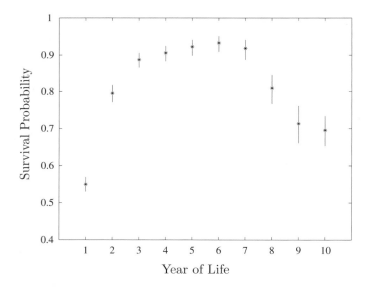

Figure 1.3 Plot of estimated probability of annual survival for female sheep, as a function of age, plotted against age. Note that year of life 10 also includes years of life greater than 10. Also shown are estimates of one estimated standard error, above and below the point estimate (on a logistic scale). Reproduced with permission from Catchpole et al. (2000, p. 461) published by Wiley-Blackwell.

□

There have been several studies of the effect of ageing on both the survival probability and the reproductive success of wild animals. The papers by Gaillard et al. (1994), Loison et al. (1999) and Nichols et al. (1997) emphasise the need for such studies to be based on long-term data sets, such as that on the Soay sheep, and we shall encounter further examples of this later. In-

SUMMARY

evitable difficulties arise as a result of only relatively small numbers of animals surviving in the oldest age groups.

1.6 Summary

The aim of this chapter is to describe the fundamental concerns of population ecology, involving the estimation of vital demographic parameters, in order that they may be used to understand better how populations of wild animals behave. There are parallels with human demography, though there are often differences in how data may be collected. An abiding concern of the 21st century lies with global warming, and its effect on all forms of life on earth. At a local scale, changes in farming practice can also have a major influence on wildlife. Before one can manage wild populations of living organisms, it is first necessary to be able to measure population sizes and biological diversity, and project how these will change in the future. The approaches of statistical ecology are to collect suitable data, devise probability models for underlying processes and, by fitting the models to the data, obtain the necessary estimates and understanding. The resulting statistical inference can be conducted in classical or Bayesian ways. The latter approach is richer than the former, and importantly may allow more realistic models to be fitted to data. Additionally, the Bayesian approach allows the incorporation of information that is separate from the data collection exercise.

1.7 Further Reading

A wider ecological perspective is provided by references such as Morgan et al. (1997), Furness and Greenwood (1993), Clutton-Brock (1988) and Clutton-Brock et al. (1982). The book by Seber (1982) is the classical text in statistical ecology, and contains a wealth of applications and analyses of challenging and interesting data sets, as well as much historical material. A review is provided by Pollock (1991). More recent work can be appreciated from Pollock (2000) and Schwarz and Seber (1999), Morgan and Thomson (2002) and Senar et al. (2004). The volume by Williams et al. (2002) is encyclopedic, and essential reading for statistical ecologists. The article by Morgan and Viallefont (2002) gives a number of Web sites which are rich sources of data on birds, including discussion of motivation and sampling methods. See for example, http://www.bto.org/ and http://www.rspb.org.uk/ (BTO stands for British Trust for Ornithology, and RSPB refers to the Royal Society for the Protection of Birds). The estimation of wild animal abundance is the subject of the book by Borchers et al. (2002). The book by Caswell (2001) is the classical reference for population dynamics modelling. Regarding the estimation of survival, analogies can be drawn with life-table analysis and other methods that are used in the analysis of human populations; for example, see Cox and Oakes (1984). The book by Hinde (1998) provides a good introduction to human demography. Illustrative fisheries applications are to be found in McAllister and Kirkwood (1998a) and Rivot et al. (2001). The paper by Barry et al. (2003) provides a detailed Bayesian analysis of ring-recovery data on lapwings, which emphasises the need to include random effects in models of this kind. The paper by Catchpole et al. (2000) provides a detailed classical analysis of Soay sheep data, and a Bayesian analysis of the same data, involving extensive model averaging, is provided by King et al. (2006). Capture-recapture methods are also used in epidemiological and medical studies – see Chao et al. (2001).

A forthcoming Bayesian revolution in population ecology has long been anticipated, for instance by Pollock (1991) and by Seber in several publication; see for example Schwarz and Seber (1999). More recent advocates include Brooks et al. (2000a), Brooks et al. (2002) and Link et al. (2002). Illustrations of applications in a variety of areas of application are given in Congdon (2003).

1.8 Exercises

1.1 Use a computer search engine to locate information for the key words: EURING, Patuxent. In both cases, read the motivation material presented for work in population ecology.

1.2 Consult the Web site for the Royal Society for the Protection of Birds. Read the material available there on conservation science, with particular reference to conservation concerns and management successes.

1.3 Explore the Web site of the Zoological Society of London, and read the material there on the Serengeti Cheetah Project.

1.4 Read selectively from Williams et al. (2002) and Seber (1982).

1.5 The following quotation is reproduced with permission from Nichols et al. (1997, p. 1009).

> Despite the existence of many reports discussing senescence in animal populations ... the pattern of age specific mortality is well known only for *Homo sapiens*. One of the reasons for this absence of knowledge stems from the small number of older individuals from which inferences about age specific mortality in natural populations must be drawn. Age-specific mortality rates are typically estimated for natural animal populations using marked individuals. Age is typically known for older animals only because they were marked at some young age at which age can be assigned unambiguously. Mortality reduces the numbers of known-age animals from cohorts marked as young, so that the numbers of animals available for study decrease with age. Thus, investigations should use large samples of marked individuals and should be relatively long in duration. Another factor limiting our knowledge of age specific mortality in natural animal populations involves estimation methodology. Mortality estimates used in the vast majority of published reviews and summary analyses of senescence in animal populations are based on models requiring extremely restrictive, yet untested, assumptions related to both population dynamics and

sampling. Capture-recapture models designed to permit reasonable inferences about age specific survival have existed for nearly two decades, yet few long-term field studies have taken advantage of these analytic methods to study senescence.

Consider the issues raised here, and consult the source paper for further information. Age can affect reproductive performance as well as survival. If reproduction reduces the survival of a wild animal the year following giving birth, suggest a reproduction strategy that might optimise fitness, in terms of ensuring the largest number of offspring.

1.6 The annual survival probability estimates for the two oldest age categories in Figure 1.3 appear to be similar. Discuss how this might arise.

1.7 Use a computer search engine to discover more about the Soay sheep research program.

CHAPTER 2

Data, Models and Likelihoods

2.1 Introduction

In this chapter we describe several ways in which population ecology data are collected. We present a variety of illustrative data sets, to which we shall return later in the book. For convenience, the data sets are mainly chosen to be small, but in practice of course real data sets can be extensive. Both classical and Bayesian methods take as their starting point the likelihood, and so we provide several examples of likelihood construction. In these examples the likelihood is the joint probability of the data. The classical method of forming maximum-likelihood estimates of model parameters proceeds by regarding the likelihood as a function of the model parameters alone, for fixed data, and then maximising the likelihood with respect to the parameters.

The current concern regarding species decline that we discussed in Chapter 1 often stems from some kind of counting or censusing method producing time-series of data which indicate a decline in numbers. One example of this has already been given in Example 1.3 (and Figure 1.2). We therefore start with a discussion of population data of this general kind.

2.2 Population Data

2.2.1 Count and Census Data

Example 2.1. Heronry Census

We start with an example of a British bird species which has been growing in numbers in recent years, in contrast to the lapwing of Example 1.3. The grey heron, *Ardea cinerea*, was, until recently, the only heron to live and breed in Britain. It is non-migratory, and birds live and breed colonially, in heronries. Each year there is a census of these heronries, which is used to produce an estimate of the total number of breeding pairs of herons in Britain. The British census of heronries was initiated by Max Nicholson, and is the longest census of its kind in the world. Census figures are shown in Figure 2.1, taken from Besbeas et al. (2001).

The heron's diet includes fish, amphibians, and other species to be found in water. If ponds freeze over then it becomes harder for herons to find the food that they need, and this is especially true of young birds. This explains the big dip in the graph of Figure 2.1, corresponding to the severe British winter of 1962; shown in Figure 2.2 is a graph of the number of days below freezing at a location in Central England, which clearly identifies the 1962

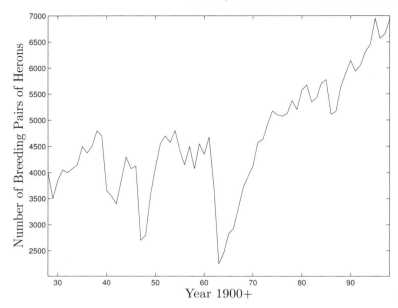

Figure 2.1 Estimated numbers of breeding pairs of grey herons in England and Wales, from 1928–1998, inclusive.

winter. However, since 1962 the grey heron has recovered well from the crash in numbers.

In this example, estimates of abundance are obtained from a national census, which is aimed at estimating a total population size. This is in contrast to the Common Bird Census (CBC) index for lapwings, from Example 1.3. For the heron, national abundance is a simple consequence of survival and productivity, as the heron does not migrate. Survival is estimated from national ring-recovery data, and an illustration of the data that result for the heron is given in Exercise 2.3.

□

Example 2.2. The CBC Index for Lapwings

As has already been discussed in Example 1.3, the CBC is not a census in the same way that the heronries census is. However, it may be regarded as an estimate of the number of birds present on the set of sites in the CBC. The CBC has recently been replaced by a different national census method in Britain. Both are used to examine trends in abundance of common British birds. The lapwing's behaviour is more complicated than that of the heron in that it is a partial migrant – in winter certain lapwings may migrate, while others may disperse locally.

□

POPULATION DATA 17

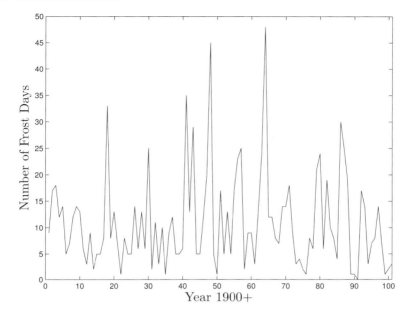

Figure 2.2 Number of days in the year when the temperature fell below freezing at a central England location, from 1900–2001, inclusive. We call this the number of frost days in a year.

2.2.2 Statistical Methods Based on Marked Animals

The two examples given previously for determining the abundance of wild animals estimate national population numbers and typically depend upon the work of large numbers of volunteers. The animals are observed at a distance and are not physically caught. An alternative approach, which is often employed for both small and large populations, involves capturing some of the animals and marking them in some way. In certain cases the marks identify the animals uniquely, while in others they do not. Subsequent observations on these animals may then be used in order to estimate population sizes, as well as demographic rates, and we shall see several illustrations of this throughout the book.

2.2.3 Marking

An early example of the scientific marking of wild animals is due to Hans Christian Mortensen who marked 164 starlings, *Sturnus vulgaris*, in Denmark in 1899. This is regarded as the first instance of marking of wild birds for scientific purposes; his objective was to study movement, and he had to experiment with different types of metallic rings. There is now much interest in the marking and subsequent study of wild animals. For instance, over a million birds are ringed each year in North America alone. At one end of the scale, insects may be marked by means of specks of fat, while at the other end of the scale birds and animals may be marked by radio transmitters and tracked by satellite. The terminology of marking is varied, with marks being sometimes described as bands, tags, or rings. Fish may be marked by PIT tags, and then the later movement of such fish over antennae in rivers automatically provides information on where and when fish are observed alive. PIT stands for Passive Integrated Transponder and these tags are very small passive identification chips, essentially like bar codes, which provide unique identification. These tags provide permanent identification, as they require no battery. Bird rings may contain an address, so that when a dead ringed bird is found, the location and timing of the death can be recorded if the ring is returned to the address on the ring. Marked animals may be seen again alive, as well as found dead, as was true of the dippers in Example 1.1. In some cases, animals can be identified through their natural markings, as is true for instance of the belly patterns of great crested newts, *Triturus cristatus*, or of the spots on a cheetah, *Acinomyx jubatus*. A modern development involves DNA matching, so that, for instance, hairs or faeces collected in the field can be used to identify individuals.

In some cases, animals are marked shortly after birth, while in others animals are marked as adults, often of unknown age, which was also true of the dippers of Example 1.1. The marking of animals shortly after birth is often convenient, and it also provides a way of estimating mortality in the first year of life. Activities such as breeding or moulting provide opportunities for scientists to capture or recapture adult wild animals.

2.2.4 The Lincoln-Petersen Estimate of the Size of a Closed Population

Once a number of animals in a population have been marked, then observing the numbers of marked animals in subsequent samples provides information about the total number of animals present in the area. An illustration is provided by a study of Nuttall's cottontail rabbits, *Sylvilagus nuttallii*, in Oregon; see Skalski et al. (1983). In this case a sample of $n_1 = 87$ rabbits were captured and had their tails and back legs dyed. A later sample resulted in $n_2 = 14$ captured animals, $m_2 = 7$ of which were marked. This allows us to form the Lincoln-Petersen estimate of the number of animals present in the area.

In the two-sample study leading to the Lincoln-Petersen estimate, we suppose that the two samples are taken sufficiently close together in time that there is no movement of animals in or out of the region sampled, and that there is no mortality or birth. This is what is regarded as a *closed* population, and of interest is the unknown size, N, of that population. Let $p_1(p_2)$ denote the probability of an animal being captured at the first (second) sample. Making appropriate assumptions of independence then allows us to derive the likelihood for this two-sample situation, which we denote $L(N, p_1, p_2 | n_1, n_2, m_2)$,

POPULATION DATA 19

and is given by,

$$L(N, p_1, p_2 | n_1, n_2, m_2) = \frac{N!}{(N-r)! m_2! (n_1 - m_2)! (n_2 - m_2)!} (p_1 p_2)^{m_2}$$
$$\times \{p_1(1-p_2)\}^{n_1 - m_2} \{(1-p_1)p_2\}^{n_2 - m_2}$$
$$\times \{(1-p_1)(1-p_2)\}^{N-r},$$

where $r = n_1 + n_2 - m_2$. It is intuitively sensible to estimate N by equating the two sample proportions of marked animals, namely:

$$\frac{n_1}{N} = \frac{m_2}{n_2},$$

and this can be shown to be approximately the *maximum-likelihood estimate* of N; see Exercise 2.1. In fact the properties of this estimator of N can be improved by means of slight modifications, resulting in the Chapman estimator, which is given by

$$\hat{N} = \frac{(n_1 + 1)(n_2 + 1)}{m_2 + 1} - 1.$$

Applying this simple formula to the rabbit data results in:

$$\hat{N} = 164 \ (35.82).$$

The number in parentheses after the estimate above is the estimated standard error of the estimator, and we discuss in general how this may be calculated, and how it can be used, in Chapter 3. In this simple example, the likelihood has what is called a *multinomial* form, and later in the chapter we shall encounter several other likelihoods that are also based on multinomial distributions; see Appendix A.

Interesting historical information regarding the Lincoln-Petersen estimate is given by Goudie and Goudie (2007). There has been statistical research in this general area for many years, with an early application to estimating the population of France undertaken by Laplace (1786). In that case the substitute for the marking of individuals was the occurrence of individuals on a birth register. The register of births for the whole country corresponds to the marked individuals n_1, a number of parishes of known total population size corresponds to n_2, and m_2 was the number of births recorded there. The simple scheme of two sampling occasions described here can be extended to several sampling occasions, which is then known as the Schnabel census.

Example 2.3. Schnabel Census Data

Examples of real data resulting from the Schnabel census are given in Table 2.1, where f_j denotes the number of animals that are caught j times. The first two examples involve wild animals, the third involves groups of golf tees, and the last involves taxi-cabs in Edinburgh. The sources of the data are as follows: Meadow voles, *Microtus pennsylvanicus*, Pollock et al. (1990); Snowshoe hares, *Lepus americanus*, collected by Burnham and Cushwa, and recorded in Otis

Table 2.1 Four Examples of Real Data Resulting from the Schnabel Census

	f_1	f_2	f_3	f_4	f_5	f_6	f_7	f_8	f_9	f_{10}
Hares	25	22	13	5	1	2	-	-	-	-
Voles	18	15	8	6	5	-	-	-	-	-
Golf tees	46	28	21	13	23	14	6	11	-	-
Taxi-cab	142	81	49	7	3	1	0	0	0	0

Note: The taxi-cab study had 10 sampling occasions, the hare study had 6 sampling occasions, etc.

et al. (1978, p. 36); golf tees, Borchers et al. (2002); and taxi-cabs, Carothers (1973).

□

Given a set of real data from a Schnabel census, we can fit a range of alternative models, using the method of maximum likelihood, and then choose between these models in some standard way. This might be done in terms of comparing maximised likelihoods or appropriate information criteria, for example, and we shall see illustrations of this in Chapter 3. The first step therefore is to construct a likelihood, and we illustrate this for the simple homogeneous case, in which there is a constant probability of recapture p, which is assumed not to vary over animals, or time, or in any other fashion. When we have t sampling occasions, then if we assume independence between those occasions, the probability that any animal is caught j times is given by the simple binomial probability:

$$\binom{t}{j} p^j (1-p)^{t-j}.$$

For distributional information see Appendix A. When a random experiment has two outcomes, and it is repeated independently several times, then the total number of occurrences of one of those outcomes is said to have the binomial distribution. The multinomial distribution has already been encountered in Section 2.2.4 and results when we have more than two outcomes; it is encountered frequently in ecology, as we shall see later. In order to estimate p, we can condition on the fact that we are modelling just those animals that have been counted. This, together with assuming independence of animals, results in the following expression for the conditional likelihood:

$$L(p|\boldsymbol{f}) = \prod_{j=1}^{t} \left[\frac{\binom{t}{j} p^j (1-p)^{t-j}}{\{1-(1-p)^t\}} \right]^{f_j}. \tag{2.1}$$

The denominator in each of the terms in the likelihood expression of Equa-

MODELLING SURVIVAL

tion (2.1) is the sum of the probabilities in the numerators, and is the probability of being caught on one of the t sampling occasions. Of course, we normally collect data from the Schnabel census in order to estimate the unknown population size, N. In order to include N in the expression for the likelihood, then we use the following likelihood:

$$L(N,p|\boldsymbol{f}) = \binom{N}{D} \prod_{j=0}^{t} \left\{ \binom{t}{j} p^j (1-p)^{t-j} \right\}^{f_j},$$

were $D = \sum_{j=1}^{t} f_j$. See for example Morgan and Ridout (2008).

We can see that this likelihood expression involves the binomial distribution, encountered above, as it provides the probability of j captures out of t occasions. We continue discussion of the Schnabel census in Example 3.3. A Bayesian approach to estimating population size is presented in Chapter 11.

2.2.5 Transect Data

Transect sampling is a way of estimating the density of an animal population by walking a transect line, and recording the locations of individuals from the population seen from that line. Care must be taken that the transect line is randomly oriented with regard to the area surveyed. For example, observations taken from a car travelling along a fenced road could easily produce biased estimates if one was surveying birds which used the fence for song posts. As with much modelling in population ecology, it is necessary to make a range of assumptions, and conclusions can depend crucially on the assumptions made. We would expect it to be easier to detect animals close to the line than animals far from the line, and density estimates can result based on a function which reflects this drop-off in detection with distance. However, estimates of density are sensitive to the assumptions made regarding the form of this function, and so more recently methods have been devised which do not make strong assumptions about the shape of the detection function. General theory is given by Buckland et al. (2001), and a Bayesian approach is provided by Karunamuni and Quinn (1995). We provide a Bayesian analysis in Section 11.5.

2.3 Modelling Survival

One of the most important aspects of demography is the estimation of survival. In human demography, statistical methods deal with times to death, which are sometimes censored, but usually known accurately. In dealing with marked wild animals, there are certainly data on times of death, following the discovery and reporting of a dead animal. However, there is typically far less precision, compared with the human case. Wild animals experience a natural yearly life cycle, and survival is often reported as annual survival, corresponding to whether or not an animal survived a particular year. In some studies, there is

2.3.1 Ring-Recovery Data

Ring-recovery data can result from studies on any spatial scale. The example of Table 2.2, taken from Catchpole et al. (1999), is an illustrative subset of a much larger data set resulting from a national study of lapwings, the subject of Example 1.3.

Table 2.2 Recovery Data for British Lapwings Ringed Shortly after Birth during the Years 1963–1973

Year of Ringing	Number Ringed	Year of Recovery – 1900+										
		64	65	66	67	68	69	70	71	72	73	74
1963	1147	13	4	1	2	1	0	0	1	0	0	0
1964	1285		16	4	3	0	1	1	0	0	0	0
1965	1106			11	1	1	1	0	2	1	1	1
1966	1615				10	4	2	1	1	1	0	0
1967	1618					11	1	5	0	0	0	1
1968	2120						9	5	4	0	2	2
1969	2003							11	9	4	3	1
1970	1963								8	4	2	0
1971	2463									4	1	1
1972	3092										8	2
1973	3442											16

Note: Ringing took place before 1st May in each year. Data provided by the British Trust for Ornithology.

Thus, for example, in the first year of this study, 1963, a total of 1147 lapwings were ringed in Britain and released. Of these birds, 14 died in their first year of life, were found and reported as dead to the British Trust for Ornithology; 4 birds survived their first year of life, but died and were reported as dead in their second year of life, and so on. If we assume independence of birds both in and between years, then the appropriate probability distribution to describe how the dead birds are distributed over the years, from any one cohort, is the multinomial distribution, the generalisation of the binomial distribution to the case of more than two outcomes, and already encountered in the likelihood for the Lincoln–Petersen procedure of Section 2.2.4. The binomial distribution has n individuals distributed over two outcomes with probabilities p and $1-p$, and it is generalised by the multinomial distribution in which n individuals are distributed over $k > 2$ outcomes, with probabilities $\{\theta_j\}$, constrained so that $\sum_{j=1}^{k} \theta_j = 1$ (see also Appendix A.1.8). Of course what we want to do is to

MODELLING SURVIVAL

estimate probabilities, ϕ, of annual survival from ring-recovery data, and this involves writing the θ_j in terms of annual survival probabilities, which means that we also have to introduce probabilities of dead animals being found and reported dead. We use λ to model these recovery probabilities. There are many different ways of making this reparameterisation, each resulting in a different model. A striking feature of Table 2.2 is the relatively large numbers of birds that are found and reported dead in their first year of life. It is a common finding that for wild animals there is appreciable mortality in the first year of life, and so one would certainly want to include in a model for the data of Table 2.2, and for similar data, some form of age dependence in the survival probabilities. The simplest way of doing this is to have a survival probability ϕ_1 for animals in their first year of life, and an annual survival probability ϕ_a for all older animals. For such a model, if we have a constant recovery probability λ, the multinomial probabilities for the case with four years of recovery, for any cohort are:

$$(1-\phi_1)\lambda, \quad \phi_1(1-\phi_a)\lambda, \quad \phi_1\phi_a(1-\phi_a)\lambda, \quad \phi_1\phi_a^2(1-\phi_a)\lambda, \quad 1-(1-\phi_1\phi_a^3)\lambda.$$

The expression $1 - (1 - \phi_1\phi_a^3)\lambda$ is the probability that an animal is not found dead after the end of the study; such animals are either alive or dead and not reported. This term can be calculated by noting that the sum of the multinomial probabilities must sum to unity. Thus, we calculate the final term as one minus the sum of the other terms.

If there are $T-1$ years of recovery and m_t animals reported dead in the tth year of the study, then the likelihood corresponding to a single cohort of marked animals of size R will be given by

$$L(\phi_1,\phi_a,\lambda|\boldsymbol{m}) \propto \{(1-\phi_1)\lambda\}^{m_1} \left[\prod_{t=2}^{T-1}\{\phi_1\phi_a^{t-2}(1-\phi_a)\lambda\}^{m_t}\right]$$
$$\times \{1-(1-\phi_1\phi_a^{T-2})\lambda\}^{m_T}, \qquad (2.2)$$

where we let $m_T = R - \sum_{t=1}^{T-1} m_t$ and denote the data by $\boldsymbol{m}$. The missing constant of proportionality does not involve the model parameters, and so is omitted for simplicity. For each cohort we form a separate multinomial likelihood, of the form of Equation (2.2), and then the likelihood for the entire recovery table is simply the product of the multinomial likelihoods for each of the cohorts.

More generally, and an example of the likelihood for an entire recovery table, suppose that we consider ring-recovery data, where there are releases in years $1, \ldots, T-1$ and recoveries in years $2, \ldots, T$ and we assume that the survival and recovery probabilities are both time dependent. We let ϕ_t denote the probability an animal alive at time t survives until time $t+1$ and λ_t the probability that an individual that dies in the interval $(t-1, t]$ is recovered at time t. The model parameters are then $\boldsymbol{\phi} = \{\phi_1, \ldots, \phi_{T-1}\}$ and $\boldsymbol{\lambda} = \{\lambda_2, \ldots, \lambda_T\}$. We let R_i denote the number of animals released at time $i = 1, \ldots, T-1$ and $m_{i,j}$ the number of animals that are released in year i and recovered dead in the interval $(j, j+1]$, for $i = 1, \ldots, T-1$ and $j = 1, \ldots, T-1$. The corresponding likelihood is given by,

$$L(\boldsymbol{\phi}, \boldsymbol{\lambda}|\boldsymbol{m}) \propto \prod_{i=1}^{T-1} \left\{ \prod_{j=i}^{T-1} \left[(1-\phi_j)\lambda_{j+1} \prod_{k=i}^{j-1} \phi_k \right]^{m_{i,j}} \chi_i^{m_{i,T}} \right\}, \quad (2.3)$$

where we use the convention that if the upper limit of a product is less than the lower limit, the product is equal to unity. The term χ_i denotes the probability that an individual released at time i is not observed again in the study and $m_{i,T} = R_i - \sum_{j=1}^{T-1} m_{i,j}$, the corresponding number of individuals released at time i and not observed again. The probability χ_i can be calculated using the sum to unity constraint of the multinomial probabilities, so that it can be expressed in the form,

$$\chi_i = 1 - \sum_{j=i}^{T-1} \left[(1-\phi_j)\lambda_{j+1} \prod_{k=i}^{j-1} \phi_k \right].$$

This product multinomial likelihood for ring-recovery data is maximised to produce the maximum-likelihood estimates of the model parameters. Typically, though not always, this involves numerical optimisation of the log-likelihood, which we cover in detail in Chapter 3.

2.3.2 Capture-Recapture Data

We have already encountered capture-recapture data in Table 1.1. As observed earlier, a display of data in this form is often called an m-array. The form of Table 1.1 is essentially the same as that of Table 2.2, and the modelling in this case is basically the same as in the last section, involving a likelihood which is a product of multinomial likelihoods, one from each cohort. In the case of dippers, we are dealing with birds which are all adults, and so we may suppose that they share the same common annual survival probability, ϕ, ignoring in this instance the possible effect of the flood year of 1983. With p denoting the probability of recapture, then for a model with just these two parameters, the multinomial probabilities corresponding to any cohort of marked birds for just four years of recapture for a single cohort of marked animals are of the form:

$$\phi p, \quad \phi^2(1-p)p, \quad \phi^3(1-p)^2 p, \quad \phi^4(1-p)^3 p, \quad \chi,$$

where χ denotes the probability an individual is not observed again in the study, following its release, and is given by,

$$\chi = 1 - \sum_{t=1}^{4} \phi^t p(1-p)^{t-1}.$$

The likelihood, $L(\phi, p|\boldsymbol{m})$ can then be formed and maximised to produce the maximum-likelihood estimates, where we again use $\boldsymbol{m}$ to denote the data. Numerical optimisation is needed, as in the application of the last section. For illustration, if we have just one cohort, of size R, of marked animals, released at time $t = 1$ and there are m_t recaptures at time $t + 1 = 2, \ldots, T$, then the likelihood has the form,

$$L(\phi, p | \bm{m}) \propto \prod_{t=1}^{T-1} [\phi^t p(1-p)^{t-1}]^{m_t} \chi^{m_T}, \qquad (2.4)$$

where $m_T = R - \sum_{t=1}^{T-1} m_t$, and now $\chi = 1 - \sum_{t=1}^{T-1} \phi^t p(1-p)^{t-1}$. There is again a missing constant term in Equation (2.4), which is once more omitted for simplicity as it does not involve the model parameters.

More generally, suppose that there releases in years $i = 1, \ldots, T-1$, with recaptures occurring at times $j = 2, \ldots, T$. We let R_i denote the number of individuals released at time $i = 1, \ldots, T-1$ (i.e. this corresponds to the total number of individuals released at this time, either captured for the first time or subsequently those that are recaptured at this time). We assume that the survival and recapture probabilities are time dependent, so that ϕ_t denotes the probability that an individual alive in year t survives until year $t + 1$, for $t = 1, \ldots, T-1$ and p_t denotes the recapture probability at time $t = 2, \ldots, T$. The model parameters are $\bm{\phi} = \{\phi_1, \ldots, \phi_{T-1}\}$ and $\bm{p} = \{p_2, \ldots, p_T\}$. We let $m_{i,j}$ denote the number of individuals released at time i and next recaptured at time $j+1$, for $i = 1, \ldots, T-1$ and $j = 1, \ldots, T-1$. Note that, by definition, $m_{i,j} = 0$ for $j < i$. The corresponding likelihood can then be expressed in the form,

$$L(\bm{\phi}, \bm{p} | \bm{m}) \propto \prod_{i=1}^{T-1} \left\{ \prod_{j=i}^{T-1} \left[\phi_j p_{j+1} \prod_{k=i}^{j-1} \phi_k (1 - p_{k+1}) \right]^{m_{i,j}} \chi_i^{m_{i,T}} \right\}, \qquad (2.5)$$

where χ_i denotes the probability an individual released at time i is not observed again in the study (either as a result of dying, or surviving to the end of the study and not being observed), and $m_{i,T}$ the corresponding number of individuals. The χ_i term can once more we obtained using the sum to unity constraints for the multinomial probabilities, given by,

$$\chi_i = 1 - \sum_{j=i}^{T-1} \left[\phi_j p_{j+1} \prod_{k=i}^{j-1} \phi_k (1 - p_{k+1}) \right].$$

When recapture and recovery data are presented, as in Tables 1.1 and 2.2 respectively, then it is usual to be given the cohort sizes, corresponding to the numbers of marked animals released at each time (i.e. the R_i, for $i = 1, \ldots, T-1$). However, when we encounter computer code written in R and WinBUGS, in Chapter 7 and Appendices B and C, then we shall see that in those cases the cohort sizes are not given, but instead, and equivalently, we are provided with the numbers of animals that are either not recovered dead, or recaptured alive, during the periods of the studies. (i.e. the $m_{i,T}$ for $i = 1, \ldots, T-1$).

2.3.3 Models for Life History Data

It would be a mistake to assume that the probability models for data arising in population ecology are all simple binomials or multinomials, or products of these. One elaboration which results in data which need a more complex model arises when recovery and recapture data are recorded on the same individuals.

Example 2.4. Shags

In some cases records are made which potentially include both recovery and recapture of the same individuals. An example of this is provided in Table 2.3, corresponding to observations on shags, *Phalacrocorax aristotelis*, observed by Mike Harris on the Isle of May in Scotland. In this case the data take the form of individual life-histories. A '1' indicates that the corresponding bird was captured or recaptured, a '2' indicates when a bird was recovered dead (after which the record for that bird only contains 0s) and a '0' denotes neither a recovery or a recapture.

Table 2.3 Illustrations of Life-Histories for 5 Different Shags

Cohort	Recapture Occasions					
	t_1	t_2	t_3	t_4	t_5	t_6
1	1	0	1	0	2	0
	1	1	0	0	1	1
2	0	1	2	0	0	0
	0	1	0	0	1	0
	0	1	0	0	0	0

Source: Reproduced with permission from Catchpole et al. (1998, p. 34) published by the International Biometric Society.

Note: In the first cohort, both birds are marked at time t_1, while in the second cohort, the three birds are all marked at time t_2.

□

Suppression of 1s (2s) in the life history (or capture-history) data results in recovery (recapture) data alone. However, we want to use the combined data, in order to estimate model parameters with the greatest precision. Let A be the maximum age for an animal, and let the years of releases be denoted by $t = 1, \ldots, T-1$ and years of recaptures by $t = 2, \ldots, T$, as in the last section.

MODELLING SURVIVAL

Then, for $j = 1, \ldots, A$ and $t = 1, \ldots, T-1$, we define,

- $\phi_{j,t}$ – the annual survival probability for an animal in its jth year of life at time t;
- $p_{j+1,t+1}$ – the capture probability of a live animal in its $(j+1)^{th}$ year of life at time $t+1$; and
- $\lambda_{j,t}$ – the recovery probability of a dead animal in its jth year of life in the time interval $[t, t+1)$.

We use $\boldsymbol{\phi}, \boldsymbol{p}$ and $\boldsymbol{\lambda}$ to denote vectors/matrices containing all the survival, recapture and recovery probabilities, as usual. It is shown by Catchpole et al. (1998) that we can specify the likelihood in the form,

$$L(\boldsymbol{\phi}, \boldsymbol{p}, \boldsymbol{\lambda}|\boldsymbol{m}) \propto \prod_{j=1}^{A} \prod_{t=1}^{T-1} \left\{ \phi_{j,t}^{w_{j,t}+z_{j,t}} p_{j+1,t+1}^{w_{j,t}} (1-p_{j+1,t+1})^{z_{j,t}} [(1-\phi_{j,t})]^{d_{j,t}} \lambda_{j,t}^{d_{j,t}} \chi_{j,t}^{v_{j,t}} \right\},$$
(2.6)

where the constant of proportionality does not involve the model parameters, and so can be ignored. Here the term $\chi_{j,t}$ is the probability that an animal observed at time t in its jth year of life is not observed after this time, either alive or dead, and it is given by the recursion,

$$\chi_{j,t} = 1 - (1-\phi_{j,t})\lambda_{j,t} - \phi_{j,t}[1-(1-p_{j+1,t+1})\chi_{j+1,t+1}],$$

with,

$$\chi_{j,T} = 1, \qquad \text{for all } j = 1, \ldots, A.$$

The likelihood is determined by four matrices constructed from the raw data, and defined below, for $j = 1, \ldots, A$:

- $w_{j,t}$ – the number of animals that are recaptured at time $t+1 = 2, \ldots, T$, in their $(j+1)^{th}$ year of life;
- $z_{j,t}$ – the number of animals that are not recaptured at time $t+1 = 2, \ldots, T$, that would be in their $(j+1)^{th}$ year of life at this time, but known to be alive, since they are observed later in the study, either alive or dead;
- $d_{j,t}$ – the number of animals recovered dead in the interval $[t, t+1)$, in their j^{th} year of life for $t = 1, \ldots, T-1$;
- $v_{j,t}$ – the number of animals that are observed for the last time at time $t = 1, \ldots, T-1$ in their j^{th} year of life.

These matrices form a set of *sufficient* statistics, since they are all that we need to retain from the data in order to evaluate the likelihood – it is not necessary to retain all the individual life-histories, and this results in a computational saving when the likelihood is formed. It is also attractive to see how simple multinomial expressions for the likelihood, which arise when there are only recovery or only recapture data, generalise to the case of integrated data. The formulation given here is completely general, and allows a wide range of complex models to be specified, making different assumptions about how various model parameters may vary with time as well as age. In fact, it was used to produce the survival probability estimates in Figure 1.3 of Example 1.4 for Soay sheep, and will be encountered again later in the book. The notation for life-histories given above can be generalised to describe animal encounters are more than one site. Formulation of a likelihood in terms of sufficient matrices extends to the case of multi-site models, which we consider next.

2.4 Multi-Site, Multi-State and Movement Data

In Section 2.2.4 we defined a closed population as one in which there is no birth, death, or migration. Of course many populations are open, as is true of most of the examples of this chapter, and one way in which populations are open is by experiencing migration. Estimates of survival probability obtained from capture-recapture studies often do not estimate survival as such, but an apparent survival, which is a product of probabilities of survival and fidelity to the study area. A characteristic of the analysis of red deer capture-recapture data by Catchpole et al. (2004) was that there it was possible to estimate the probability of movement as well as survival. Global warming can result in fragmented habitat, resulting in a particular type of movement, with individuals living and reproducing in a small number of colonies which occasionally exchange individuals. In this book we shall consider multi-site models, in which individuals move between sites; recovery and recapture of marked animals may take place at a number of different sites, and the resulting data may then be used to estimate probabilities of movement of animals between the sites, as well as probabilities of recovery, recapture and survival in each site. An example of one such set of data, recording re-sightings at three different sites, is presented in Table 2.4 for a population of Hector's dolphins, *Cephalorhynchus hectori*, around Banks Peninsula off the coast of Akaroa (near Christchurch) on the Southern Island of New Zealand between 1985 and 1993.

An immediate consequence of multi-site models is a potential explosion in the number of parameters to be estimated, because as well as the usual issues of possibly incorporating time and age varying parameters, one has also to consider whether parameters vary between sites, in addition to having parameters of movement between sites. However, whatever models one might decide to fit and compare in the context of multi-site data, the classical approach is basically the same as when one has data from just the one site. A particularly interesting aspect of the probability modelling of multi-site data is how in effect scalar probabilities for single-site data are replaced by matrices of probabilities in the multi-site generalisation, in what are called Arnason-Schwarz models (Brownie et al. 1993, Schwarz et al. 1993).

One of the concerns of ecology is to study how wild animals regulate their breeding. For example, as discussed in Exercise 1.5, for a relatively long-lived animal, is it advantageous, in terms of overall fitness, to attempt to reproduce each year, or should one skip years, taking a sabbatical from breeding, which

Table 2.4 The Number of Hector's Dolphins Observed and Then Re-sighted Off Banks Peninsula, New Zealand Between 1985–1993

Year	Number Released	1986	1987	1988	1989	1990	1991	1992	1993
1985	3	0 1 0	0 0 0	0 0 0	0 0 0	1 0 0	0 0 0	0 0 0	0 0 0
	0	0 0 0	0 0 0	0 0 0	0 0 0	0 0 0	0 0 0	0 0 0	0 0 0
	4	0 2 0	0 0 1	0 1 0	0 0 0	0 0 0	0 0 0	0 0 0	0 0 0
1986	1		0 1 0	0 0 0	0 0 0	0 0 0	0 0 0	0 0 0	0 0 0
	30		1 20 1	1 0 0	0 0 0	0 0 0	1 1 0	1 0 0	0 0 0
	5		2 2 0	0 0 0	0 0 0	0 0 0	0 0 0	0 0 0	0 0 0
1987	9			0 5 0	0 1 0	0 0 0	2 0 0	0 0 0	0 0 0
	34			0 24 0	0 1 0	0 0 1	0 1 0	1 0 0	0 0 0
	5			0 2 1	0 0 0	0 1 1	0 0 0	0 0 0	0 0 0
1988	5				0 0 0	0 1 0	0 0 0	0 0 0	0 1 0
	41				0 8 0	0 4 0	3 3 0	0 4 0	0 0 0
	2				0 1 0	0 0 0	0 0 0	0 0 0	0 0 0
1989	0					0 0 0	0 0 0	0 0 0	0 0 0
	15					0 5 0	0 2 1	0 0 1	0 1 0
	0					0 0 0	0 0 0	0 0 0	0 0 0
1990	1						0 0 0	0 0 0	0 0 0
	13						2 5 1	1 0 0	0 1 0
	5						0 0 2	0 0 0	0 0 0
1991	10							3 1 0	0 0 0
	12							1 2 3	1 0 0
	9							0 0 7	0 0 1
1992	12								1 1 1
	10								0 3 1
	14								0 2 4

Source: Reprinted with permission from *The Journal of Agricultural, Biological, and Environmental Statistics*. Copyright 1999 by the American Statistical Association. All rights reserved. Published in Cameron et al. (1999, p. 129).

may substantially increase the chance of dying in a breeding year? Thus in some studies it is important to be able to record whether or not an animal is breeding that year. The theory of multi-site models also applies to multi-state models, in which animals exist in one of several states, such as breeding and non-breeding. King and Brooks (2003a) show how the general likelihood formulation for integrated recovery and recapture data, given in Equation (2.6), extends naturally to the case of multiple sites. A Bayesian approach to multi-site modelling is provided in Chapter 9.

2.5 Covariates and Large Data Sets; Senescence

We expect demographic parameters to vary with time, as well as with age. Thus the potential number of models may be large. A convenient way to describe time variation can be to try to regress relevant parameters on time varying covariates. Biologists often have measurements on a range of covariates, at both the population level and individual level, but it is only relatively recently that their incorporation into models has become feasible.

The use of covariates in mark-recapture-recovery studies dates from North and Morgan (1979). In that paper the authors modelled the survival of grey herons, with the annual survival probability of birds in their first year of life (denoted by ϕ) being a logistic function of the number of frost days in the year (W) when the temperature in central England was below freezing: that is,

$$\text{logit } \phi = \log\{(\phi/(1-\phi))\} = \beta_0 + \beta_1 W,$$

where β_0 and β_1 are parameters to be estimated from the data. Such regressions are valuable, as they both suggest causes for mortality and reduce the overall number of parameters in models, often resulting in more parsimonious models being fitted to the data.

There are now many examples of the use of covariates in mark-recapture-recovery work. For instance, Clobert and Lebreton (1985) provided a mark-recapture application for starlings and Catchpole et al. (1999) modelled the survival of Northern lapwings in terms of a range of different winter weather measures; they also considered possible covariate dependence of both adult survival probability and first-year survival probability, as well as the probability of reporting of a dead bird. A striking finding of the latter paper is the decline of reporting probability of dead birds over time. Barry et al. (2003) have included random effects by the simple addition of a random component to the right-hand side of the above equation, and then used a Bayesian analysis; this has already been mentioned in Example 1.3.

The only limits to the number and type of covariates to be included in models lie in the imagination of the ecologist and the availability of data. A popular covariate used in ecology is the North Atlantic Oscillation (NAO), which is a measure of the pressure gradient between Stykkisholmur, Iceland and Lisbon (Wilby et al. 1997, see `www.met.rdg.ac.uk/cag/NAO`). The average NAO over the winter months has been found to be a good measure of overall winter severity, and is used, for example, by King and Brooks (2003a) in their modelling of the survival of Soay sheep – see Example 1.4. Food availability can be an important determinant of survival, and recent declines in the survival of British seabirds may be attributed to a corresponding decline in customary foodstocks. This can have a consequent knock-on for conservation

COVARIATES AND LARGE DATA SETS; SENESCENCE   31

if the birds in question then prey on other seabirds in order to survive. In some cases it can be difficult to obtain appropriate measures of a desired covariate. It may also be necessary to consider lagged covariates, if it is anticipated that covariates might have a delayed effect. As a general rule, it is a good idea to undertake a sensible preliminary selection of covariates before starting an analysis.

Age may be an important covariate. Two prime examples result from long-term studies of Soay sheep encountered earlier in Example 1.4 and red deer, *Cervus elaphus*. Statistical analyses of mark-recapture-recovery data from these studies are given in Catchpole et al. (2000) and Coulson et al. (2001), for the sheep, and Catchpole et al. (2004) for the deer. In these papers the authors proceed incrementally, by first of all determining appropriate age classes, within which survival does not vary with age, with the exception of the oldest class for the deer, when it was possible to model senescence explicitly as a logistic function of age. To these age dependent models, the authors then added covariates, taking each age class separately. The covariates in this case were not just environmental, but also included population density, individual covariates such as coat type, horn type, genetic covariates and birth weight.

In both of these applications the regressions are logistic, as was true also for the original grey herons analysis. Different link functions could be considered, and it is possible that if there were enough data, then there could be a preference for using one link rather than another – see, for example, the discussion in Morgan (1992, Chapter 4). While much of this work could be done in the computer package MARK, which we discuss in Section 3.6.1, (White and Burnham 1999, see: www.cnr.colostate.edu/~ gwhite/), the authors used the MATLAB programs developed by Ted Catchpole specifically for the purpose, and available from http://condor.pems.adfa.edu.au/~s7301915/Emu/. The following example illustrates the results from two such analyses.

Example 2.5. Red Deer on Rum

Red deer studied on the Scottish island of Rum have detailed life history data, dating from 1971. The age dependent survival probability graph in this case is shown in Figure 2.3. Shown are results for both males and females, as well as two fitted models. In one model there is a separate survival probability for each age, whereas in the other model a particular structure, involving both age groups and a logistic regression for senescence, is assumed. It is interesting that although the two sexes were analysed separately, senescence appears to occur at the same age for males and females. Males senesce faster than females. The evidence for senescence is stronger for deer than for sheep, as deer live longer overall. Once again, as was also true for Soay sheep in Example 1.4, the highest age category includes all deer of that age and older.

After the introduction and selection of a range of covariates, a model for female deer that includes age and covariate regressions is given below:

$$\phi_1(P+N+B),\ \phi_2,\ \phi_{3:8}(R),\ \phi_{9+}(age+N+R)\ /\ \nu(P+Y)\ /\ \lambda(t).$$

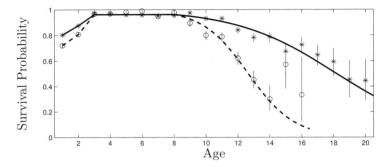

Figure 2.3 Plot of estimated probability of annual survival for male (dashed line) and female deer (solid line), as a function of age, plotted against age. Also shown are estimates assuming a separate survival probability for each age for males (○) and females (*) with associated one standard error above and below the point estimate (on a logistic scale). Reprinted with permission from *The Journal of Agricultural, Biological, and Environmental Statistics*. Copyright 2004 by the American Statistical Association. All rights reserved. Published in Catchpole et al. (2004, p. 9).

This population is so well studied that, unusually, the recapture probability p is unity, and so it does not feature in the model. The interpretation of this model is that there is an annual survival probability, ϕ_1, for animals in their first year of life, and this is logistically regressed on P, the population size, and N, the average winter NAO index, as well as on birth weight B. A separate, constant probability, ϕ_2, of annual survival applies to animals in their second year of life. Animals aged 2 to 7 share the same annual survival probability $\phi_{3:8}$, which is logistically regressed on the binary covariate R, denoting the breeding status of the hind. Animals aged at least 8 years have an annual survival probability ϕ_{9+} which is a logistic function of the age of the animal, NAO index, and reproductive status. In addition, the model allows for an annual dispersal probability ν, which is a logistic function of the population density and a measure Y of the location of the animal. The reporting probability of dead animals is a general time varying parameter, $\lambda(t)$, which is not related to any covariates. The corresponding model for the stags is similar, but differences include, for instance, an interaction between P and N in the regression for ϕ_1, suggesting that a combination of poor weather and high population density is especially life-threatening for animals in their first year of life.

□

It is interesting to note that, for the Soay sheep and red deer data, it is possible to include density dependence in addition to other types of covariate. This has also been done by Jamieson and Brooks (2004b), who present a Bayesian analysis of North American census data on 10 species of ducks. They show that in those cases inclusion of national counts of the numbers of

May ponds in their state-space modelling improves the estimation of density dependence effects.

The integrated modelling of Besbeas et al. (2002), described in Section 2.6, demonstrates how combining information from different studies may increase the significance of covariates in logistic regressions.

2.6 Combining Information

Biologists often have multiple sources of information about a species that initially may appear to be rather separate, but that can be combined to provide more precise estimators of survival. For example, sometimes time series of population size estimates exist along with resighting histories.

An appealing way to combine information is through the Bayesian paradigm, discussed in Chapter 4. An example is provided by Thomas et al. (2005), who analyse data on the annual census of grey seal, *Halichoerus grypus*, pups in Scotland. Another obvious way to combine information is through the use of suitable covariates, as described in Section 2.5. This is made easier by the publication of relevant data sets on the World Wide Web.

2.6.1 Same Animals, but Different Data

As well as combining information from different studies, one may be able to perform integrated analyses of mark-recapture-recovery-resighting data measured on the same animals, and relevant work is given by Burnham (1993), Barker (1997) and Barker (1999). Catchpole et al. (1998) analysed integrated mark-recapture-recovery data to produce a general likelihood which demonstrated the existence of sufficient data matrices, containing all the information needed for modelling and analysis, and the result has already been seen earlier, in Equation (2.4). They also considered the effects of analysing different subsets of the data separately, and Table 2.5 presents their results for an illustrative analysis of the survival of shags. We can see from this table that, in this example, the recapture data are more informative than the recovery data for the estimation of the adult annual survival probability ϕ_a, for which the data from ringing adults are also more informative than the data from marking the pulli (a term used to denote nestlings) alone. However, we note that we need to mark the pulli (i.e. birds in their first year of life) in order to estimate the juvenile survival probability ϕ_{imm}.

2.6.2 Different Animals and Different Data: State-Space Models

Many countries undertake regular national censuses of wild animals, without considering how the resulting information might be analysed to estimate survival. Additionally, much national effort is devoted to investigating productivity. For birds, for example, there is a range of methods available; see the papers by Pollock and Cornelius (1988), Heisey and Nordheim (1990), Heisey

Table 2.5 Parameter Estimates for the Shag Data Set

	Parameter	Recaptures (live)	Recoveries (dead)	Recaptures and Recoveries
Pulli	ϕ_{imm}	0.602 (0.026)	0.769 (0.087)	0.712 (0.026)
	ϕ_a	0.873 (0.034)	0.866 (0.093)	0.879 (0.020)
Adults	ϕ_a	0.838 (0.017)	0.916 (0.092)	0.843 (0.016)
Pulli and adults	ϕ_{imm}	0.661 (0.196)	0.822 (0.042)	0.698 (0.021)
	ϕ_a	0.864 (0.014)	0.918 (0.024)	0.866 (0.012)

Source: Reproduced with permission from Catchpole et al. (1998, p. 41) published by the International Biometric Society.

Note: The table provides maximum-likelihood estimates, with estimated asymptotic standard errors in parentheses, of annual survival probabilities ϕ_{imm}, for the 2nd and 3rd years of life, and ϕ_a, from the 4th year onwards. The columns show estimates from the live recaptures only, from the dead recoveries only, and from both. The rows show data from birds ringed as pulli only (ringed as young birds after hatching), from birds ringed as adults only, and from both.

and Nordheim (1995) and Aebischer (1999). Various authors have shown how one can model census data, using state-space models based on Leslie matrices which include parameters for productivity as well as survival (for example Millar and Meyer 2000; Newman 1998; Besbeas et al. 2003). An illustration is given below from this last paper.

A state-space model comprises two stochastic equations, a transition equation, relating the evolution of the underlying process from time $t-1$ to time t, and a measurement equation, relating what is measured to the underlying process. A typical transition equation is

$$\begin{pmatrix} N_1 \\ N_a \end{pmatrix}_t = \begin{pmatrix} 0 & \rho\phi_1 \\ \phi_a & \phi_a \end{pmatrix} \begin{pmatrix} N_1 \\ N_a \end{pmatrix}_{t-1} + \begin{pmatrix} \epsilon_1 \\ \epsilon_a \end{pmatrix}_t, \qquad (2.7)$$

where N_1 and N_a denote the numbers of one-year-old female birds and (adult) female birds aged $\geq$ 2 years, respectively. Similarly ϕ_1 and ϕ_a denote the annual survival probabilities of birds in their first year of life and of older birds, respectively. The parameter ρ here denotes the annual productivity of females per female, and the ϵ terms are errors, with variances which are given by suitable Poisson and binomial expressions. Thus $\mathrm{Var}(N_{1,t}) = N_{a,t-1}\rho\phi$

MODELLING PRODUCTIVITY

and $\text{Var}(N_{a,t}) = (N_{1,t-1} + N_{a,t-1})\phi_a(1 - \phi_a)$. We use binomial distributions when we want to represent survival or death, and Poisson distributions in order to represent birth. In both of these cases we assume that there is no overdispersion, though the model could be made more complex by the addition of overdispersion in either or both of these instances. We assume no sex effect on survival and that breeding starts at age 2. The matrix above is an example of a Leslie matrix. If we observe (with error) only the number of adults, then the measurement equation is,

$$y_t = \begin{pmatrix} 0 & 1 \end{pmatrix} \times \begin{pmatrix} N_{1,t} \\ N_{a,t} \end{pmatrix} + \eta_t.$$

One possibility is to assume that the observation error η_t is normally distributed, with zero mean and constant variance.

It is often the case in practice that the binomial and Poisson distributions of the state-space model can be well approximated by normal distributions. If normal approximations are made to the binomial and Poisson distributions, then conveniently the Kalman filter can be used to provide a likelihood. This likelihood is parameter redundant, due to the fact that the parameters ρ and ϕ_1 only occur together, as a product, and Besbeas et al. (2003) combine the likelihood with a corresponding likelihood for ring-recovery data in order to provide an integrated analysis in which productivity can be estimated. This work has recently been extended to include models for productivity. In a combined analysis, it is assumed that different component data sets are independent, so that likelihoods may be multiplied together. We provide a Bayesian analysis of state-space data in Chapter 10.

2.7 Modelling Productivity

The annual productivity of wild animals is generally more easily estimated than either survival or movement. For mammals, for example, one can count numbers of young born in the spring, and for birds one can observe clutch sizes, the sizes of resulting broods that hatch and fledging rates. Many bird species have characteristic clutch sizes that show little variation over time. For example, Cory's shearwaters, *Calonectris diomedea*, lay just one egg, whereas a blue tit (*Parus caeruleus*) can have a clutch size as large as 17. However, there can be complications. For example, account may need to be taken of animals skipping breeding in certain years, depending on a variety of conditions; some birds may have multiple broods, perhaps depending on the availability of food, or the outcome of earlier broods in the same season; certain animals may sometimes have twinned offspring. An illustration of an interesting data set on productivity is the one obtained by J. D. Nichols and colleagues, and modelled in Pollock and Cornelius (1988).

Example 2.6. Mourning Dove

In this case the target species is the mourning dove, *Zenaida macroura*, studied in Maryland. The mourning dove is similar in appearance to the extinct Passenger Pigeon, but smaller. In a particular area, 59 nests were located, each containing either 1 or 2 eggs. Future observation of these nests revealed that 25 of the nests were successful in producing nestlings; 11 corresponded to nests first located no more than 8 days after the eggs were laid; 8 were found when the eggs were 9 to 16 days old, and 6 were found when at least 17 days old. The remaining 34 nests failed, and did not produce any nestlings; 16 failed within 8 days of discovery, 9 failed within 9 to 16 days after discovery, and 9 after the 16th day. Typically, nests are not located until after incubation of the eggs has started, so that the located nests are of unknown "age." However, a useful feature is that the combined egg incubation and fledging time is essentially constant for mourning doves. Thus, for a nest that succeeds, one can work back from the time of success in order to gauge the time of initiation, whereas for failed nests, times of initiation are unknown.

□

2.8 Parameter Redundancy

In Section 2.3.1 we saw that probability models in ecology can involve a transformation from a set of well-defined parameters, such as multinomial probabilities $\{\theta_j\}$, to a derived set involving parameters of interest. When we build models involving such parameters of interest, such as age and time varying survival probabilities, then we might be guided by ecological considerations (such as the timing of features such as a flood, or the fact of high mortality in the early years of life), and possibly include too many parameters in the model. A simple example of this could occur with recapture data, when it might not be possible to estimate separately probabilities of survival and site fidelity, as discussed in Section 2.4. There is nothing to stop us forming likelihoods that contain more parameters than can be estimated from data, but the likelihoods will be flat, and will not have a single point in the parameter space that will uniquely maximise the likelihood. As we saw in Section 2.6.2, and as we shall see also below, there are very simple examples when it can be spotted immediately, from looking at the multinomial probabilities, that it is not possible to estimate all of the model probabilities. However, there are many cases when it is not clear whether one can or cannot estimate all the parameters of a model. Also, even if one can estimate all model parameters in principle, in practice missing data, such as a zero in an m-array, can mean that in practice not all parameters can be estimated for that particular data set. As we shall discuss in Chapter 3, in classical model selection, information criteria, are often used to select the best classical model for a data set, and in order to construct information criteria we need to know how many parameters have been estimated from a data set. We start with a simple example.

PARAMETER REDUNDANCY

Example 2.7. The Cormack-Jolly-Seber Model
The Cormack-Jolly-Seber model (hereafter denoted CJS) extends the model for capture-recapture data in Section 2.3.2, by allowing both survival and recapture probabilities to vary with time. There is no age variation of parameters in this model, and it might therefore be useful for situations in which animals have been marked as adults of unknown age. In practice, one might well consider such time variation to be potentially important, and worth investigating, and indeed the CJS model is an important tool in the armoury of the statistical ecologist. The multinomial probabilities for a study involving 3 years of marking and 3 years of recovery are shown in Table 2.6(a); for each row we omit the final multinomial probability, which corresponds to the animals that are not recaptured during the study. We now adopt the notation that ϕ_i is the probability that an animal survives from occasion i to occasion $i+1$, and p_i is the probability that an animal is recaptured, if alive, on occasion i.

Table 2.6 Multinomial Cell Probabilities for (a) Cormack-Jolly-Seber Model for Capture-Recapture Data and (b) Cormack-Seber Model for Capture-Recovery Data

(a)

Release Occasion	Recapture Occasion		
	2	3	4
1	$\phi_1 p_2$	$\phi_1(1-p_2)\phi_2 p_3$	$\phi_1(1-p_2)\phi_2(1-p_3)\phi_3 p_4$
2		$\phi_2 p_3$	$\phi_2(1-p_3)\phi_3 p_4$
3			$\phi_3 p_4$

(b)

Year of Birth	Year of Recovery		
	1–2	2–3	3–4
1	$(1-\phi_1)\lambda$	$\phi_1(1-\phi_2)\lambda$	$\phi_1\phi_2(1-\phi_3)\lambda$
2		$(1-\phi_1)\lambda$	$\phi_1(1-\phi_2)\lambda$
3			$(1-\phi_1)\lambda$

We can see immediately that the multinomial probabilities (and hence the likelihood - equivalent to that given in Equation 2.5) can be represented by a parameter set that is smaller by one than the original parameter set, since the parameters ϕ_3 and p_4 occur only as the product $\phi_3 p_4$, and so they can be replaced by a single parameter $\beta = \phi_3 p_4$, say. The likelihood can now be maximised to obtain unique maximum-likelihood estimates of members of the new parameter set (including β). What we have observed and done here is not a feature of there having been 3 years of release and recapture.

Exactly the same result applies if the number of years of study increases to $k > 3$, but does so for both release and recapture simultaneously. It would have been desirable to have estimated the parameters ϕ_k and p_{k+1}, but without additional information we cannot. It is, however, useful to be able to estimate all of the other parameters in the original parameter set. What makes this an interesting area for study, which also has repercussions for Bayesian analysis, is that slight changes to the experimental paradigm can change the picture. For example, if there were only 2 years of release but still 3 years of recaptures, so that the multinomial probabilities comprise just the first two rows of Table 2.6(a), how many parameters can we now estimate in principle?

□

The paper by Catchpole and Morgan (1997) introduced the idea of parameter redundancy: a model based on a parameter vector $\boldsymbol{\theta}$ of dimension p is said to be *parameter redundant*, with deficiency d, if it can be reparameterised in terms of a vector of dimension $p - d$. Another simple, but subtler, example of parameter redundancy occurs in the Cormack-Seber model, which was developed independently by Cormack (1970) and Seber (1971) for ring-recovery data.

Example 2.8. Cormack-Seber Model
In this model there is a constant reporting probability λ, and fully age dependent survival probabilities, with ϕ_i now being the probability of surviving from age $i-1$ to age i. Unlike the CJS model, there is no time variation in this model, just age variation in survival. If the animals were ringed when young, shortly after birth, then that could motivate the use of this model, because of the need for at least some age variation in the set of survival probabilities. If there are 3 years of release of marked animals, and 3 subsequent years of recovery, then the probabilities of recovery in year j, given release on occasion i, are given by the $(i,j)^{th}$ entry of Table 2.6(b). As with the illustration of the CJS model probabilities, we do not present the probabilities corresponding to animals that have not been recovered dead during the study.

This model has 4 parameters; however, we cannot estimate them all because if we look carefully at the multinomial probabilities, then we can see that the data can be rewritten in terms of the new parameters $\beta_1 = (1-\phi_1)\lambda$, $\beta_2 = \phi_1(1-\phi_2)\lambda$, and $\beta_3 = \phi_1\phi_2(1-\phi_3)\lambda$. Thus the deficiency is 1. As with the CJS model when the number of years of release equals the number of years of study, the deficiency remains 1 for a k-year study, for any $k > 3$, as is easily checked. However, a clear difference, compared with the CJS model, is that now *none* of the original parameters are estimable.

□

For a given model we need to answer the questions: how many estimable parameters are there, and which of the parameters can we estimate? Catchpole and Morgan (1997) developed a technique for answering these questions, and we illustrate their method by considering again the CJS model comprising the first two rows of Table 2.6(a). A slightly modified version of their

PARAMETER REDUNDANCY

technique requires the calculation of a derivative matrix D, with $(i,j)^{th}$ entry $\theta_i \partial \log \mu_j / \partial \theta_i$, where θ_i is the ith parameter and μ_j is the jth overall multinomial cell probability, both of which can be taken in any order. When there are no missing data, or no cells with zero entries, it is not necessary to include in the differentiation multinomial cell probabilities corresponding to missing birds, that have not been recaptured during the study; see Catchpole and Morgan (1997). Here we have $\boldsymbol{\theta} = (\phi_1, \phi_2, \phi_3, p_2, p_3, p_4)$ and

$$\boldsymbol{\mu} = \{\phi_1 p_2,\ \phi_1(1-p_2)\phi_2 p_3,\ \phi_1(1-p_2)\phi_2(1-p_3)\phi_3 p_4,\ \phi_2 p_3,\ \phi_2(1-p_3)\phi_3 p_4\}.$$

Thus, we obtain,

$$D = \begin{bmatrix} 1 & 1 & 1 & 0 & 0 \\ 0 & 1 & 1 & 1 & 1 \\ 0 & 0 & 1 & 0 & 1 \\ 1 & -p_2^* & -p_2^* & 0 & 0 \\ 0 & 1 & -p_3^* & 1 & -p_3^* \\ 0 & 0 & 1 & 0 & 1 \end{bmatrix}$$

where we denote $p^* = p/(1-p)$ for any probability p. A computer algebra package such as *Maple* or *Mathematica* may be used both for the above differentiation and for each of the following steps. (For examples of Maple code see Catchpole et al. (2002); www.ma.adfa.edu.au/eac/Redundancy/Maple/ and ftp://ftp.ceef.cnrs-mop.fr/biom/PRM). Next we find the symbolic row rank (that is the number of independent rows) of D. The result is 4, two less than the number of rows, and so the deficiency is now 2. (We can see this from inspection of D: the 3rd and last rows are identical, and we can obtain the 2nd row by suitably combining rows 5 and 6.) Thus leaving off the last year of release has resulted in the deficiency increasing from 1 to 2.

To find out which parameters are estimable, Catchpole et al. (1998) show that it is sufficient to solve the equation $\boldsymbol{\alpha}^T D = 0$. The solution is

$$\boldsymbol{\alpha} = a\,(0, -(1-p_3), 1, 0, -(1-p_3), 0) + b\,(0, -(1-p_3), 0, 0, -(1-p_3), 1)$$

where a and b are arbitrary. According to the theory in Catchpole et al. (2002), since $\boldsymbol{\alpha}$ has zero entries only in the first and fourth positions, then the only estimable parameters, from amongst the original set, are the first and fourth, corresponding to the order in which the differentiation took place in forming D. Thus we can only estimate the original parameters ϕ_1 and p_2. The full set of estimable quantities can be found by solving a system of differential equations (see Catchpole et al. 1998 and Gimenez et al. 2003); Maple code can be obtained for solving these equations can be downloaded from ftp://ftp.ceef.cnrs-mop.fr/biom/PRM), although in this particular case it is easy in this example to see by inspection that the other two estimable quantities are $\beta_1 = \phi_2 p_3$ and $\beta_2 = \phi_2(1-p_3)\phi_3 p_4$.

The work of this section deals exclusively with classical maximum-likelihood estimation; however, it is will be important also for Bayesian analysis, as we shall see in later chapters.

2.9 Summary

Ecological data are collected in a wide variety of ways. Censusing methods provide information on population trends, often without any physical interaction with the animals concerned. By contrast, information on the survival of wild animals is often obtained through observations of previously marked individuals, making the assumption that the marking does not interfere with behaviour. A substantial body of statistical methodology has been developed over the past 100 years for analysing data collected on marked animals. A relatively modern development extends existing methods to cope with animals that may move between different states and sites. Also of current interest is the combination of data from different sources. It is possible to estimate population sizes as well as survival, although population size estimation is not robust with respect to violations of assumptions made in the modelling. Likelihood formation is fundamental to inference in both the classical and Bayesian frameworks. As models become more complex and realistic it is important to be able to check that likelihood surfaces do not possess flat regions, such as ridges, which is the result of parameter redundancy. This can be done using symbolic algebra.

2.10 Further Reading

With regard to data collection, we can distinguish among national surveys, which rely on the work of large numbers of volunteers, and small-scale scientific studies of communities which are not widely distributed, by dedicated observers. An example of a small-scale scientific study is to be found in Catchpole et al. (2001), who analysed mark-recapture-recovery data on abalone, *Haliotis rubra*, and modelled survival as a function of size, estimated from a fitted growth curve. They also extended the standard ring-recovery methodology to allow for the possibility that the shells from dead animals might be recovered substantially later than the time of death.

Regarding the estimation of survival, analogies can be drawn with life-table analysis and other methods that are used in the analysis of human populations; for example, see Cox and Oakes (1984). What is different about modelling the survival of wild animals, as compared with modelling survival in human populations, is the difficulty of keeping track of marked animals in the wild. Even with Soay sheep, for example, involving the detailed study of a relatively small population in a restricted area, it is still difficult to account for all deaths, and possible to miss seeing live animals. Radio-tracking data can correspond closely to human life-time data, as long as batteries do not fail and have reasonably long life times. Cost is a limiting factor in radio-tracking work, and there remains the possibility that radio-marking can interfere with the behaviour of wild animals, and then bias estimation of their demographic rates. Because the life styles of wild animals are closely regulated by the seasons of the year, it is natural to consider annual measures of survival, productivity and movement, though in some cases data are available to allow

FURTHER READING

modelling on a finer time scale. The long-term behaviour of some wild animals is determined by attempts to optimise lifetime productivity. For example, in terms of optimising life-time breeding success it may be advantageous for animals to skip breeding in some years, rather than breed every year. We shall not consider detailed models for such processes, but we will refer to these features when appropriate.

Likelihoods may be more complex than those considered in this chapter. See for example Morgan (2008) and the references therein. The work of Besbeas et al. (2003) demonstrated how one can use multivariate normal approximations in order to form combined likelihoods. This can greatly simplify the combination different sources of information, and provides a simple prescription for the reporting of model-fitting, in order that results might, in the future, be readily included in such an integrated modelling exercise. The work of Brooks et al. (2004) has provided a flexible Bayesian analysis for combined data, and demonstrated the robustness of the normal approximation approach to the presence of small counts. Besbeas et al. (2005) have used the combined modelling technique to investigate the effects of habitat on productivity for Northern lapwings.

The paper by Brownie et al. (1985) showed that the parameters of models for recoveries with age dependence in both survival and reporting probabilities were estimable only if both young and adults were marked and released. Mixing recoveries and recaptures of animals marked as young is one way to achieve appropriate data in this instance, since the release after live recapture of birds after their first year will correspond to data from birds marked as adults. Note also that in similar vein, Freeman et al. (1992) investigated the combination of ring-recovery data with information on survival obtained from radio-marking in order to obtain estimable parameters. With regard to heterogeneity of recapture probability, Figure 6.4, on p. 115 of Borchers et al. (2002) provides a graphical demonstration of negative bias due to heterogeneity.

Questions of parameter redundancy first appeared in linear models, in the context of confounding; see for example Fisher (1935) and Yates (1937); in factor analysis, see for example Thurstone (1947); and in structural relations, see for example Reiersøl (1950). The first general treatment was by Rothenberg (1971). Other applications occur in compartment models (Bellman and Åström 1970), contingency tables (Goodman 1974) and Bayesian networks (Geiger et al. 1996). A natural question to ask is whether one can extend parameter redundancy results, such as those for the CJS model, to larger studies with more years of data? Catchpole and Morgan (1997) and Catchpole and Morgan (2001) provide a simple method for extending results from small-scale examples to arbitrarily large studies. This permits general results to be found for classes of models. See also Catchpole et al. (1996), who prove general theorems.

We have dealt above exclusively with parameter redundancy for classical maximum-likelihood estimation. Bayesian analyses are usually concerned with posterior means rather than modes and unique parameter estimates are obtained even for parameter-redundant models (see Carlin and Louis 1996, p. 203 and Brooks et al. 2000b), though prior sensitivity is often a problem in such cases. Carlin and Louis (1996) also warn that in such cases there are often high correlations between the parameters, which cause univariate marginal posterior summaries to be misleading. We return to a discussion of such issues in Chapters 5 and 6. As explained above, model deficiency can also be estimated numerically, for any given model/data combination, by evaluating the eigenvalues of an appropriate matrix evaluated at the maximum-likelihood estimate. This method is used in, for example, program MARK. One problem with this method is determining when an eigenvalue that is almost zero should be counted as zero. See Viallefont et al. (1998) for more details. The computer package M-SURGE uses the Catchpole-Morgan method, calculating the derivative matrix D numerically, in several points in a neighbourhood of the maximum-likelihood estimate. This has been found to work much better than using the Hessian. Other methods are discussed in Gimenez et al. (2004).

2.11 Exercises

2.1 In the notation of Section 2.2.4, show that the maximum-likelihood estimate of the number of animals in a region, N, is given approximately by: $\hat{N} = \frac{n_1 n_2}{m_2}$.

2.2 The data of Table 2.7 are simulated Schnabel census data. For the simulated data of Table 2.7(a), we obtain the maximum-likelihood estimate, $\hat{p} = 0.28$. For the simulated data of Table 2.7(b), we get $\hat{p} = 0.31$. Discuss whether you think that these are sensible estimates. For case (b), we can fit a mixture model in which $p = \alpha p_1 + (1-\alpha)p_2$, to correspond to the data-generating mechanism. In this case the likelihood becomes a function of the three parameters, α, p_1 and p_2). When we maximise this three-parameter likelihood, then we get the maximum-likelihood estimates, $\hat{p_1} = 0.09, \hat{p_2} = 0.40$, and $\hat{\alpha} = 0.45$. Discuss these results.

For the data in Table 2.7(b), if one uses a simple binomial model, then the maximum-likelihood estimate of N is given by $\hat{N} = 102$ (96.9, 109.9), while for the model which is a mixture of two binomial distributions, the estimate of N is $\hat{N} = 132$ (101.3, 62139). The values in parentheses after each of these estimates in this case correspond to 95% profile likelihood confidence intervals for N; see Section 3.5. Discuss these results.

2.3 Extend the single-cohort likelihoods of Equations (2.2) and (2.4) to the case of the release of several cohorts of marked animals, one for each of several years.

2.4 Given in Table 2.8 are a subset of grey heron ring-recovery data for the years 1975 through 1985, supplied by the British Trust for Ornithology. If the data are described by a model with the three parameters, ϕ_1, ϕ_a, λ, by

EXERCISES

Table 2.7 Two Sets of Simulated Schnabel Census Data

(a)

Occasion	1	2	3	4	5	6	7
Caught	32	40	35	42	23	41	31
New	32	30	17	12	8	8	6
Recaptures	0	10	18	30	15	33	25
$\{f_j\}$	36	39	24	12	2	0	0

(b)

Occasion	1	2	3	4	5	6	7
Caught	38	31	32	27	31	32	34
New	38	20	12	2	9	7	10
Recaptures	0	11	20	25	22	25	24
$\{f_j\}$	34	21	23	10	5	2	0

Source: This table was published in Williams et al. (2002, pp. 308, 309), copyright Elsevier.

Note: In case (a) the data are simulated under the assumption that all 120 animals in the population share the same capture probability of $p = 0.3$, whereas in case (b), half of the animals have recapture probability $p = 0.15$, and the other half have recapture probability $p = 0.4$. The first three rows refer, in order, to the total number of animals caught at each occasion, the animals that are caught for the first time at each occasion, and the animals that are caught that have been caught before. The second and third row totals clearly sum to give the first row. The last row gives the $\{f_j\}$, corresponding to the number of animals caught j times.

considering the geometric progression of cell probabilities with regard to ϕ_a, working across rows, devise an estimate of ϕ_a.

The data of Table 2.9 give average central England temperatures (Centigrade) for December, January, February and March in each bird year (corresponding to April–March). Propose a way of regressing the model parameters on these 4 weather covariates.

2.5 The lizard, *Lacerta vivipara*, has been observed on one of three sites, denoted by the integers 1 through 3, for 6 time periods. Thus, in this instance, there are only live recaptures (no recoveries of dead animals), and when

Table 2.8 Recovery Data of Grey Herons from a British National Ringing Study

Year of Release	Number Ringed	Year of Recovery – 1900+									
		76	77	78	79	80	81	82	83	84	85	86
1975	450	29	7	3	0	1	0	0	1	1	0	1
1976	589	0	50	7	4	1	1	0	1	1	0	0
1977	464	0	0	32	13	4	2	1	1	0	0	0
1978	237	0	0	0	20	4	0	2	2	0	0	0
1979	434	0	0	0	0	32	13	12	0	3	4	0
1980	648	0	0	0	0	0	35	13	7	0	2	1
1981	740	0	0	0	0	0	0	43	5	6	1	0
1982	1014	0	0	0	0	0	0	0	41	12	3	4
1983	777	0	0	0	0	0	0	0	0	31	7	7
1984	1100	0	0	0	0	0	0	0	0	0	63	21
1985	814	0	0	0	0	0	0	0	0	0	0	49

Source: Reproduced with permission from Freeman and North (1990, p. 141) published by the Polish Zoological Society.

Table 2.9 Average Central England Temperatures for the Months December, January, February and March in Each Bird Year (April-March)

Bird Year	Dec	Jan	Feb	Mar
1975	5.3	5.9	4.5	4.8
1976	2.0	2.8	5.2	6.9
1977	6.1	3.4	2.8	6.7
1978	3.9	−0.4	1.2	4.7
1979	5.8	2.3	5.7	4.7
1980	5.6	4.9	3.0	7.9
1981	0.3	2.7	4.8	6.1
1982	4.4	6.8	1.8	6.5
1983	5.7	3.9	3.4	4.8
1984	5.3	0.9	2.2	4.8
1985	6.3	3.6	−1.0	4.9

there is a recapture, then that is denoted by the number corresponding to the site on which the lizard is seen; a 0 indicates that the individual was not observed at that time. The data shown in Table 2.10 have been *parsed*, so that next to each observed life history is a number in parentheses, indicating the number of times that life history occurs in the data set. Discuss

EXERCISES

Table 2.10 Lizard Life History Data, Provided by Jerome Dupuis

101000 (1)	101100 (1)	110000 (9)	110100 (4)	111000 (1)	111001 (1)
111100 (3)	111101 (1)	111110 (1)	111111 (1)	120000 (9)	200222 (1)
202002 (1)	202200 (1)	210200 (1)	211200 (1)	220000 (9)	220220 (1)
221100 (1)	222000 (1)	222200 (5)	222300 (5)	223200 (1)	233300 (1)
302200 (2)	303000 (1)	332000 (3)	330300 (1)	330303 (1)	330000 (14)
333000 (1)	333300 (5)	333303 (1)	332000 (2)		

the values of parsing in this way. Provide an evaluation of the parsed data set.

2.6 The data in Table 2.11 give the population figures for red deer stags on the island of Rum for each year from 1957 to 1965 (data supplied by Dennis Mollison). The figures were estimated by examining the teeth of all deer that died from 1957 up to June 1966, and thus only include deer which died before June 1966. As most deer are born around May to June, it is relatively easy to allocate them to distinct age groups. Thus we can use this table to follow each cohort: for instance, of the 133 stag calves from 1957, 104 were still alive the following summer. Provide an analysis of the data.

2.7 A model for what are called *timed species count data* has the multinomial probabilities given below, where ν is the probability that the species is present, and $(1-p)$ is the probability that a bird species which is present is detected in each 10-minute interval. Here n_i denotes the number of surveys when a bird is first encountered in the $(7-i)th$ time interval, for $i = 1, \ldots, 6$, and n_0 is the number of surveys when a bird is never encountered.

n_6	n_5	n_4	n_3	n_2	n_1	n_0
$\nu(1-p)$	$\nu(1-p)p$	$\nu(1-p)p^2$	$\nu(1-p)p^3$	$\nu(1-p)p^4$	$\nu(1-p)p^5$	$\nu p^6 + (1-\nu)$

Write down the likelihood. Maximum-likelihood parameter estimates result from maximising the likelihood as a function of the two parameters, p and ν, using numerical optimisation, and searching the two-dimensional parameter space for the maximum. An alternative approach is to reparameterise. Show that when the model is reparameterised in terms of $\theta = \nu(1-p^6)$ and p then there exists an explicit maximum-likelihood estimator for θ. Interpret θ, and provide its estimates for the data above. Use this result to explain why the expected frequency corresponding to n_0 equals the observed frequency.

Table 2.11 Population Figures for Red Deer Stags

Age	1957	1958	1959	1960	Year 1961	1962	1963	1964	1965
0	133	136	111	78	94	39	25	16	12
1	107	104	128	103	65	44	24	14	2
2	75	98	99	124	97	54	38	23	11
3	79	74	96	98	118	80	44	33	19
4	70	78	71	96	92	108	64	36	28
5	86	69	73	66	80	74	69	41	20
6	78	82	67	52	47	46	53	46	17
7	79	66	62	45	40	36	22	29	9
8	59	71	55	46	32	34	24	16	10
9	30	42	39	29	29	23	25	17	2
10	11	20	25	23	21	18	14	15	4
11	9	8	11	13	14	8	11	9	6
12	9	5	3	3	6	3	3	4	0
13	4	4	2	0	2	3	2	2	0
14	6	2	4	1	0	1	3	2	0
15	2	4	2	4	0	0	1	3	0
16	1	1	1	1	3	0	0	1	3
17	0	0	1	1	0	1	0	0	0
18	0	0	0	0	0	0	1	0	0

2.8 One way to estimate population size, N, using data from the Schnabel census of Example 2.3, is to estimate the probabilities of recapture from the likelihood given in Equation (2.1), and then set

$$\hat{N} = \frac{\sum_{j=1}^{t} f_j}{1 - (1-p)^t}.$$

Discuss whether you think that this might be a sensible way to proceed.

2.9 In Exercise 2.2, a model is fitted to simulated data in which the probability of detection is written as a mixture of two binomial distributions. Give an illustration of when you think such a model would be appropriate. Write down the likelihood for the case of a general mixture, for k component distributions. An alternative approach to dealing with variation in detection probability is to suppose that it varies continuously over animals. Suggest a way of incorporating such variation in a model.

2.10 A multi-site recovery data set is provided by Schwarz et al. (1993), relating

Table 2.12 Numbers of Herrings Tagged and Recovered Off the West Coast of Vancouver Island, at Two Sites, N and S

| | | | Number of Tags Recovered ||||||||
| | | Number | 1946 || 1947 || 1948 || 1949 ||
Year	Stratum	Released	S	N	S	N	S	N	S	N
1946	S	14,921	120	26	69	12	0	4	3	0
	N	13,227	30	128	5	36	1	33	0	2
1947	S	21,763			1,117	106	15	92	53	3
	N	8,638			48	126	3	115	2	2
1948	S	14,798					39	96	78	5
	N	17,149					4	588	7	16
1949	S	10,686							197	3
	N	11,170							39	14

Source: Reproduced with permission from Schwarz et al. (1993, p. 186) published by the International Biometric Society.

to the British Columbia herring, *Clupea harengus pallasi*. Recovery data of herrings from two sites is given in Table 2.12.

Study the data and discuss how you would construct a suitable probability model.

2.11 Compare and contrast two different ways in which the "frost days" covariate of Figure 2.2 might be included as a covariate for the annual survival probability of grey herons.

CHAPTER 3

Classical Inference Based on Likelihood

3.1 Introduction

Although the classical and Bayesian paradigms are quite different, both take the likelihood as their starting point. As we have seen in the last chapter, classical inference proceeds by forming likelihoods, and regarding them as functions of the model parameters. The estimates of those parameters are then obtained by finding the values of the parameters that maximise the likelihoods. This typically requires numerical analysis methods in order to perform the optimisation. The reason for the enduring success of maximum-likelihood methods is because the estimators that result have excellent properties, which we shall discuss later. The Bayesian paradigm is quite different, as explained in Chapter 1. As we shall appreciate in more detail in Section 4.1, we form the posterior distribution of the parameters by effectively multiplying the likelihood by the prior distribution. In order to make appropriate comparisons between classical and Bayesian methods, we devote this chapter to aspects of likelihood optimisation, and the properties of the resulting estimators.

3.2 Simple Likelihoods

Every day, many statisticians use specialised computer packages for forming and maximising likelihoods in statistical ecology. Usually these likelihoods are functions of many parameters, and their optimisation therefore takes place in many dimensions, when it is impossible to visualise the likelihood surfaces that result. It is, however, useful to be able to see what likelihood surfaces look like in simple examples, and we shall do that in this chapter. We start by returning to the Schnabel likelihood of Equation (2.1).

Example 3.1. Schnabel Likelihood: Estimating p
We start with the Schnabel census discussed in Example 2.3. Here the likelihood is given below, as the function of the single recapture probability, p;

$$L(p|\boldsymbol{f}) = \prod_{j=1}^{t} \left\{ \frac{\binom{t}{j} p^j (1-p)^{t-j}}{(1-(1-p)^t)} \right\}^{f_j}.$$

For any data set resulting in the values $\{f_j\}$, we can plot the likelihood as a function of p. The resulting graphs are shown in Figure 3.1, for two of the data sets from Table 2.1.

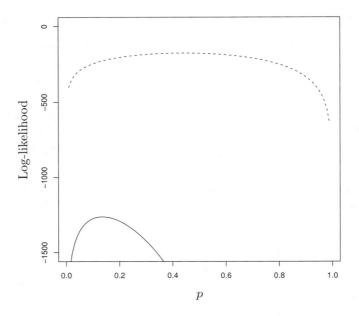

Figure 3.1 The log-likelihood graphs for two data sets from Table 2.1. The dotted line corresponds to the voles data set, and the solid line corresponds to the taxi-cab data set.

An attractive feature of both of the graphs in Figure 3.1 is that they indicate a unique maximum. Furthermore, the smaller data set has the flatter likelihood, suggesting that when data sets are small then it is more difficult to identify precisely the parameter estimates that will maximise the likelihood than if the data sets are larger. This point is simply formalised in maximum-likelihood theory, and we continue this discussion later in the chapter. Even though in this example we have an explicit expression for the maximum-likelihood estimate, it is still interesting to be able to see the shape of the log-likelihood.

□

Of course it is rare that we have a likelihood that is a function of a single parameter. The next step up in complexity is to consider a two-parameter likelihood, and therefore we now return to the capture-recapture model of Section 2.3.2.

Example 3.2. Capture-Recapture Likelihood
The log-likelihood in the simplest example of a capture-recapture model has just the two parameters, ϕ and p. In this case the likelihood is a surface which

MODEL SELECTION

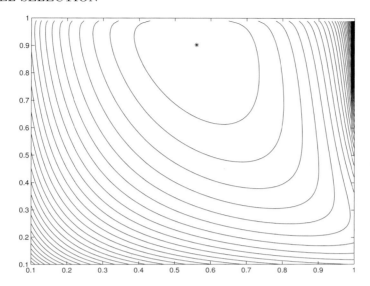

Figure 3.2 The contours of the log-likelihood surface when the two-parameter capture-recapture model is fitted to the dipper data of Table 1.1. The location of the optimum is denoted by *.

may be plotted as a function of the two parameters, ϕ and p, given in Figure 3.2 for the dipper data in Example 1.1. The maximum-likelihood parameter estimates are given by $\hat{p} = 0.5602$ and $\hat{\phi} = 0.9026$.

□

3.3 Model Selection

The model of Example 3.2 is only one of several that might be proposed for the dipper data; for instance, an alternative is the CJS model, with time variation in both of the parameters. We now need to decide on which models to fit as well as a notation for the various alternative models that might be fitted to the data. Then, having fitted the models to the data, we need to decide how to choose between them.

One way of doing this is in terms of information criteria, two of which are defined below:

$$\text{AIC} = -2 \log L_{max} + 2n_p,$$

$$\text{BIC} = -2 \log L_{max} + n_p \log(n),$$

where $\log L_{max}$ denotes the maximised log-likelihood, n_p is the number of parameters in a model, and n is the sample size. The first criterion is the Akaike information criterion (AIC), and the second is the Bayesian information criterion (BIC) - see Morgan (2008, p. 96).

As we shall discuss below, we can also compare models in terms of their maximised log-likelihoods, and so each of these information criteria may be regarded as ways of penalising models with large numbers of parameters: for any model, its maximised log-likelihood value may be made larger by the addition of more parameters, but the addition of too many parameters may result in both over fitting the data and loss of simplicity. Thus we can see that for each information criterion, we want to select a model with a low value for the information criterion, corresponding to a large maximised log-likelihood, achieved by means of a relatively small number of parameters. The difference between the two information criteria is that the BIC imposes a smaller penalty on the number of parameters than the AIC.

Example 3.3. Alternative Models for Schnabel Census Data on Hares

We illustrate the use of the AIC for model selection by considering five alternative models for the snowshoe hare data of Table 2.1. The alternative models are binomial (given in Section 2.2.4), beta-binomial, logistic-normal-binomial (LNB), a mixture of two binomial distributions (2 Bins) and a mixture of a binomial with a beta-binomial distribution (Bin + Beta-bin). These models are discussed in detail by Morgan and Ridout (2008). The estimate of the total number of snowshoe hares and corresponding AIC statistic for each model are provided in Table 3.1.

Table 3.1 Comparison of Models for the Snowshoe Hare Data of Table 2.1

Model	n_p	$-\log L_{max}$	χ^2	$\hat{N}$	se	AIC
Binomial	2	15.33	44.56	74.4	3.40	34.66
Beta-bin	3	12.55	5.09	90.8	16.10	31.1
LNB	3	12.42	-4.63	88.8	12.46	30.8
2 Bins	4	10.43	0.50	76.6	4.13	28.9
Bin + Beta-bin	5	10.41	0.42	77.0	4.56	30.8

Note: For the chi-square goodness-of-fit statistics, the number of cells is 6. Also shown are the number of parameters in each model, n_p, the maximised log-likelihood, the estimate of N, the population size, and the estimated standard error (se).

In this case the AIC suggests that the best model is the mixture of two binomials, though there is not a great difference between the AIC values for several models. See also Exercises 3.9 and 3.10.

□

MODEL SELECTION 53

3.3.1 Model Notation

Throughout the book we use a convenient shorthand notation for describing models. We have already described the CJS model, but more generally, for capture-recapture experiments involving animals ringed as adults of unknown age, following Brooks et al. (2000a), we describe recapture models by means of two components. The first component describes the recapture probabilities and the second the survival probabilities. Recapture and survival may be constant (C) or time dependent (T), and in addition, with particular reference to the dipper data of Table 1.1, we also consider a model with two constant survival probabilities, allowing the survival probability in flood years to differ from the survival probability in any other year; this is denoted by $C2$. We therefore consider the following five models for the dipper data: C/C, $C/C2$, T/C, C/T and T/T. Note that the T/T model is the CJS model already encountered in Chapter 2, and mentioned above.

Models for *ring-recovery* data obtained from adult animals may also be described by the same X/Y notation. In this case, the first letter describes the survival probabilities and the second letter the reporting probabilities. When ring-recovery data are obtained from animals ringed shortly after birth, models can contain age dependence, and nearly always require a separate description of first-year survival, compared with survival of later years, due to the typically high mortality of wild animals in their first year of life. The model notation that we use in this instance has three components: $X/Y/Z$, where X describes first-year survival, Y describes adult survival, and Z describes the reporting probability of dead animals. As above, we can use C and T, but in addition we may use Ak to denote k age classes of adult survival or of reporting probability. Thus for example the model of Section 2.3.1, with the three parameters, λ, ϕ_1 and ϕ_a can be denoted by $C/C/C$. This notation was introduced by Catchpole et al. (1996), and is readily extended, for instance to the case of data arising from recovery and recapture studies.

Example 3.4. Choosing a Best Model for the Dipper Data Using Information Criteria

Selecting an initial set of models is something which would benefit from biological knowledge, and in this case the primary biological input is that there was a flood in 1983 which might have affected the probability of surviving the two years, 1982–1983 and 1983–1984.

We show in Table 3.2 the maximised log-likelihoods $\log L_{max}$ for each of these models, and also the AIC, as well as the BIC criterion. Also shown in Table 3.2 are the ΔAIC and ΔBIC values: the ΔAIC is just the AIC minus its smallest value for the set of models considered, and this aids model selection using the AIC, as the best model is readily identified as the one with a zero ΔAIC, and similarly for the BIC. There is some evidence, resulting from specific simulation studies, that favours using the BIC rather than the AIC, but usually there is little difference in their overall performance. If several models have similar information criterion values, then it would indicate that there is little difference between them; a rough rule of thumb is to regard models as essentially indistinguishable if their information criteria differ by less than about 2, though we note that higher values are also in use.

Table 3.2 Information Criteria and Maximised Log-Likelihoods for Five Models for the Dipper Data

Model	AIC	BIC	n_p	ΔAIC	ΔBIC	$-\log L_{max}$
C/C	670.87	668.23	2	4.49	6.43	333.42
$C/C2$	**666.38**	662.58	3	0	0.78	330.19
T/T	679.71	**661.80**	11	13.33	0	328.50
T/C	678.75	668.65	7	12.37	6.85	332.38
C/T	674.00	663.89	7	7.62	2.09	330.00

Note: The term $\log L_{max}$ denotes the maximised log-likelihood. The smallest AIC and BIC values are shown in bold.

In this example, the AIC selects the $C/C2$ model as the best for the data, whereas the BIC, imposing a smaller penalty for the CJS model with 11 parameters, indicates two best alternative models for the data, namely the $C/C2$ and the CJS model (which is the same as the T/T model). As discussed in Chapter 2, the CJS model is parameter redundant, and so although the model contains 12 parameters, only 11 can be estimated from the data. If we had incorrectly set $n_p = 12$ in Table 3.2, that would have resulted in $BIC = 661.97$, even closer to the corresponding value for the $C/C2$ model, though the difference is very small in this instance.

□

Another way of comparing models using classical statistical inference is to use asymptotic tests, such as likelihood-ratio or score tests. See for example Morgan (2008, Chapter 4). Asymptotic theory tells us that for a pair of nested models, that is to say models when one model is a special case of the other one, if we take twice the difference of the maximised model log-likelihoods, then the result can be referred to an appropriate chi-square distribution, as a way of judging whether the difference is significant or not. We would expect the difference to depend to some extent on the numbers of parameters in the two models, and in fact the degrees of freedom for the chi-square distribution are the difference between the numbers of parameters in the two models. This very neat theory is easily illustrated using the dipper data of Example 3.4.

Example 3.5. Selecting a Model for the Dipper Data Using Likelihood-Ratio Tests

For models of the dipper data, model C/C is nested within model $C/C2$. The second of these two models has an extra parameter, and so it is not

MODEL SELECTION

surprising that it has a larger value for $\log L_{max}$. In this case the test statistic for the likelihood-ratio test has the value 6.46, which is highly significant when referred to chi-square tables on 1 degree of freedom. The conclusion is that the extra survival probability corresponding to the flood in 1983 substantially increases the likelihood, and so we prefer the more complex model. Similarly, we can compare separately each of the models C/T and T/C with the CJS model, as they are both nested inside the CJS model, and similarly we can compare the C/C model with the CJS model. As the 5% value for a χ^2_4 distribution is 9.49, then we can see from Table 3.2 that neither of the models C/T and T/C is significantly improved by the CJS model, and also neither of them improves significantly on the C/C model, which in turn is not improved by the CJS model. This basically tells us we have no evidence in the data to indicate general time variation in either of the basic model parameters, relating to recapture and survival. However, there is evidence for elaborating the model to take account of the 1983 flood, and we draw the same conclusion as when using information criteria in the last example. As was observed in Chapter 1, the dipper data set is a sparse one, and so there will be little power for likelihood-ratio tests. Non-significant tests do not necessarily mean that more complex models are inappropriate when there are sparse data.

□

Score tests are asymptotically equivalent to likelihood-ratio tests when the null hypothesis being tested holds. Score tests have the advantage that when a simple model is compared to a more complex model, then it is only necessary to fit the simpler of the two models. See Morgan et al. (2007a) for further discussion. Likelihood-ratio and score tests make the assumption that asymptotic results can be applied, and this is certainly questionable for the dipper data set. Additionally, when several models are being compared using such models, then one should adjust significance levels to take account of the fact that multiple tests are being carried out. We note finally that the asymptotic theory for likelihood-ratio and score tests only applies to nested models. Simply comparing models in terms of information criteria is clearly far simpler, especially when there are very many models being considered.

Example 3.6. Recovery Data on Herring Gulls, *Larus argentatus*
Model selection procedures based on likelihood-ratio tests for the herring-gull ring-recovery data set of Exercise 3.4 result in the information in Table 3.3. There is much information in this table. Here we just focus on the likelihood-ratio test results. For each model comparison we compare a simpler model with a more complex model, within which the simpler model is nested. The result is a likelihood-ratio (LR) statistic, which is referred to chi-square tables on the relevant, shown, degrees of freedom (d.f.). We can see, for example, that the $T/C/C$ model provides a better fit to the data than the $C/C/C$ model, but that the $T/C/C$ model is not appreciable better than the $T/C/T$ model. A similar application to this one is given in Exercise 3.2.

Table 3.3 Information for Assessing Relative and Absolute Goodness-of-Fit of Models to a Particular Set of Ring-Recovery Data on Herring Gulls Ringed as Young

Model Comparison	Score Statistic	LR Statistic	d.f.	p(%) (score)	Model	AIC	X^2	d.f.	p(%) (X^2)
					$C/C/C$	173.9	63.4	18	0.0
$C/C/C : C/C/T$	15.96	15.46	5	0.70	$C/C/T$	168.4	48.8	13	0.0
$C/C/C : T/C/C$	31.71	30.11	5	0.0007	$T/C/C$	153.8	32.7	13	0.2
$C/C/C : C/A_2/C$	12.23	12.53	1	0.05	$C/A_2/C$	163.4	48.1	17	0.0
$T/C/C : T/C/T$	4.53	4.42	5	48	$T/C/T$	159.4	26.6	8	0.1
$T/C/C : T/A_2/C$	11.20	10.44	1	0.08	$T/A_2/C$	145.3	19.7	12	7.3
$T/A_2/C : T/A_3/C$	0.30	0.21	1	58	$T/A_3/C$	145.1	19.5	11	5.3
$T/A_2/C : T/A_2/T$	6.19	6.27	5	29	$T/A_2/T$	149.0	12.3	7	9.1

Source: Reproduced with permission from Catchpole and Morgan (1996, p. 668) published by the International Biometric Society.

□

3.4 Maximising Log-Likelihoods

Only very rarely in statistical ecology can one obtain explicit expressions for maximum-likelihood estimates. What this means is that classical inference usually involves understanding and using numerical optimisation procedures, designed for optimising general functions of many variables. Although modern numerical optimisation procedures are very sophisticated, they iteratively seek local optima, from a given starting point in the parameter space. Classical inferential procedures rely on finding global optima, such as that in Figure 3.2 and in order to guard against finding a local optimum, it is a good idea to run numerical optimisation procedures from a number of different starting points. Even then, there is no guarantee that a global optimum can be obtained. In the next sections we describe some of the features of numerical optimisation procedures used in statistics.

3.4.1 Deterministic Procedures

The optimisation procedures that are typically used in maximising likelihoods may be described as deterministic, which means that they iterate from a given starting point in the parameter space, and whenever they are repeated, if they start from the same starting point, then they always end at the same end-point. We may distinguish three different methods that are currently used, namely simplex search methods, quasi-Newton methods, and dynamic search methods. Of these, the last are the most complicated, allowing users to specify constraints on parameters. An implementation in the computer package MATLAB, is provided by the function fmincon, and in R through the facilities in optim. Simplex methods involve evaluating the function at selected points on the likelihood surface, corresponding to a particular simplex in the parameter space. Depending on the relative magnitudes of the function values at the points of the simplex, the simplex is changed, and the hope is that the simplex contracts to the point corresponding to the maximum. Quasi-Newton methods are based on the Newton-Raphson iterative method. In their simplest forms, these methods require expressions for first- and second-order derivatives of the surface being optimised, but it is possible either to evaluate these derivatives by means of computer packages for symbolic algebra, or to use difference approximations to the derivatives. Methods that make use of the shape of surfaces, through exact or approximate first- and second-order derivatives can be expected to converge faster than simplex methods. However, convergence of any of these methods is not guaranteed, and they may diverge. Numerical optimisation routines are typically written by numerical analysts for general function *minimisation*. Statisticians then simply need to submit negative log-likelihoods, when using such routines. Statisticians find it easier to try to maximise log-likelihoods, as these involve sums, rather than the products of likelihoods. Additionally, log-likelihoods may be easier to maximise than likelihoods, as they change more gradually as functions of the parameters. This is true, for example, of the log-likelihood and corresponding likelihood corresponding to the dipper example.

3.4.2 Stochastic Procedures

A clear disadvantage of deterministic methods is that from a given starting point their progress is determined and unvarying. Stochastic search methods involve random searches of a surface, which means that different runs from the same starting point can proceed quite differently. Methods based on simulated annealing can explore a range of different optima, and so have the potential of finding a global optimum, whatever the starting value; for more details, see Brooks and Morgan (1995) and Morgan (2008, Chapter 3). A routine for simulated annealing is provided in optim in the statistics package R, which is used in Chapter 7 and described in Appendix B. An extension of methods for classical inference that allow simulated annealing to be used to explore different models, as well as the parameter space of individual component models is called trans-dimensional simulated annealing, and is described by Brooks et al. (2003a). This relates to the work of Chapter 6.

3.5 Confidence Regions

We have already seen in Figure 3.1 the effect of sample size on the shape of a likelihood at its maximum. This feature is formally included in the asymptotic properties of maximum-likelihood estimators. If a model parameter $\boldsymbol{\theta}$, of length k, has maximum-likelihood estimator $\hat{\boldsymbol{\theta}}$, then asymptotically, $\hat{\boldsymbol{\theta}}$ has a multivariate normal distribution, given by:

$$\hat{\boldsymbol{\theta}} \sim \mathcal{N}(\boldsymbol{\theta}, I(\boldsymbol{\theta})^{-1}),$$

where $I(\boldsymbol{\theta})$ is the expected information matrix given by,

$$I(\boldsymbol{\theta}) = -\mathbb{E}\left[\frac{\partial^2 log(L)}{\partial \boldsymbol{\theta}^2}\right],$$

in which $\mathbb{E}$ denotes expectation, taken with regard to the members of the sample. For more detail, see Morgan (2008). Thus for scalar θ, the greater the curvature of the log-likelihood surface, then the greater the expected information, and the greater the precision (that is the smaller the variance) of the estimator. When maximum-likelihood estimates are presented, they are usually presented along with their estimated standard error, obtained from the appropriate term in the inverse of the expected information matrix. We have seen an example of this in Section 2.6.1.

An alternative way to construct confidence regions is by taking appropriate sections of profile log-likelihoods. This method can result in asymmetric intervals and regions that reflect the shape of the log-likelihood surface away from the maximum. How this is done is explained in Morgan (2008, Section 4.4), and illustrations were given in Exercise 2.2.

Example 3.7. Site Occupancy

Site occupancy models are similar to those for timed species counts – see Exercise 2.7. They result in a vector $\boldsymbol{m}$ of data, in which m_y is a number of sites where a particular species of animal has been detected y times. As an illustration from Royle (2006), observations on the common yellow-throat *Geothlypis trichas* resulted in $T = 11$ and

$$\boldsymbol{m} = \begin{pmatrix} 14 & 6 & 7 & 5 & 3 & 1 & 4 & 5 & 3 & 2 & 0 & 0 \end{pmatrix}.$$

In this case we have a three-parameter model, and are able to construct a two-parameter profile log-likelihood surface with respect to two of the model parameters, as shown in Figure 3.3. The maximum-likelihood point estimates of the parameters in this case are given by $\hat{\mu} = 0.3296$ (0.0596), and $\hat{\tau} = 0.2851$ (0.1272). For more discussion of this example, including how one constructs confidence intervals and regions from profile log-likelihoods, see Morgan et al. (2007b) and Morgan (2008).

□

3.6 Computer Packages

The paper by Catchpole (1995) advocates using the computer package MATLAB for programming model-fitting in statistical ecology; the same argument can be used for programming in the statistical package R, which we use in

COMPUTER PACKAGES

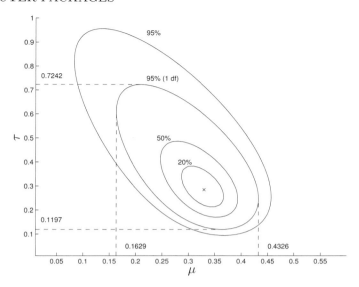

Figure 3.3 Profile log-likelihood contours and confidence regions resulting from the common yellow-throat data. Shown are the 95% confidence region for (μ, τ) together, as well as the marginal 95% interval for each parameter. The location of the maximum-likelihood estimate is denoted by a ×. Reproduced with permission from Morgan et al. (2007b, p. 620) published by Wiley-Blackwell.

Chapter 7 and is described in Appendix B. An alternative approach is to use specialised computer packages; for instance, for Bayesian methods the package WinBUGS is introduced in Chapter 7 and described in Appendix C. In this section we mention three of several that are now available for analysing various kinds of capture-recapture data. Computer packages need some system for communicating models and model structure, and the three computer packages considered below do this in different ways. These packages are all freely available through the Web.

3.6.1 MARK

The package MARK is probably the most widely used computer program for analysing data from observations on previously marked animals. This freely available computer program can be downloaded from:

`http://welcome.warnercnr.colostate.edu/~gwhite/mark/mark.htm`

The package allows users to estimate abundance as well as survival. One reason for MARK's success is its use of a database management system in order for users to store the results from fitting many different models. Although traditionally its emphasis has been to provided classical analyses, it now also includes Bayesian facilities, which we shall describe later in Chapter 7. The facilities provided by MARK are extensive. See www.cnr.colostate.edu/~gwhite. It has an excellent online manual and help system.

Example 3.8. Using MARK to Analyse the Dipper Data

MARK uses parameter index matrices (PIMs) to communicate model structure. For the C/C and CJS models these are shown in Table 3.4 for the survival parameter only. Exactly the same approach is used for the reporting parameter (but using a different integer value, otherwise MARK will read the recapture and survival parameters to be equal). Thus in the case of the two-parameter model, having a 1 in each position in the PIM for the survival parameter simply means that the corresponding parameter is the same in each place in the m-array. For the corresponding PIM for the recapture parameters, we would typically set each value to be equal to 2, once more setting each recapture parameter to be equal. In contrast, for the CJS model, we can see that there is a time effect, and each year has its own parameter.

Table 3.4 Examples of PIMs for the Survival Parameters for the Dipper Data

1	1	1	1	1	1
	1	1	1	1	1
		1	1	1	1
			1	1	1
				1	1
					1
1	2	3	4	5	6
	2	3	4	5	6
		3	4	5	6
			4	5	6
				5	6
					6

Note: The first example is for the C/C model, and the second is for the CJS model. Analogous PIMs are defined for the recapture parameters, but using different parameter values.

Standard models, like those illustrated in Table 3.4, are available as defaults, thus reducing further the possibility of input errors. However, it can be nontrivial to set up the PIMs for complex examples. MARK deals with parameter redundancy by considering particular model/data combinations, and judging which of a set of appropriate eigenvalues are zero. As they involve subjective judgment, numerical procedures are not as good as symbolic ones. However, they do allow the user to investigate the combined effect of model and data,

SUMMARY

and it may be that full rank models might perform poorly in practice, for particular data sets. See Catchpole et al. (2001) for an illustration of this.

□

3.6.2 Eagle

The package *eagle* is restricted to model-fitting and comparison for single-site data from ring-recovery experiments. However, it is unusual in that it communicates models through a convenient, transparent shorthand, and it also selects models through score tests (see Section 3.3); *eagle* moves between alternative models by means of a step-up system, starting with the simplest possible model. The idea of using score tests in this way in statistical ecology is provided by Catchpole and Morgan (1996), and provides an attractively simple way of steering a path through a large number of alternative possible models. The model notation is that established in Section 3.3.1, namely, a triple of letters, $X/Y/Z$, where the first letter indicates the assumptions made regarding the first-year survival of the animals, the second letter does the same for older animals, and the third letter relates to the reporting probability of dead animals. The letters used include, for example, C for constant and T for time dependent.

3.6.3 M-SURGE

Details of this package are to be found at:
>   http://www.cefe.cnrs.fr/biom/En/softwares.htm

This important package is designed to deal with multi-site and multi-state data. A novel feature is that it uses exact log-likelihood derivatives in the numerical optimisation for obtaining maximum-likelihood estimates. A new development, E-SURGE, accounts for when the state of an animal may not be known exactly.

3.7 Summary

Methods of classical inference based on the construction and maximisation of likelihoods are well developed and in frequent use in population ecology. Likelihoods are regarded as functions of model parameters. Fisher's information is minus the expected Hessian matrix evaluated at the likelihood maximum, and its inverse is used to estimate standard errors. Maximisation of likelihoods results in maximum-likelihood parameter estimates, along with their associated measures of standard error. Profile log-likelihoods may also be used to produce confidence intervals and regions. One can discriminate between alternative models, involving different degrees of complexity, by means of score and likelihood-ratio tests. Score tests are asymptotically equivalent to likelihood ratio tests under the null hypothesis, but have the advantage that when one is comparing two models, it is only necessary to fit the simpler of the two models. If a score test is significant, indicating that the more complex model should be fitted, then it will be necessary to fit the more complex model. However, this can then be done with confidence that the data support the more complex model. Models may also be compared through different information criteria, such as the Akaike and Bayes information criteria. These may be regarded as penalising models for including too many parameters. Likelihood maximisation is usually done using appropriate deterministic methods of numerical analysis implemented on computers; packages such as R provide a flexible range of alternative optimisation algorithms. Stochastic optimisation methods may be used to explore complex likelihood surfaces, which may possess multiple optima. Specialised computer packages such as MARK and M-SURGE exist for classical methods of model fitting to data in population ecology.

3.8 Further Reading

The use of statistics in ecology is a continuing success story, due to good communications between ecologists and statisticians, the basic classical theory being made accessible through publications like Lebreton et al. (1992), and the availability of user-friendly computer packages like MARK and DISTANCE. An important feature that has not been considered in this chapter is how to judge whether a model provides a good fit to the data. General classical methods are described in Morgan (2008, Section 4.7); however, these often need adapting in the analysis of ecological data, when observed numbers may be small. For a detailed explanation of score tests, see Morgan (2008, Section 4.6) and Morgan et al. (2007a). Relevant discussion and ideas are to be found in Catchpole et al. (1995), Freeman and Morgan (1990) and Freeman and Morgan (1992). The general use of MATLAB for modelling mark-recapture-recovery data is described by Catchpole et al. (1999); note that both *eagle* and M-SURGE are written in MATLAB. A classical approach to model averaging, using AICs is described by Buckland et al. (1997).

3.9 Exercises

3.1 An analysis of ring-recovery data on male mallards, *Anas platyrhynchos*, ringed as young produces the maximum-likelihood estimates and standard errors of Table 3.5. Explain the meaning of the model parameters and discuss the results. Specify the model, both using the Catchpole-Morgan */*/* notation, and also through appropriate PIMs.

3.2 We wish to use methods of classical inference in order to select a model to describe ring-recovery data on blue-winged teal, *Anas discors*. A subset of results is given in Table 3.6, where the models are specified using the notation of Section 3.3.1. Consider how the score tests differ from the likelihood-ratio (LR) tests. The X^2 values denote the Pearson goodness-of-

EXERCISES

Table 3.5 Maximum-Likelihood Parameter Estimates for a Model Fitted to Recovery Data on Mallards Ringed as Young

	$\widehat{\phi}_a = 0.69(0.02)$	
i (year)	$\widehat{\lambda}_i$	$\widehat{\phi}_{1i}$
1	0.15 (0.02)	0.44 (0.05)
2	0.28 (0.03)	0.48 (0.06)
3	0.18 (0.02)	0.58 (0.05)
4	0.22 (0.02)	0.46 (0.04)
5	0.17 (0.01)	0.42 (0.04)
6	0.24 (0.02)	0.60 (0.04)
7	0.27 (0.02)	0.59 (0.05)
8	0.24 (0.02)	0.56 (0.06)
9	0.15 (0.02)	0.29 (0.14)
Arithmetic average	0.21	0.49

Source: Reproduced with permission from Freeman and Morgan (1992, p. 222) published by the International Biometric Society.

Table 3.6 Information for Assessing Relative and Absolute Goodness-of-Fit of Models to a Particular Ring-Recovery Data Set on Teal Ringed as Young

Model Comparison	Score Statistic	LR Statistic	d.f.	p (%) (score)	Model	AIC	X^2	d.f.	p (%) (X^2)
					$C/C/C$	224.3	40.23	25	2.8
Stage 1									
$C/C/C : C/C/T$	21.16	23.53	11	3.2	$C/C/T$	222.8	23.24	19	22.7
$C/C/C : T/C/C$	5.39	5.71	6	50	$T/C/C$	230.6	32.59	19	2.7
$C/C/C : C/A_2/C$	2.40	2.42	1	12	$C/A_2/C$	223.9	39.14	24	2.6
Stage 2									
$C/C/T : T/C/T$	5.00	5.22	6	54	$T/C/T$	229.6	19.11	13	12.0
$C/C/T : C/A_2/T$	1.82	1.84	1	18	$C/A_2/T$	222.9	22.76	18	20.0

Source: Reproduced with permission from Catchpole and Morgan (1996, p. 669) published by the International Biometric Society.

fit statistics. Explain how the AIC and X^2 values are to be used. Suggest a suitable model for the data, giving your reasons.

3.3 Soay sheep were encountered in Example 1.4. The annual survival probabilities of female Soay sheep studied in the Village Bay area of Hirta, the largest island of the St. Kilda archipelago, can be categorised into age classes as follows:

ϕ_1: first-year animals;

ϕ_2: second-year animals;

ϕ_a: third- through seventh-year animals;

ϕ_s: animals aged ≥ 7 years.

When these parameters are logistically regressed on two environmental covariates, the coefficient estimates and standard errors are shown in Table 3.7. Population size is measured in hundreds of animals, and March rainfall in hundreds of millimeters, with both the covariates mean-centered; this means that each covariate value is replaced by that value minus the covariate mean, a procedure that often improves model fitting. Provide the full mathematical description of the survival probabilities.

Table 3.7 Maximum-Likelihood Estimates (and Standard Errors) of Logistic Regression Coefficients of Covariates When a Particular Age Structured Survival Model is Fitted to Soay Sheep Data

Parameter	Covariate	
	Population Size	March Rainfall
ϕ_1	−1.1641 (0.1349)	−1.2015 (0.2499)
ϕ_2	−0.3823 (0.2110)	−1.7834 (0.5291)
ϕ_a	−0.2106 (0.1320)	−1.4531 (0.3087)
ϕ_s	−0.7838 (0.2239)	−1.3089 (0.4202)

Source: Reproduced with permission from Catchpole et al. (2000, p. 466) published by Wiley-Blackwell.

On the basis of the results given, explain which variables you would retain in the model, and consider whether your selected model might be simplified further.

3.4 A ring-recovery data set on herring gulls, *Larus argentatus*, is analysed in the computer package SURVIV, using the code below. Note that SURVIV is the precursor of the package MARK.

```
cohort=3646;
    62:(1-s(1))*s(12);
    15:s(1)*(1-s(2))*s(12);
    9:s(1)*s(2)*(1-s(3))*s(12);
    7:s(1)*s(2)*s(3)*(1-s(4))*s(12);
    13:s(1)*s(2)*s(3)*s(4)*(1-s(5))*s(12);
    8:s(1)*s(2)*s(3)*s(4)*s(5)*(1-s(6))*s(12);
```

EXERCISES

```
cohort=10748;
   169:(1-s(7))*s(12);
   67:s(7)*(1-s(2))*s(12);
   29:s(7)*s(2)*(1-s(3))*s(12);
   22:s(7)*s(2)*s(3)*(1-s(4))*s(12);
   20:s(7)*s(2)*s(3)*s(4)*(1-s(5))*s(12);

cohort=6665;
   92:(1-s(8))*s(12);
   64:s(8)*(1-s(2))*s(12);
   27:s(8)*s(2)*(1-s(3))*s(12);
   16:s(8)*s(2)*s(3)*(1-s(4))*s(12);

cohort=4652;
   70:(1-s(9))*s(12);
   21:s(9)*(1-s(2))*s(12);
   15:s(9)*s(2)*(1-s(3))*s(12);

cohort=2983;
   29:(1-s(10))*s(12);
   25:s(10)*(1-s(2))*s(12);

cohort=3000;
   72:(1-s(11))*s(12);
```

Although not fully explained here, the model notation is straightforward, since the program is given cohort sizes, the numbers of birds ringed in each year, the numbers of birds reported dead in each year after ringing, and then expressions for the corresponding probabilities. Thus $s(1)$ is the probability of surviving the first year of life, $s(12)$ is the reporting probability, and so on. Write down the complete model description using the model notation of Section 3.3.1. Write down the parameter index matrices (PIMs) for the model. Compare and contrast the three different ways of specifying the model, in SURVIV, *eagle* and MARK.

3.5 The data in Table 3.8 relate to the white stork, *Ciconia ciconia*, population in Baden Württemberg, Germany. The data consist of 321 capture histories of individuals ringed as chicks between 1956 and 1971. From the 1960's to the 1990's, all Western European stork populations were declining; see Bairlein (1991). This trend is thought to be due to the result of reduced food availability caused by severe droughts observed in the wintering ground of storks in the Sahel region; see Schaub et al. (2005). This hypothesis has been examined in several studies, see Kanyamibwa et al. (1990) and Barbraud et al. (1999). Analyse the data, and suggest an appropriate probability model. Note that we see frequent use of this example in later chapters, especially when we consider the use of measures of rainfall in order to describe mortality.

3.6 The data in Table 3.8 are evidently sparse. Suggest an alternative analysis,

Table 3.8 Capture-Recapture Data for White Storks

Year of Ringing	Number Ringed	Year of first recapture – 1900+														
		57	58	59	60	61	62	63	64	65	66	67	68	69	70	71	72
1956	26	19	2	0	0	0	0	0	0	0	0	0	0	0	0	0	0
1957	50		33	3	0	0	0	0	0	0	0	0	0	0	0	0	0
1958	53			35	4	0	0	0	0	0	0	0	0	0	0	0	0
1959	69				42	1	0	0	0	0	0	0	0	0	0	0	0
1960	73					42	1	0	0	0	0	0	0	0	0	0	0
1961	71						32	2	1	0	0	0	0	0	0	0	0
1962	64							46	2	0	0	0	0	0	0	0	0
1963	64								33	3	0	0	0	0	0	0	0
1964	66									44	2	0	0	0	0	0	0
1965	55										43	1	0	0	1	0	0
1966	60											34	1	0	0	0	0
1967	53												36	1	0	0	0
1968	51													27	2	0	0
1969	38														22	0	0
1970	33															15	1
1971	23																15

Source: Reproduced with permission from Gimenez et al. (2009a, p. 887) published by Springer.

involving only the information on the numbers of birds released each year and those captures in the main diagonal of the table.

3.7 Indicate how you would construct the likelihood for the ring-recovery data and model of Section 2.3.1.

3.8 Consider how you would model the nest-record data of Example 2.6.

3.9 Construct the BIC values for the comparison of models for snowshoe hare data in Table 3.1. and compare your conclusion based on the BIC with those based on the AIC.

3.10 Consider how you would produce a model-averaged estimate of N from the results of Table 3.1. See Buckland et al. (1997) for suitable discussion.

PART II

Bayesian Techniques and Tools

In recent years there has been a huge increase in the number of publications of Bayesian statistical analyses across all application areas. This is particularly true for the area of statistical ecology. The potential and expected increase in the application of Bayesian methods to this area has been foreseen (see for example Schwarz and Seber 1999 and Ellison 2004). The recent publication of a number of books focussing on Bayesian applications to statistical ecology reflects the increase in the application of Bayesian techniques within this field (McCarthy 2007; Royle and Dorazio 2008; Link and Barker 2009). Some of the early journal publications generally providing initial work in this field considered the application to closed populations (Castledine 1981; Smith 1991), ring-recovery data (Vounatsou and Smith 1995; Brooks et al. 2000a), capture-recapture data (Brooks et al. 2000b) and multi-state capture-recapture data (Dupuis 1995). We particularly focus on capture-recapture data and ring-recovery data in this part of the book. These and other forms of data, along with additional examples, are considered in Part III).

Bayesian analyses provide a flexible approach for analysing complex data, often permitting intricate models to be fitted, and hence the identification of fine detail contained within the data. We aim to provide the underlying principles behind Bayesian statistics and demonstrate how these are applied to problems in statistical ecology. The Bayesian principles are illustrated by analyses of real data and the interpretation of the results obtained are discussed.

CHAPTER 4

Bayesian Inference

4.1 Introduction

The Bayesian approach to statistical inference is quite different from the classical approach, though they both depend essentially upon the likelihood. As stated in Section 1.4, in the classical approach the model parameters are regarded as fixed quantities that need to be estimated, and this is typically done by finding the values of the parameters that maximise the likelihood, considered as a function of the parameters. As we have seen, this usually reduces to a problem in numerical optimisation, often in a high dimensional space. In the Bayesian approach, parameters are seen as variable and possessing a distribution, on the same footing as the data. Before data are collected, it is the *prior* distribution that describes the variation in the parameters. After the collection of data the prior is replaced by a *posterior* distribution. No optimisation is involved. As we shall see, the approach follows from a simple application of Bayes' Theorem, however the problem with the use of Bayes' Theorem is that it provides a posterior distribution for *all* of the model parameters jointly, when we might be especially interested in just a subset of them. For example, if we are interested in only a particular parameter, we want to estimate the marginal posterior distribution for that parameter alone. In order to obtain the marginal posterior distribution of the parameters of interest, we have to do what we always do when seeking a marginal distribution, which is integrate the joint posterior distribution. It is here that the difficulties typically arise when a Bayesian analysis is undertaken. The optimisation of classical analysis has been replaced by integration for the Bayesian approach. The modern approach to Bayesian analysis is not to try to integrate the posterior joint distribution analytically, but instead to employ special simulation procedures which result in samples from the posterior distribution. Having simulated values from the posterior joint distribution means that one then naturally has simulated values from the posterior marginal distribution of the parameters of interest. This approach is called *Markov chain Monte Carlo* (MCMC), and it will be described in detail in Chapter 5.

4.1.1 Very Brief History

The Bayesian approach to statistical inference, in which prior information is combined with the information provided by the data in order to produce a consensus, was the dominant approach for over 150 years, from its introduction in the late 18th century. However, at the turn of the 20th century, the Bayesian approach began to fall out of favour due mainly to two distinct issues:

- The computational complexity of the Bayesian approach became too great as more and more complex statistical problems began to be addressed.

- Researchers at the time became increasingly concerned about the subjectivity of the approach, which stemmed from the inclusion of the prior in the analysis.

Staunch opponents of the Bayesian approach began to dominate the field and, looking for alternatives, they developed the classical approach to statistical inference (Kotz and Johnson 1992).

With the popularity of the classical alternative, much of the 20th century was bleak for the Bayesian. The revival of Bayesian inference was due to a combination of new statistical innovations and the computer revolution which suddenly made the Bayesian approach tractable once more. In particular, the MCMC algorithm was introduced to the literature (see Chapter 5), allowing realisations to be simulated from the posterior distribution and permitting the necessary numerical integration. These new statistical methods made possible the analysis of far more complex processes than was previously possible under either the Bayesian or classical paradigms. Thus, for many the choice of Bayesian versus classical is much less a philosophical one than it is a pragmatic one. The tools available to the Bayesian provide a far more powerful and flexible framework for the analysis of complex stochastic processes than the corresponding traditional ones. Thus, by adopting a Bayesian approach:

- Complex processes, several of which we consider in Part III, can be analysed with little additional effort.

- Unrealistic assumptions and simplifications can be avoided.

- Both random effects and missing values are easily dealt with.

- Model choice and model averaging are relatively easy even with large numbers of models.

Many application areas, including ecology, require these characteristics of statistical inference. This has been the main driving force behind the enormous upsurge of interest in the Bayesian approach.

4.1.2 Bayes' Theorem

Bayesian methods date to the original paper by the Rev. Thomas Bayes (Bayes 1763) which was read to the Royal Society in 1763. Known then as *inverse probability*, the approach dominated statistical thinking throughout the 19th century until the classical methods of Neyman and Fisher began to gain prominence during the 1930s (Savage 1954; Stigler 1986; Kotz and Johnson 1992).

The Bayesian approach is based upon the idea that the experimenter begins with some prior beliefs about the system under study and then updates these

INTRODUCTION

beliefs on the basis of observed data, x. This updating procedure is based upon what is known as Bayes' Theorem:

$$\pi(\boldsymbol{\theta}|\boldsymbol{x}) = \frac{f(\boldsymbol{x}|\boldsymbol{\theta})p(\boldsymbol{\theta})}{f(\boldsymbol{x})},$$

in which $\boldsymbol{x}$ denotes the set of observed data, as before, and $\boldsymbol{\theta}$ is the vector of model parameters. The function f is either the probability density function, for continuous data, or the probability mass function for discrete data, associated with the observed data under the chosen model (conditional on the parameter values), and p and π denote what are called, respectively, the prior and posterior distributions of the parameters $\boldsymbol{\theta}$. For simplicity, we shall refer to f as a density below. We emphasise here that we obtain a posterior *distribution* for the parameters, rather than simply a single point estimate.

Note that the term $f(\boldsymbol{x})$ is a function of *only* the observed data (i.e. it is independent of the parameters, $\boldsymbol{\theta}$). Typically the term $f(\boldsymbol{x})$ is omitted from Bayes' Theorem which is expressed in the form:

$$\pi(\boldsymbol{\theta}|\boldsymbol{x}) \propto f(\boldsymbol{x}|\boldsymbol{\theta})p(\boldsymbol{\theta}). \tag{4.1}$$

The omitted term $f(\boldsymbol{x})^{-1}$ is essentially simply the normalising constant, which is needed so that the posterior distribution $\pi(\boldsymbol{\theta}|\boldsymbol{x})$ is a proper distribution (and integrates to unity).

In order to undertake a Bayesian analysis, we must first define an appropriate probability model, which describes the relationship between the observed data, $\boldsymbol{x}$, and the set of model parameters, $\boldsymbol{\theta}$, (i.e. specify $f(\boldsymbol{x}|\boldsymbol{\theta})$) and in addition an appropriate prior distribution for the model parameters (i.e. specify $p(\boldsymbol{\theta})$). Plugging the $f(\boldsymbol{x}|\boldsymbol{\theta})$ and $p(\boldsymbol{\theta})$ terms into Bayes' Theorem, we obtain the posterior distribution $\pi(\boldsymbol{\theta}|\boldsymbol{x})$. This is a new distribution for the model parameters, corresponding to our new beliefs, which formally combine our initial beliefs represented by p with the information gained from the data through the model. This posterior distribution is the basis for all subsequent inference within the Bayesian paradigm. The specification of prior distributions is discussed in Section 4.2.

As described, the Bayesian approach combines the model for the data with the prior for the parameters to form the posterior distribution of the parameters. This is in contrast to the classical approach which simply does away with the prior and focuses on the model term $f(\boldsymbol{x}|\boldsymbol{\theta})$. Rather than inverting $f(\boldsymbol{x}|\boldsymbol{\theta})$ through Bayes' Theorem, (which requires a prior) to obtain a function of the model parameters, the classical approach simply sets

$$f(\boldsymbol{\theta}|\boldsymbol{x}) \equiv f(\boldsymbol{x}|\boldsymbol{\theta}),$$

where $f(\boldsymbol{\theta}|\boldsymbol{x})$ is referred to as the likelihood, as we have seen in the last two chapters. Unlike $\pi(\boldsymbol{\theta}|\boldsymbol{x})$, the likelihood is not a probability distribution and, rather than reflecting the experimenter's beliefs, the likelihood refers to the relative likelihood of the observed data under the model, given the associated parameters, $\boldsymbol{\theta}$. As we know from Part I, the classical approach typically seeks to identify the value of the parameters that maximises this likelihood. Thus a point estimate is obtained for the parameters that maximises the likelihood. Conversely, in the Bayesian framework the parameters are assumed to have a distribution rather than a single value. We note that the term *likelihood* is (perhaps a little confusingly) also generally used for the probability density function $f(\boldsymbol{x}|\boldsymbol{\theta})$ within the Bayesian framework, since it is of the same mathematical form.

We now provide two examples of the application of Bayes' Theorem. We begin with the simplest example, with just a single parameter, before extending to a two-parameter example.

Example 4.1. One-Parameter Example – Radio-Tagging Data

We consider radio-tagging data collected on individuals over a period of time. We assume that there is no radio failure, so that at each time point we know whether each individual in the study is alive or dead. The data corresponding to each individual are then of the form of life histories, such as that illustrated below:

$$0\ 0\ 1\ 1\ 1\ 2\ 0\ 0.$$

This particular history indicates that the individual is initially tagged at time 3, is then alive at times 4 and 5, but dies between times 5 and 6, denoted by 2. Thus the general life-histories of Section 2.3.3 are simplified here. There is no live recapture, and all dead birds are recovered. The data can be summarised in the form of an m-array, as in Table 1.1, where the entries in the array correspond to the numbers of animals that die. Since death is always recorded, all other animals are known to be alive (contrary to the case in ring-recovery data).

We consider a subset of the data presented by Pollock et al. (1995) relating to canvasbacks, *Aythya valisineria*. In particular, we consider only those ducks with known fate within the study area (i.e. survival to the end of the study or recorded death). We also omit from the study any individuals that are observed outside of the study area. This is an oversimplification of this particular data set in order to provide a simple illustrative example. A summary of the corresponding data is provided in Table 4.1.

Table 4.1 The Number of Radio-Collared Canvasbacks Released and Corresponding Number Recovered Dead at Each Capture Time

Number Released	Capture Time					
	1	2	3	4	5	6
36	3	2	3	3	2	1

The likelihood is a function of only survival probabilities. For simplicity we assume a constant survival probability, ϕ, with a single release year. Then the

INTRODUCTION

likelihood can be expressed in the multinomial form,

$$f(\boldsymbol{x}|\phi) = \frac{(\sum_{i=1}^{7} x_j)!}{\prod_{j=1}^{7} x_j!} \prod_{j=1}^{7} p_j^{x_j},$$

where x_j denotes the number of deaths at times $j = 1, \ldots, 6$; x_7 corresponds to the number alive at the end of the study, and p_j denotes the corresponding cell probability of the m-array. In particular, we have,

$$p_j = \begin{cases} 1 - \phi & j = 1 \\ (1-\phi)\phi^{j-1} & j = 2, \ldots, 6 \\ \phi^{j-1} & j = 7, \end{cases}$$

where the last term is obtained by subtraction, and corresponds to individuals surviving until the final capture time. This expression for $\{p_j\}$ corresponds to a right-truncated geometric distribution; see Exercise 4.5.

We now need to specify a prior on the parameter ϕ. Without any prior information, we specify a $U[0, 1]$ prior on ϕ, so that $p(\phi) = 1$, for $0 \leq \phi \leq 1$. The corresponding posterior distribution for ϕ is given by,

$$\begin{aligned}
\pi(\phi|\boldsymbol{x}) &\propto f(\boldsymbol{x}|\phi)p(\phi) \\
&= \frac{(\sum_{j=1}^{7} x_j)!}{\prod_{j=1}^{7} x_j!} \prod_{j=1}^{7} p_j^{x_j} \times 1, \quad \text{for} \quad 0 \leq \phi \leq 1, \\
&\propto (1-\phi)^{x_1} \{\phi(1-\phi)\}^{x_2} \{\phi^2(1-\phi)\}^{x_3} \{\phi^3(1-\phi)\}^{x_4} \times \\
&\quad \{\phi^4(1-\phi)\}^{x_5} \{\phi^5(1-\phi)\}^{x_6} (\phi^6)^{x_7} \\
&= \phi^{\sum_{j=2}^{7}(j-1)x_j} (1-\phi)^{\sum_{j=1}^{6} x_j}.
\end{aligned}$$

Now if the parameter $\phi \sim Beta(\alpha, \beta)$, then the probability density function for ϕ is given by

$$p(\phi) = \frac{\Gamma(\alpha+\beta)}{\Gamma(\alpha)\Gamma(\beta)} \phi^{\alpha-1}(1-\phi)^{\beta-1},$$

(see Appendix A for further details and discussion of this distribution). We can see that $\pi(\phi|\boldsymbol{x})$ is of this form. Thus, we have as the posterior distribution for ϕ

$$\phi|\boldsymbol{x} \sim Beta\left(1 + \sum_{j=2}^{7}(j-1)x_j, 1 + \sum_{j=1}^{6} x_j\right).$$

For the data given above, this gives the posterior distribution,

$$\phi|\boldsymbol{x} \sim Beta(163, 15).$$

This is plotted in Figure 4.1, along with the corresponding uniform prior specified on the parameter, ϕ.

Note that, more generally, if we specify the prior, $\phi \sim Beta(\alpha, \beta)$, the

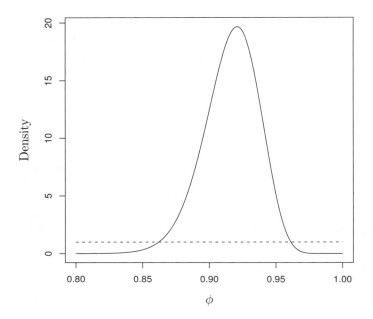

Figure 4.1 The prior and posterior distributions of ϕ for the canvasback radio-collared data. Note that the x-axis is only part of the range for ϕ.

posterior distribution for ϕ is,

$$\phi|\boldsymbol{x} \sim Beta\left(\alpha + \sum_{j=2}^{n+1}(j-1)x_j, \beta + \sum_{j=1}^{n} x_j\right),$$

where n is the number of capture events (note that $Beta(1,1) \equiv U[0,1]$). In this case, the posterior distribution of the parameter is from the same family as the prior distribution. This is a special form of prior, called a *conjugate* prior (see Section 4.2.1).

□

The above example was atypical, as the model contained just the one parameter, and furthermore it was possible to write down explicitly, and identify, the posterior distribution. In the next example numerical analysis is needed in order to obtain posterior marginal distributions.

Example 4.2. Example 1.1 Revisited – Two-Parameter Model for Dipper Data

We have already seen in Figure 3.2 the log-likelihood contours for a two-parameter capture-recapture model applied to the dipper data of Table 1.1.

PRIOR SELECTION AND ELICITATION

In order to conduct a corresponding Bayesian analysis for this application, the likelihood plays the role of the density f. In Figure 4.2 we present the *likelihood* contours, the two posterior marginal distributions, one for each of the two model parameters, and also histograms from MCMC simulations (see Chapter 5), used for estimating the two marginal distributions. The prior distribution in this example was a product of two uniform distributions (i.e. independent $U[0,1]$ distributions), so that the two parameters were independent in the prior distribution. In order to obtain the two marginal distributions in this case it was necessary to perform numerical integration of the posterior distribution.

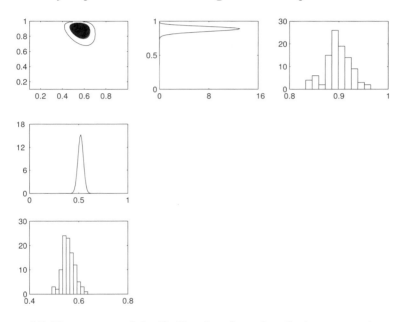

Figure 4.2 The contours of the likelihood surface when the two-parameter capture-recapture model is applied to the data of Table 1.1. Also shown are the two marginal posterior distributions of the model parameters, obtained by numerical integration of the bivariate posterior distribution, and simulations from these two marginal distributions using MCMC.

□

4.2 Prior Selection and Elicitation

As explained above, the Bayesian analysis begins with the specification of a model and a prior on the corresponding parameters. The specification of this prior, and the elicitation process that determines its precise form, can be something of a mystery for those not already acquainted with the idea, and so we deal with this process here.

Though modern scientific thinking is much more at home with the subjective interpretation and analysis of data than was previously the case, the incorporation of subjective information via the prior remains a focus of much debate, especially when little or no knowledge is available. In ecological applications, the prior is a convenient means of incorporating expert opinion or information from previous or related studies that would otherwise be ignored. With sparse data, the role of the prior can be to enable inference to take place on parameters that would otherwise be impossible, though with sufficiently large and informative data sets the prior typically has little effect on the results.

When one specifies priors, there are basically two situations that can be encountered:

- there is no prior information; or
- there is prior information that needs to be expressed in the form of a suitable probability distribution.

We begin with a discussion of a particular type of prior that is often used for simplicity – the *conjugate* prior, already encountered in Example 4.1.

4.2.1 Conjugate Priors

When a posterior distribution for a model parameter has the same distributional form as the corresponding prior, then we say that the prior is *conjugate*. For example, suppose that we observe data $x_1, \ldots, x_n$, such that each $X_i \stackrel{iid}{\sim} N(\mu, \sigma^2)$, where μ is unknown and σ^2 is known. We specify a normal prior on μ, such that, $\mu \sim N(0, \tau)$. Then the posterior distribution for μ is given by,

$$\mu | \boldsymbol{x} \sim N\left(\frac{\tau n \bar{x}}{\tau^2 n + \sigma^2}, \frac{\sigma^2 \tau^2}{\tau^2 n + \sigma^2}\right).$$

Thus, the prior and posterior distributions for μ are both normal. In other words a normal prior distribution on the mean of a normal distribution is a conjugate prior. Examples of additional common conjugate priors are:

- An inverse gamma (Γ^{-1}) distribution on the variance of a normal distribution. Note that this is equivalent to a gamma (Γ) distribution on the precision of a normal distribution, as we define precision = 1/variance;
- A gamma (Γ) distribution on the mean of a Poisson distribution;
- A beta distribution on the probability parameter of a binomial distribution, or of a geometric distribution, as we have seen in Example 4.1;
- A Dirichlet distribution on the probabilities of a multinomial distribution (a generalisation of the previous beta conjugate prior for the binomial distribution);
- A multivariate normal distribution on the mean vector of a multivariate normal distribution (the multivariate generalisation of the univariate normal distribution);
- An inverse Wishart distribution on the covariance matrix of a multivariate normal distribution.

PRIOR SELECTION AND ELICITATION

Conjugate priors have historically been used to simplify a Bayesian analysis. However, with modern computational algorithms, conjugate priors are no longer necessary in order to obtain posterior estimates. Despite this, conjugate priors remain very common within Bayesian analyses, and can improve computational efficiency. We reiterate that a prior distribution should be specified that represents expert opinion, before collecting any data. This may be able to be represented by a conjugate prior, but the prior should not in general be restricted to only be of conjugate form.

We now consider a range of *uninformative* or *vague* priors that are commonly used when there is no prior information, before we discuss informative priors.

4.2.2 "Flat" Priors

In the absence of any prior information on one or more model parameters, we would normally aim to ensure that this lack of knowledge is properly reflected in the prior. Typically, this is done by choosing a prior distribution with a suitably wide variance (Carlin and Louis 2001; Gelman et al. 2003). At first, it may seem that taking a flat prior that assigns equal probability to all possible parameter values is a sensible approach. However, flat priors are rarely invariant to reparameterisations of the model. What may appear flat for parameters under one parameterisation may be far from flat under another. (Note that this is similar to the classical problem of the invariance of the concept of unbiasedness to reparameterisations of the model.) In ecological applications many parameters of interest are probabilities (for example, of survival, recapture, etc.). We often express these as logistic regressions such as $\text{logit}(\phi_t) = \alpha + \beta t$, where for instance t might denote time. Placing a flat prior on parameters α and β does not induce a flat prior on ϕ_t (see Section 8.2.1 for further discussion of this). In such circumstances, the statisticians must decide whether they want a prior that is flat on the ϕ-space or (α, β)-space.

Another problem with flat priors is that for continuous variables these priors are *improper* distributions unless bounds on the parameter space are either naturally available (as is the case for probabilities, for example, since they are contained in the interval [0,1]) or are imposed. Improper priors can lead to improper posteriors which, in turn, can mean, for example, that the posterior mean simply does not exist. On the other hand, the imposition of bounds which are too restrictive may result in an unrealistically restricted posterior.

4.2.3 Jeffreys Prior

One solution to the reparameterisation problem is to use the Jeffreys prior (Lindley 1965; Robert 1994). This prior attempts to minimise the amount of influence of the prior on the posterior and is invariant to reparameterisations of the model. It is based upon estimation of Fisher's expected information matrix

– the same matrix used to estimate classical standard errors (see Section 3.5). Mathematically, suppose that we are interested in the parameter θ, and specify $\phi = h(\theta)$, where h is a bijective function. Then the prior for θ is the same as for ϕ, when the scale is transformed. Jeffreys prior is given by,

$$p(\theta) \propto \sqrt{I(\theta|\boldsymbol{x})},$$

where $I(\theta|\boldsymbol{x})$ is the Fisher Information, given in Section 3.5 as

$$I(\theta|\boldsymbol{x}) = -\mathbb{E}\left[\frac{d^2 \log L(\theta|\boldsymbol{x})}{d\theta^2}\right].$$

However, we note that Jeffreys prior is an improper prior.

4.2.4 Hierarchical Prior

In practice most people choose vague priors that are both proper and have large variance, see Gilks et al. (1996), and this is the default choice in computer packages such as WinBUGS, for example (see Chapter 7 and Appendix C). An alternative approach is to adopt a hierarchical prior structure in which the prior parameters are assumed to be unknown and are themselves given prior distributions. Such prior distributions are called hyper-priors. For example, rather than writing $\alpha \sim N(0, 1000)$, we might instead write $\alpha \sim N(0, \tau^{-1})$ with $\tau \sim \Gamma(a, b)$, for suitable a and b (this is a conjugate prior for τ, as stated above). This dilutes the influence on the posterior of any prior assumptions made and essentially creates random effects of the model parameters as we shall see in Example 4.12 and Section 8.5. The influence of the prior can and always should be assessed via a sensitivity analysis and we shall also return to this issue in Section 4.3.

4.2.5 Informative Priors

The aim of informative priors is to reflect information available to the statistician that is gained independently of the data being studied. A prior elicitation process is often used and involves choosing a suitable family of prior distributions for each model parameter, and then attempting to find parameters for that prior that accurately reflect the information available. This is often an iterative process and one might begin by obtaining some idea of the range of plausible model parameter values and then in turn using this to ascribe values to suitable prior parameters. Plots can be made and summary statistics calculated, which can then be shown to the expert to see if they are consistent with his/her beliefs. Any mismatches can be used to alter the prior parameters and the process is repeated (O'Hagan 1998).

For example, Dupuis et al. (2002) consider multi-site capture-recapture-recovery data relating to a population of mouflon. They analyse the data using informative priors. These priors are constructed from independent radio-tagging data of a smaller population of mouflon. Essentially, an initial Bayesian

PRIOR SELECTION AND ELICITATION

analysis is performed on the radio-tagging data, assuming independent uniform priors on the parameters. These uninformative priors are combined with the radio-tagging data to form the posterior distribution of survival and movement probabilities. This posterior distribution is then used as the prior distribution for the multi-site capture-recapture data. An uninformative prior was specified on the recapture probabilities, for which there was no information contained in the radio-tagging data.

Note that this underlying idea of using independent data to construct an informative prior can be viewed in another way. Suppose that we have independent data, denoted by x_1 and x_2. Further, we are interested in the parameters $\boldsymbol{\theta}$. Following the idea above, we initially assume an uninformative prior for $\boldsymbol{\theta}$, denoted by $p(\boldsymbol{\theta})$, and calculate the corresponding posterior distribution of the parameters, using only data x_1. Thus, we have,

$$\pi(\boldsymbol{\theta}|x_1) \propto f(x_1|\boldsymbol{\theta})p(\boldsymbol{\theta}).$$

We then use this posterior distribution to be the corresponding prior distribution in the Bayesian analysis of data x_2. The corresponding posterior distribution of the parameters is given by,

$$\begin{aligned}\pi(\boldsymbol{\theta}|x_1, x_2) &\propto f(x_2|\boldsymbol{\theta})\pi(\boldsymbol{\theta}|x_1) \\ &\propto f(x_2|\boldsymbol{\theta})f(x_1|\boldsymbol{\theta})p(\boldsymbol{\theta}) \\ &= f(x_1, x_2|\boldsymbol{\theta})p(\boldsymbol{\theta}),\end{aligned}$$

since the data x_1 and x_2 are independent. In other words, the building up of an informative prior from independent data is equivalent to jointly analysing all of the data, assuming the same initial prior specified on the parameters. Typically, it is generally easier to perform a single Bayesian analysis of all of the data simultaneously than take the additional step of constructing the informative prior. See, for example, Section 10.2.1 for integrated data analyses, i.e. jointly analysing independent data sets.

Example 4.3. Example 4.1 Revisited – Radio-Tagging Data
We reconsider the radio-tagging data given in Example 4.1. Recall that there is a single parameter, the survival probability, ϕ. As this parameter is a probability, it is constrained to the interval $[0, 1]$. Thus an obvious prior distribution to specify on ϕ is a beta distribution, which is defined over the interval $[0, 1]$. We might then elicit values for the parameters α and β by considering what sets of values we believe are most likely for ϕ. In the absence of any relevant information (i.e. expert prior opinion), we might simply wish to assume a flat prior over the interval $[0,1]$ by setting $\alpha = \beta = 1$.

We combine the prior specified on ϕ, with the corresponding likelihood (see Example 4.1) to obtain the posterior distribution,

$$\phi|x \sim Beta(\alpha + m, \beta + n),$$

where,

$$m = \sum_{i=2}^{7}(i-1)x_i = 162; \quad \text{and} \quad n = \sum_{i=1}^{6} x_i = 14.$$

Under the uniform prior with $\alpha = \beta = 1$, we obtain the following posterior statistics for ϕ:

- posterior mean = 0.916;
- posterior mode = 0.920;
- posterior standard deviation = 0.021.

The corresponding MLE of ϕ is readily seen to be equal to,

$$\hat{\phi} = \frac{m}{m+n} = 0.921.$$

The MLE of ϕ is equal to the posterior mode, since, in this case, the posterior distribution is simply proportional to the likelihood function, assuming a uniform prior. Note that there is some discrepancy between the posterior mean and mode, due to a small amount of skewness in the posterior distribution. In examples where the this skewness is more pronounced, or where the prior is not flat, the posterior mean and the MLE may be considerably different.

For a general $Beta(\alpha, \beta)$ prior, the posterior mean of ϕ is given by,

$$\frac{\alpha + m}{\alpha + m + \beta + n} = (1-w)\left(\frac{\alpha}{\alpha+\beta}\right) + w\left(\frac{m}{m+n}\right),$$

where,

$$w = \frac{m+n}{\alpha + m + \beta + n}.$$

In other words, the posterior mean is a weighted average of the prior mean $(\alpha/(\alpha+\beta))$ and the MLE $(m/(m+n))$. Note that as the number of individuals within the study increases, the value $m+n$ increases, and the weight w tends to unity, so that the posterior mean tends to the MLE. The posterior mean converges to the MLE irrespective of the prior parameters α and β, though the speed at which convergence occurs will vary with respect to the prior specification. In addition, it is straightforward to show that the posterior variance also converges to zero as the sample size increases; see Exercise 4.6. Thus, in the limiting case, the posterior distribution converges to a point mass at the MLE.

□

4.3 Prior Sensitivity Analyses

As discussed earlier, it is important to assess the sensitivity of key results to the prior assumptions made. In the presence of reasonably informative data, there should be little prior sensitivity, as the information from the data should dominate that in the prior. Prior sensitivity in itself is not a problem,

PRIOR SENSITIVITY ANALYSES

but extreme prior sensitivity often points to problems such as parameter redundancy, discussed in Section 2.8, (see Section 4.3.1), or overly restrictive prior assumptions.

A simple sensitivity analysis might involve increasing and decreasing the prior variances to see what, if any, difference this makes to the posterior statistics of interest. This can be done in a fairly structured manner with perhaps three levels of variance for each parameter and a separate analysis conducted for each combination.

Example 4.4. Example 4.1 Revisited – Radio Tagging Data
For the radio-tagging data in Example 4.1, we can explicitly calculate the posterior distribution for ϕ. In Example 4.1, we showed that assuming a $Beta(\alpha, \beta)$ distribution on ϕ, the corresponding posterior distribution for ϕ is given by,

$$\phi|\boldsymbol{x} \sim Beta(\alpha + 162, \beta + 14).$$

The corresponding posterior mean is,

$$\mathbb{E}_\pi(\phi) = \frac{\alpha + 162}{\alpha + 162 + \beta + 14} = \frac{\alpha + 162}{\alpha + \beta + 177}.$$

Thus, we can derive an explicit expression for the posterior mean, and clearly see how the location of the posterior distribution (in terms of the mean) is affected by different priors placed on the parameters. For example, placing a $Beta(1, 1)$ prior (i.e. U[0,1]) prior on ϕ we obtain the posterior mean of 0.916 with lower and upper 2.5% and 97.5% quantiles of 0.871 and 0.952, respectively (this corresponds to a 95% symmetric credible interval; see Section 4.4 for further description of posterior summary statistics). Alternatively, suppose that we have strong prior belief that the value of ϕ is located around 0.8, with a specified 95% interval of (0.7, 0.9). We could represent these prior beliefs by specifying a $Beta(50, 12)$ distribution, which has the specified distributional characteristics (i.e. prior mean of 0.806; prior median of 0.810; 2.5% quantile of 0.700, and 97.5% quantile of 0.894). This provides a posterior distribution for ϕ of $Beta(212, 26)$. The corresponding posterior mean is 0.891, with lower and upper 2.5% and 97.5% quantiles of 0.848 and 0.927, respectively. We note that even though the prior distribution specified did not contain the posterior mean of the parameter under a flat prior (or the MLE of the parameter), within the central 95% interval, the corresponding posterior mean is still only marginally less than the posterior mean under the flat prior. In addition, the posterior mean is at the upper 97.5% prior bound. The posterior distributions for the different priors (and prior distributions) are plotted in Figure 4.3.

Clearly, from Figure 4.3 and prior summary statistics, the $Beta(50, 12)$ prior is quite an informative prior specification. The corresponding posterior distribution for ϕ is then "pulled" towards the values with high prior mass, with the posterior distribution essentially providing a compromise between the information contained in the likelihood concerning ϕ and the prior specified.

□

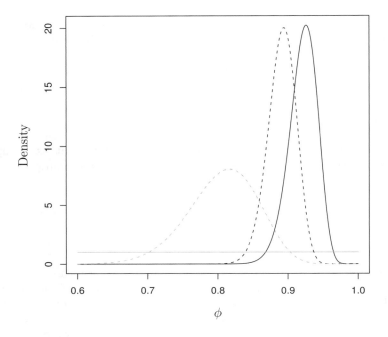

Figure 4.3 Posterior distribution for ϕ (in black) for $Beta(1,1)$ prior (solid line), $Beta(50,12)$ prior (dashed line) and corresponding prior distributions (in grey). Note that the x-axis is only part of the range for ϕ.

Example 4.5. Dippers – Three-Parameter Model

To demonstrate how we might perform a sensitivity analysis for more complex data, we consider the annual capture-recapture data set relating to dippers introduced in Example 1.1. There are a total of 6 release events and 6 recapture events. For simplicity, we consider the model $C/C2$ with 3 parameters: a constant recapture probability, p; survival probability for non-flood years (at capture times 1, 4, 5 and 6) denoted by ϕ_n, and survival probability for flood years (capture times 2 and 3), denoted by ϕ_f. In the absence of any expert prior information, we might specify a $Beta(1,1) (\equiv U[0,1])$ prior on each of the model parameters. To assess the sensitivity of the posterior distribution to the priors, we might consider two alternative priors: a left-skewed and right-skewed distribution. For example, we consider the prior $Beta(1,9)$ on each parameter with mean 0.1 and standard deviation 0.09; and a $Beta(9,1)$ prior on each parameter with mean 0.9 and standard deviation 0.09. Clearly, we could consider the prior sensitivity by simply changing the prior on a single parameter. However, here we consider the more extreme case, changing all the

PRIOR SENSITIVITY ANALYSES

priors simultaneously. If the posterior distribution is insensitive to the prior change on all the parameters, then the posterior distribution is likely to be insensitive to changes in a single prior specification.

In this example we use an MCMC algorithm, to be described in detail in Chapter 5, in order to obtain samples from the joint posterior distribution. The estimated posterior distribution for each parameter is shown as a histogram in Figure 4.4 under the different prior specifications, and compared to the corresponding prior specified on each parameter. In addition, we present the posterior mean and standard deviation of each of the parameters for each prior in Table 4.2.

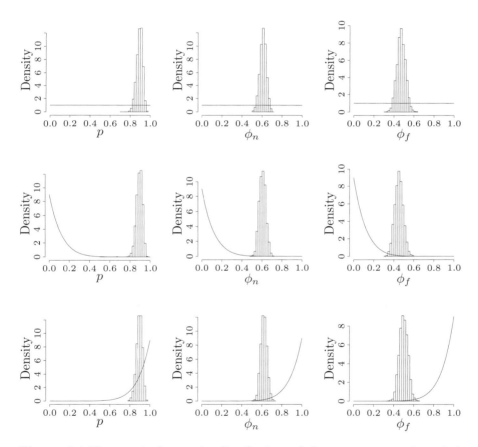

Figure 4.4 The marginal posterior distributions of the parameters p, ϕ_n and ϕ_f estimated by histograms and compared to the prior density under the different independent priors $U[0, 1]$ (top row); $Beta(1, 9)$ (middle row); and $Beta(9, 1)$ (bottom row).

Table 4.2 The Posterior Mean and Standard Deviation (in Brackets) of Each Parameter under the Three Different Priors

Parameter	Prior		
	$U[0,1]$	$Beta(1,9)$	$Beta(9,1)$
p	0.893 (0.030)	0.842 (0.032)	0.895 (0.030)
ϕ_n	0.609 (0.031)	0.605 (0.032)	0.620 (0.031)
ϕ_f	0.472 (0.043)	0.451 (0.042)	0.499 (0.042)

From both Figure 4.4 and Table 4.2 it is clear that the marginal posterior distributions of each of the parameters, p, ϕ_n and ϕ_f, are essentially insensitive to the prior specification. Under three quite different priors the corresponding posterior distributions are very similar. In other words the posterior appears to be data-driven rather than prior-driven.

□

4.3.1 Prior Sensitivity – Parameter Redundancy

The two-parameter capture-recapture model of Example 4.2 for the dippers data can be extended so that both the recapture and annual survival probabilities are functions of time. The resulting model is the CJS model, encountered earlier in Example 2.7. We saw there that this model is parameter redundant, as it is not possible to estimate two of its parameters, only their product. Because of the use of a prior distribution, it can still be possible to obtain a posterior distribution when a model is parameter redundant. An illustration of this is provided in the next example.

Example 4.6. Prior and Posterior Distributions for the CJS Model

In Figure 4.5 we display the independent uniform prior distributions assumed for all of the model parameters, and superimposed on them are the estimated posterior distributions, obtained from using an MCMC algorithm, to be discussed in detail in Chapter 5. In this case the histograms have been smoothed using kernel density estimates. Shown shaded in the figure is the amount of overlap between the prior and posterior distributions.

Due to the parameter redundancy of the CJS model, we cannot obtain separate maximum-likelihood estimates of the parameters ϕ_6 and p_7. We can see from Figure 4.5 that for these two parameters there is appreciable overlap between the marginal prior and posterior distributions, as we would expect. There is a similar amount of overlap for the parameters ϕ_1 and p_2, and this is because the year of release 1981, at the start of the study, had such a small cohort of marked birds.

□

SUMMARISING POSTERIOR DISTRIBUTIONS

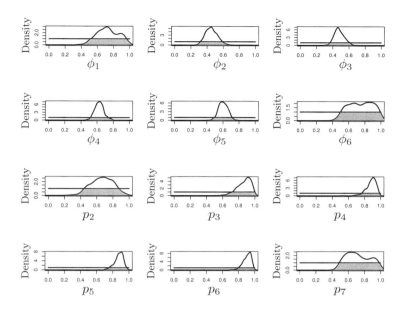

Figure 4.5 Overlap of prior and posterior distributions for the Cormack-Jolly-Seber model applied to the data of Table 1.1. Reproduced with permission from Gimenez et al. (2009b, p. 1062) published by Springer.

Further discussion of this and related examples is provided by Gimenez et al. (2009b). It is shown there that if one assumes uniform prior distributions, then an overlap of more than 35% between prior and posterior distributions is an indication that a parameter is not being estimated well, and that the posterior distribution is dependent upon the prior. As a general rule for all Bayesian work, posterior inferences should always be checked for the influence of prior distributions (i.e. a prior sensitivity analysis performed), for example by performing the analysis for a range of alternative prior distributions, and considering how conclusions might change (see Section 4.3). For further discussion, see Exercises 4.1–4.3.

4.4 Summarising Posterior Distributions

The posterior distribution is typically very complex (and often high-dimensional). Presenting the posterior distribution (up to proportionality) generally provides no directly interpretable information on the parameters of interest. Thus, the posterior distribution is often summarised via a number of summary statistics. For example, the posterior mean is the most commonly used point estimate to summarise the "location" of the posterior distribution. This is often justified as minimising the squared error loss function, with respect to the posterior distribution. However, we note that this summary statistic may not be the most appropriate point estimate in all circumstances. In particular, if the posterior distribution is skewed, then the mean can be highly affected by the skewness of the distribution. In such instances the median is often used as an alternative.

The "spread" of a distribution is often summarised via the variance or standard deviation. However, this does not provide any information regarding possible skewness, for example, of the posterior distribution. A more informative description is provided via credible intervals. Formally, suppose that we are interested in the parameter θ, which has posterior distribution $\pi(\theta|\boldsymbol{x})$. Then, the interval (a, b) is defined as a $100(1-\alpha)\%$ credible interval if,

$$\mathbb{P}_\pi(\theta \in [a,b]|\boldsymbol{x}) = \int_a^b \pi(\theta|\boldsymbol{x})d\theta = 1 - \alpha, \qquad 0 \leq \alpha \leq 1.$$

Note that a $100(1-\alpha)\%$ credible interval is not unique since, in general, there will be many choices of a and b, such that, $\int_a^b \pi(\theta|\boldsymbol{x})d\theta = 1-\alpha$. For example, see Figure 4.6, where we have two 95% credible intervals (i.e. setting $\alpha = 0.05$), denoted by $[a_1, b_1]$ and $[a_2, b_2]$.

Often, a symmetric $100(1-\alpha)\%$ credible interval (a,b) is used. This credible interval is unique, and is defined such that,

$$\int_{-\infty}^a \pi(\theta|\boldsymbol{x})d\theta = \frac{\alpha}{2} = \int_b^\infty \pi(\theta|\boldsymbol{x})d\theta,$$

i.e. the credible interval such that a corresponds to the lower $\frac{\alpha}{2}$ quantile, and b the upper $1 - \frac{\alpha}{2}$ quantile of the posterior distribution, $\pi(\theta|\boldsymbol{x})$.

Consider Figure 4.6, again. Both of the intervals contain 95% of the distribution (i.e. $\alpha = 0.05$), so that there is no objection to either of them. However, what about communicating information about θ? The interval $[a_1, b_1]$ is clearly more informative, since a shorter interval represents a "tighter" inference. This motivates the following refinement of a credible interval. The interval $[a,b]$ is a $100(1-\alpha)\%$ *highest posterior density interval* (HPDI) if:

1. $[a,b]$ is a $100(1-\alpha)\%$ credible interval; and

2. for all $\theta' \in [a,b]$ and $\theta'' \notin [a,b]$, $\pi(\theta'|\boldsymbol{x}) \geq \pi(\theta''|\boldsymbol{x})$.

In an obvious sense, this is the required definition for the "shortest possible" interval having a given credible level $1-\alpha$, and essentially centres the interval around the mode, in the unimodal case. Clearly, if a distribution is symmetrical about the mean, such as the normal distribution, the $100(1-\alpha)\%$ symmetric credible interval is identical to the $100(1-\alpha)\%$ HPDI. In the case where the posterior distribution is multimodal, the corresponding HPDI may consist of several disjoint intervals. In Figure 4.6, the interval $[a_1, b_1] = [0.35, 2.57]$ is the 95% HPDI; and the interval $[a_2, b_2] = [0.50, 3.08]$ the 95% symmetric credible interval. There is clearly quite a difference between these two 95% credible intervals. This is as a result of the skewness of the distribution.

Note that if we have a symmetric credible interval $[a,b]$ for a given parameter

SUMMARISING POSTERIOR DISTRIBUTIONS

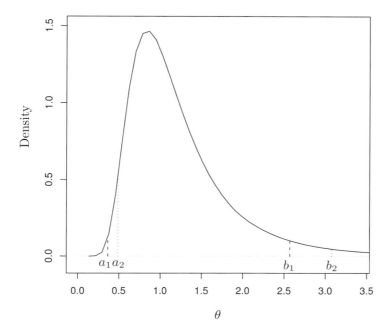

Figure 4.6 The density of a particular inverse gamma distribution, namely, $\Gamma^{-1}(5,5)$ and two alternative 95% credible intervals.

θ, then, if we consider a bijective transformation of the parameter, such as $\phi = g(\theta)$, the corresponding symmetric credible interval for ϕ is given by $[g(a), g(b)]$, However, this is not always true for HPDIs. The corresponding HPDI for ϕ is $[g(a), g(b)]$, if and only if, g is a linear transformation; otherwise, the HPDI needs to be recalculated for ϕ.

Example 4.7. Example 4.1 Revisited – Radio-Tagging Data
In Example 4.1 we derived the posterior distribution for the survival parameter, ϕ, to be a $Beta(163, 15)$ distribution. The specification of the distribution provides all the information needed concerning the parameter and is the most complete description. However, the posterior information given in this distributional form is typically not very informative to practitioners or ecologists. Alternatively, we could plot the posterior distribution as in Figure 4.1, or provide posterior summary statistics. For example, a posterior mean of 0.916 with 0.021 standard deviation and 95% symmetric credible interval (0.872, 0.952). The 95% HPDI is equal to (0.875, 0.955), which is virtually identical to the symmetric credible interval as a result of the posterior distribution being ap-

proximately symmetrical in shape. Note that credible intervals often provide more information than the standard deviation, for example, as they typically provide information on any skewness of the posterior distribution.

□

Example 4.8. Example 1.1 Revisited – Two-Parameter Model for Dipper Data

When there is more than a single parameter, the marginal posterior distributions are often reported for the different parameters. This provides information on a single parameter, integrating out over the rest of the parameter space. For the dipper data, recall that there are two parameters: ϕ (survival probability) and p (recapture probability). The marginal posterior mean and standard deviation for ϕ (integrating out over all possible values of p) are 0.56 and 0.025. Similarly, the marginal posterior mean and standard deviation for p (integrating out over all possible values of ϕ) are 0.90 and 0.029. Credible intervals can similarly be obtained and/or the marginal posterior distributions plotted, as in Figure 4.2. Providing only the posterior marginal distributions provides no information regarding the relationship between the parameters. In the case of more than one parameter, posterior credible regions can be plotted (again see Figure 4.2). Alternatively, to provide a summary point estimate on the relationship between the two parameters, the posterior correlation of ϕ and p can be calculated. For this example, we obtain a posterior correlation of -0.16. As expected there is a negative relationship between the survival and recapture probabilities (although this appears to be relatively small).

□

4.4.1 Credible Intervals versus Confidence Intervals

The difference in philosophy between the classical and Bayesian approaches to statistical inference is well illustrated by looking at the two distinct forms of uncertainty interval associated with the two approaches: confidence intervals and credible intervals.

The classical interval is known as the confidence interval, as already explained in the last chapter. Given the observed data, x, the $100(1-\alpha)\%$ confidence interval $[a(x), b(x)]$ for θ then satisfies the statement:

$$\mathbb{P}([a(x), b(x)] \text{ contains } \theta) = 1 - \alpha.$$

Since θ is assumed fixed, it is the interval constructed from the data that is uncertain, and its randomness derives from the stochastic nature of the data under theoretical replications of the data collection process. The Bayesian interval is known as the *credible interval*. Given data, the $100(1-\alpha)\%$ credible interval $[a, b]$ for θ then satisfies the statement:

$$\mathbb{P}(\theta \in [a, b] | x) = 1 - \alpha.$$

Here it is θ that is the random variable and the interval is fixed. Recall that

DIRECTED ACYCLIC GRAPHS

a credible interval is not unique; there are many such intervals $[a, b]$ such that $100(1 - \alpha)\%$ of the posterior distribution lies within bounds of the interval. Furthermore, the Bayesian statistician conditions on the data observed, rather than appealing to asymptotic properties of estimators, which may not apply for small sample sizes.

4.5 Directed Acyclic Graphs

In this section we present a diagrammatical representation often used to express the manner in which the data, model parameters and corresponding prior distributions are linked. These graphical models neatly summarise and communicate complex model structures in a convenient and straightforward way. In particular we focus on *directed acyclic graphs* (DAGs).

Within a DAG, the data, model parameters and prior parameters are each represented as graphical nodes. Square nodes typically represent a fixed value (e.g. observed data, prior parameter values) and circular nodes a parameter to be estimated (e.g. model parameters). Relationships between these variables are then represented graphically by edges, with the direction of the edge between the nodes denoting the directional relationship between the two variables. A deterministic relationship is represented by a broken line, a stochastic relationship with a solid line. Such graphs are usually presented as a hierarchical structure, with the data presented at the bottom of the graph; the variables that exert the most direct influence upon the data are placed closest to the bottom, and those of lesser influence placed in decreasing order up the graph. The prior parameter values are then at the uppermost level of the graph. Thus, square nodes (for the data and prior parameter values) are typically at the lowest and highest level of the graph. Note that the graph is *acyclic* since there are no cycles in the graph between the nodes via the directional edges. Examples of DAGs for the two examples of this chapter are provided next.

Example 4.9. Radio-Tagging Example
We consider the simple case of the one-parameter radio-tagging data, presented in Example 4.1. The data are the number of individuals radio-tagged and the corresponding number that die between each pair of subsequent capture events, denoted by x. The only model parameter is the survival probability, ϕ. Finally, we need to specify a prior distribution on the survival probability. As before, we specify,

$$\phi \sim Beta(\alpha, \beta).$$

The corresponding DAG in Figure 4.7 provides the relationship between each of the variables. The survival probability is a stochastic function of the prior parameters α and β, although we emphasise that the DAG does not specify the exact distributional form of the relationship (only whether the relationship is stochastic or deterministic). Similarly, the data are a stochastic function of the model parameter ϕ, via the likelihood $f(\boldsymbol{x}|\phi)$, which is of multinomial form.

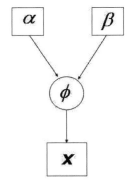

Figure 4.7 A directed acyclic graph (DAG) for the radio-tagging data.

From Figure 4.7 we can clearly see the relationships between the parameters. In addition, the shape of the nodes (square or circular) also clearly distinguishes between the fixed variables (square nodes – data and specified prior parameters) and parameters to be estimated (circular nodes – survival probability). The square nodes are in the lowest level of the graph (data) and the highest level (prior parameters) encompassing the parameters to be estimated.

□

Example 4.10. Dippers Example
We return to the dippers capture-recapture example initially presented in Example 1.1. We consider two different possible models, and their corresponding DAGs. In particular, we consider the two-parameter model, with a constant survival probability over time (as given in Example 4.2) and the three-parameter model, with a different survival probability for flood and non-flood years (as given in Example 4.5). The data are represented by the standard m-array and denoted by $\boldsymbol{m}$. Then for both models, we specify a beta distribution on the recapture probability, p. In particular,

$$p \sim Beta(\alpha_p, \beta_p).$$

For the survival probabilities there are two possible models – a constant survival probability, ϕ; or different survival probabilities relating to flood years, ϕ_f and non-flood years, ϕ_n. For the simpler model, with a constant survival probability, we specify the prior:

$$\phi \sim Beta(\alpha_\phi, \beta_\phi).$$

Similarly, for the more complex three-parameter model, we specify indepen-

DIRECTED ACYCLIC GRAPHS

dent beta distributions on the different survival probabilities, in the form,

$$\phi_f \sim Beta(\alpha_{\phi_f}, \beta_{\phi_f}); \quad \text{and} \quad \phi_n \sim Beta(\alpha_{\phi_n}, \beta_{\phi_n})$$

The corresponding DAGs are presented in Figure 4.8 for the two different models. The DAGs are clearly very similar, since the models are themselves very similar.

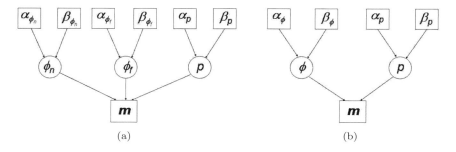

Figure 4.8 Directed acyclic graphs for the dippers data for (a) two-parameter model (constant survival probability) and (b) three-parameter model (flood dependent survival probability).

□

Example 4.11. Covariate Regression Analysis
As we have seen in Section 2.5, within the analysis of ecological data, we often wish to express the model parameters as a function of different factors, or covariates. This approach permits the model parameters to be directly associated with the given covariates of interest. For example, survival probabilities may be expressed as a function of resource availability or predator abundance. We consider a particular example relating to ring-recovery data of UK lapwings from 1963–1998. The corresponding likelihood and further description of this type of data are described in Section 2.3.1 (with the ring-recovery data provided for the lapwings for 1963–1973). The regression of demographic probabilities on covariates is also briefly described in Section 2.5, and considered in greater detail in Chapter 8 (and Section 10.2). There are three parameters within the model that we consider: ϕ_1, the first-year survival probability; ϕ_a, the adult survival probability; and λ, the probability of recovery of dead individuals. To explain the temporal variability of the demographic parameters, we consider two covariates: time (t) and and $fdays$ denoting the number of days that the minimum temperature fell below freezing in Central England during the winter leading up to the breeding season in year t given in Figure 2.2. Here, $fdays$ can be regarded as a (crude) measure of the harshness of the winter (and hence also resource availability). Each of the survival parameters are regressed on $fdays$ and the recovery probability on time. Thus,

mathematically, we have,

$$\text{logit } \phi_1 = \alpha_1 + \beta_1 f_t;$$
$$\text{logit } \phi_a = \alpha_a + \beta_a f_t;$$
$$\text{logit } \lambda = \alpha_\lambda + \beta_\lambda u_t,$$

where f_t and u_t denote the normalised values for the covariates $fdays$ and time, respectively. This model is typically denoted by:

$$\phi_1(fdays); \phi_a(fdays)/\lambda(t).$$

Note that normalising covariate values are simply defined to be,

$$\frac{f_t - \mathbb{E}(f)}{SD(f)},$$

where $\mathbb{E}(f)$ denotes the sample mean of the observed covariate values; and $SD(f)$ the sample standard deviation of the observed covariate values. Within Bayesian analyses, it is important that the covariates are normalised so that (i) all covariates are on the same scale (if there is more than one covariate); and (ii) the priors on the corresponding regression coefficient are specified on a meaningful scale (see Borchers et al. 2010 for further discission).

We place priors on each of the regression parameters. In particular, we specify independent normal priors of the form,

$$\alpha_i \sim N(\mu_{\alpha_i}, \sigma^2_{\alpha_i}); \qquad \beta_i \sim N(\mu_{\beta_i}, \sigma^2_{\beta_i}), \qquad (4.2)$$

for $i \in \{1, a, \lambda\}$.

The corresponding DAG is provided in Figure 4.9. Once more we note that the lowest and highest levels of the DAG are represented with square nodes (of known parameters), with all the other nodes circular, denoting parameters to be estimated. Note that the dashed lines represent the deterministic relationship between the demographic parameters (ϕ_1, ϕ_a and λ) and the regression coefficients ($\alpha_1, \beta_1, \alpha_a, \beta_a, \alpha_\lambda$ and β_λ) given in (4.2).

□

Example 4.12. Fixed Effects versus Random Effects
We once more consider the radio-tagging data presented in Example 4.1 and discussed above. However, we now assume that the survival probability is time dependent, so that we have parameters, $\phi = \{\phi_1, \ldots, \phi_T\}$. We could then specify independent beta distributions on each of the parameters $\phi_1, \ldots, \phi_T$. However, we consider an alternative parameterisation and specify,

$$\text{logit } \phi_t = \beta_t$$

where $\beta_t \sim N(\alpha, \sigma^2)$, and we specify $\alpha \sim N(\mu_\alpha, \sigma^2_\alpha)$. Typically in the absence of prior information, we might set $\mu_\alpha = 0$ and $\sigma^2_\alpha = 10$ (see Section 8.2.1 for further discussion of prior specification for parameters within logistic regressions)

SUMMARY

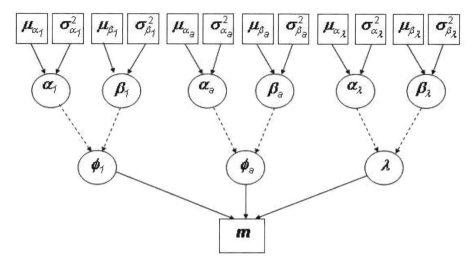

Figure 4.9 A directed acyclic graph (DAG) for the UK lapwing ring-recovery data for a logistic regression covariate analysis.

We note that we can rewrite this parameterisation in the form,

$$\text{logit } \phi_t = \alpha + \epsilon_t,$$

where $\epsilon_t \sim N(0, \sigma^2)$.

Specifying a value for σ^2 we obtain a fixed effects model with parameters α and $\epsilon = \{\epsilon_1, \ldots, \epsilon_T\}$. The corresponding DAG is given in Figure 4.10(a). Alternatively, we could specify a hierarchical prior on σ^2, for example,

$$\sigma^2 \sim \Gamma^{-1}(a, b).$$

An alternative interpretation of this hierarchical prior is that of a random effects model, with parameters α and σ^2. The corresponding DAG is given in Figure 4.10(b). Thus, in terms of the DAG, the hierarchical prior (or equivalently random effects model) adds in an extra level to the graph. See Sections 4.2.4 and 8.5 for further discussion.

□

4.6 Summary

In this chapter we have described the basis for Bayesian inference, which follows naturally from Bayes' Theorem. In contrast to classical inference, parameters and data are considered on an equal footing, and the effect of data is to modify distributions for parameters, which change from prior (beliefs concerning the parameters before observing the data) to posterior forms (updated beliefs regarding the parameters after observing the data). The Bayesian ap-

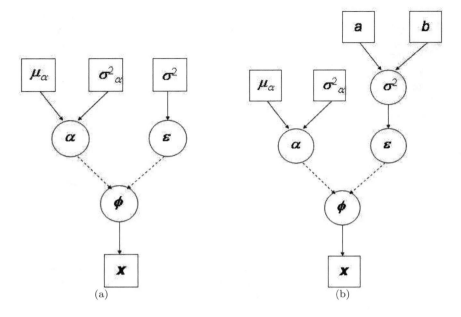

Figure 4.10 The DAG for the time dependent radio-tagging data for (a) fixed effects model; and (b) hierarchical prior (or random effects model).

proach allows prior information to be combined formally with the information to be obtained from the data via the likelihood. Formal elicitation methods exist to assist experts to translate their prior knowledge into explicit prior distribution form that can be used within Bayesian analyses. However, in many analyses, uninformative priors may be preferred. We emphasise that the posterior distribution should always be assessed for prior sensitivity, even in the case of informative priors, since their specification can sometimes be difficult to represent the prior beliefs accurately. O'Hagan (1998) discusses how informative prior distributions can be translated from expert prior beliefs using an iterative approach. A practical example of forming informative prior distributions from additional external information is provided by King et al. (2005).

Within Bayesian analyses we can calculate an expression for the corresponding posterior distribution (up to proportionality). For multi-parameter models, the posterior distribution is often high-dimensional, so that direct interpretation is difficult and inference difficult. This typically leads to marginal posterior distributions being of interest, essentially integrating out the other parameter values. Direct integration can sometimes be performed (see for example Gelman et al. 2003), allowing the exact shape of the marginal posterior distribution to be calculated. However, alternative computational techniques need to be implemented when this is not the case (see Chapter 5).

4.7 Further Reading

There are numerous publications detailing the Bayesian approach to both general statistical problems and particular applications. For example, Lee (2004) provides an excellent general introduction to Bayesian statistics. Additional books that provide a wider discussion of issues and applications include Gelman et al. (2003), Congdon (2006) and Robert (2007). The paper by Bellhouse (2004) provides a biography on Bayes. In addition, there are a number of recent books either containing a statistical ecology application, or wholly aimed at the Bayesian approach to statistical problems, including McCarthy (2007), Royle and Dorazio (2008), Link and Barker (2009) and Marin and Robert (2007). These books each focus on different aspects and ecological applications. For example, Royle and Dorazio (2008) focus on hierarchical modelling within statistical ecology, including application areas of species richness and occupancy models not considered within this book. An introductory reference to Bayesian methods is the book by Lee (2004), and a historical review is provided by Brooks (2003). More advanced topics are provided by Carlin and Louis (2001). The classical reference for modern applied Bayesian methods, with applications drawn from a range of different areas is Gelman et al. (2003), while the edition by Gilks et al. (1996) illustrates the great potential for modern Bayesian methods. Bayesian state-space models are constructed in Millar and Meyer (2000) and illustrations of hierarchical models are given by Link et al. (2002) and Rivot and Prévost (2002); these papers provide examples of complex probability models which are readily fitted using Bayesian methods. O'Hagan (1998) discusses how one can incorporate prior knowledge in a prior distribution, when that is felt to be appropriate, and a good illustration of this is provided in Gelman et al. (2003, p. 260) in the context of a hierarchical model in the area of pharmacokinetics. The paper by Millar (2004) presents an analytic approach to prior sensitivity analysis; see Exercise 4.4.

Various extensions to the CJS model have been proposed. For example, Pradel (1996) considers the estimation of recruitment into the population, in addition to the standard estimation of removal from the population (via death or permanent migration). Similar ideas applied in a Bayesian framework, that consider not only survival probabilities but also birth rates, are provided by Link and Barker (2005) and O'Hara et al. (2009). Alternatively, structured error processes have been proposed by Johnson and Hoeting (2003) for capture-recapture data, where the demographic parameters are modelled as an autoregressive process. Thus, this approach does not assume that survival probabilities are independent of each other between different capture events, but are related via a modelled process (i.e. autoregressive process). Additional ecological areas where Bayesian methods are used include fisheries stock assessment (Millar and Meyer 2000; McAllister and Kirkwood 1998b; Punt and Hilborn 1997), site occupancy models (Kéry and Royle 2007), species richness (Mingoti 1999 and Dorazio et al. 2006) and meta-population models for patch occupancy (O'Hara et al. 2002 and ter Braak and Etienne 2003)

4.8 Exercises

4.1 We continue here the discussion of overlap of prior and posterior distributions, from Example 4.6. Here and in the next exercise we use τ to indicate the proportion of overlap. We give in Table 4.3 the amounts of overlap of prior and posterior distributions, illustrated in Figure 4.5. Provide a suitable discussion of these values; comment in particular on the suggested threshold for τ of 35 %.

Table 4.3 Overlap of Priors and Posteriors Values Expressed as Percentages for the CJS Model Applied to the Dipper Data, Corresponding to Figure 4.5

Parameter	τ
ϕ_1	53.6
ϕ_2	33.4
ϕ_3	29.2
ϕ_4	27.9
ϕ_5	27.4
ϕ_6	57.2
p_2	54.7
p_3	35.9
p_4	28.4
p_5	26.4
p_6	23.4
p_7	56.5

Source: Reproduced with permission from Gimenez et al. (2009b, p. 1062) published by Springer.

Note: For all parameters $U(0,1)$ priors are used.

4.2 We return to the mallard recovery data discussed in Exercise 3.1. Shown in Figure 4.11 are the prior and posterior distributions when the model fitted has two annual survival probabilities, ϕ_1, corresponding to birds in their first year of life, and ϕ_a, corresponding to all older birds, as well as two corresponding recovery probabilities, λ_1 and λ_a. This model is parameter redundant (see Section 2.8). The overlap proportions are $\tau_{\phi_1} = 75.93, \tau_{\phi_a} = 10.78, \tau_{\lambda_1} = 60.66, \tau_{\lambda_a} = 51.33$. Provide a suitable discussion, commenting on the advisability of a Bayesian analysis of a parameter redundant model.

4.3 The herring gull data of Exercise 3.4 have been analysed by Vounatsou and Smith 1995, using a Bayesian approach, and fitting the parameter-redundant Cormack-Seber model of Example 2.8. Consider what form the

EXERCISES

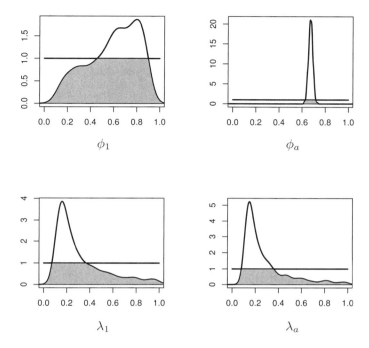

Figure 4.11 Display of the prior-posterior distribution pairs for ϕ_1, ϕ_a, λ_1 and λ_a in the parameter redundant submodel of the FM model, applied to the mallard data. The priors for all parameters were chosen as $U(0,1)$ distributions. Note that scales in panels differ. Reproduced with permission from Gimenez et al. (2009b, p. 1060) published by Springer.

overlaps of prior and posterior distributions will take in this case, and provide suitable discussion.

4.4 The random variable X has a $N(\mu, \sigma^2)$ distribution, where σ^2 is known and the parameter μ has a $N(\nu, \tau^2)$ distribution. Form the partial derivative $\partial \nu^*/\partial \nu$, where ν^* denotes the mean of the posterior distribution of μ. Consider how this might be used to determine the sensitivity of the posterior distribution to the prior. (Millar 2004).

4.5 Verify the cell probabilities of the m-array given in Example 4.1. Discuss how you might modify the model to allow the survival probability ϕ to vary between animals.

4.6 Verify the result stated in Example 4.1 that the posterior variance converges to zero as the sample size increases.

4.7 Verify the conjugate forms for prior distributions given in Section 4.2.1.

4.8 In Example 4.8 it is stated that the negative correlation obtained between the survival and recapture estimates is to be expected. Explain why this is the case.

4.9 Discuss whether the DAG of Figure 4.9 might be appropriate for a model of the ring-recovery data on herons in Example 2.1.

4.10 Prove that Jeffreys prior is invariant to a transformation of variables.

CHAPTER 5

Markov Chain Monte Carlo

At the beginning of the 1990s there were two developments that provided an enormous impetus to the development of modern Bayesian methods, namely:

- the computer revolution, which made powerful computers an affordable reality;
- the Markov chain Monte Carlo (MCMC) algorithm was introduced to the statistical literature (Smith and Gelfand 1992).

Before we describe MCMC in detail, we first consider the two components of this approach: Monte Carlo integration and Markov chains.

5.1 Monte Carlo Integration

In many circumstances, we are faced with the problem of evaluating an integral which is too complex to calculate explicitly. We recall that Bayesian inference is based upon the estimation of posterior summaries such as the mean, and that these require integration of the posterior density. For example, we may wish to estimate the posterior expectation of a function $\psi(.)$ of a parameter, θ, given observed data $\boldsymbol{x}$:

$$\mathbb{E}_\pi[\psi(\theta)] = \int \psi(\theta)\pi(\theta|\boldsymbol{x})d\theta.$$

We can use the simulation technique of *Monte Carlo integration* to obtain an estimate of a given integral (and hence posterior expected value). The method is based upon drawing observations from the distribution of the variable of interest and simply calculating the empirical estimate of the expectation. For example, given a sample of observations, $\theta^1, \ldots, \theta^n \sim \pi(\theta|\boldsymbol{x})$, we can estimate the integral

$$\mathbb{E}_\pi[\psi(\theta)] = \int \psi(\theta)\pi(\theta|\boldsymbol{x})d\boldsymbol{x}$$

by the average

$$\overline{\psi}_n = \frac{1}{n}\sum_{i=1}^{n}\psi(\theta^i). \tag{5.1}$$

In other words, we proceed by drawing samples $\theta^1, \ldots, \theta^T$ from the posterior distribution and then calculating the sample mean. This is Monte Carlo integration (Morgan 1984, p. 163)

Example 5.1. Simulating Beta Random Variables Using Rejection

In Example 4.1 the posterior distribution is beta, with known parameters, and we saw there that we have explicit expressions for the posterior mode and mean. However, it is of interest to consider how we might simulate from such a posterior distribution. One way to do this is to use the method of rejection, illustrated in Figure 5.1. Here a beta density is given a rectangular envelope, under which points are simulated uniformly and at random. We reject those points that lie above the probability density function, and retain the points that lie below the density function, as the x-coordinates of the latter points have the beta distribution; see Morgan (2008). As an illustration, for the $Beta(3, 2)$ distribution, in a simulation of 100 points in the appropriate rectangle, 48 points are accepted, with a sample average of 0.5733, when the population expectation is known to be 0.6.

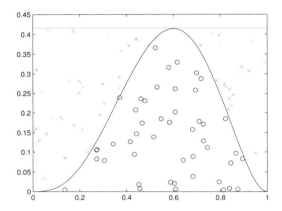

Figure 5.1 Illustration of a rejection method used to simulate beta random variables. The distribution shown is $Beta(3,2)$. The ×'s are rejected and the ○'s are accepted. The x-coordinates of the ○'s are realisations of the random variable with the $Beta(3,2)$ density function.

□

For independent samples, the Law of Large Numbers ensures that

$$\overline{\psi}_n \to \mathbb{E}_\pi[\psi(\theta)] \quad \text{as } n \to \infty.$$

Independent sampling from $\pi(\theta|\boldsymbol{x})$ may be difficult, however Equation (5.1) may still be used if we generate our samples, not independently, but via some other method.

The above example concentrates on obtaining an estimate of the posterior mean of a distribution. However, any number of posterior summary statistics may be of interest. For example, suppose that we are interested in the posterior probability that the parameter of interest, θ, has a value greater than 10.

MARKOV CHAINS

Then we can estimate $\mathbb{P}(\theta > 10|\boldsymbol{x})$ by simply calculating the proportion of the sample from the posterior distribution for which the parameter value is greater than 10. Alternatively, the sample obtained can be used to estimate the density of the (marginal) posterior distribution of θ. We have already seen this done in Figures 4.4 and 4.5.

Thus, we replace the integration problem by a sampling problem, but this creates two new problems. In general, $\pi(\theta|\boldsymbol{x})$ represents a high-dimensional and complex distribution from which samples would usually be difficult to obtain. In addition, large sample sizes are often required and so powerful computers are needed to generate these samples. So, how do we obtain a potentially large sample from the posterior distribution, when in general, this will be very complex and often high-dimensional? The answer is via the use of a Markov chain.

5.2 Markov Chains

A Markov chain is simply a stochastic sequence of numbers where each value in the sequence depends *only* upon the last (Gilks et al. 1996). We might label the sequence $\theta^0, \theta^1, \theta^2$, etc., where the value for θ^0 is chosen from some arbitrary starting distribution. In general, if we want to simulate a Markov chain, we generate the new state of the chain, say θ^{k+1}, from some density, dependent only on θ^k:

$$\theta^{k+1} \sim \mathcal{K}(\theta^k, \theta) \quad \left(\equiv \mathcal{K}(\theta|\theta^k)\right).$$

We call $\mathcal{K}$ the transition kernel for the chain. The transition kernel uniquely describes the dynamics of the chain.

Under certain conditions, that the chain is aperiodic and irreducible, the distribution over the states of the Markov chain will converge to a *stationary* distribution. Note that we shall always assume that these conditions are met. Suppose that we have a Markov chain with stationary distribution $\pi(\theta)$. The Ergodic Theorem states that,

$$\overline{\psi}_n = \frac{1}{n}\sum_{i=1}^{n} \psi(\theta^i) \to \mathbb{E}_\pi(\psi(\theta)) \qquad \text{as } n \to \infty.$$

For our purposes, $\pi(\theta) \equiv \pi(\theta|\boldsymbol{x})$ (i.e. the stationary distribution will be the posterior distribution of interest).

Combining Markov chains with Monte Carlo integration results in Markov chain Monte Carlo (MCMC).

5.3 Markov Chain Monte Carlo

MCMC methods perform Monte Carlo integration using a Markov chain to generate observations from π. Essentially, we construct a Markov chain (using standard methods described below) where the updating of the parameters in the Markov chain is performed such that the probability distribution associated with the kth observation gets closer and closer to $\pi(\theta|x)$ as k increases. We say that the distribution of the chain *converges* to π and we return to the concept of convergence later on.

MCMC therefore allows us to construct a sequence of values whose distribution converges to the posterior distribution of interest (if the chain is run for long enough). Once the chain has converged to the stationary distribution, we can then use the sequence of values taken by the chain in order to obtain empirical (Monte Carlo) estimates of any posterior summaries of interest, such as posterior means. Note that we need to ensure that the Markov chain has reached the stationary distribution *before* we can use the realisations to obtain our Monte Carlo estimates of the posterior distributions of interest. This means that we need to discard realisations from the start of the chain and only use observations once the chain has converged. This initial period of the chain is referred to as the *burn-in*. We return to the issue of determining the length of the burn-in in Section 5.4.2, once we have described the general MCMC methods. We emphasise that the beauty of MCMC is that the updating procedure remains relatively simple, no matter how complex the posterior distribution of interest. Thus we can do the required integration by sampling from the posterior distribution, and we can sample from the posterior by generating a Markov chain. We construct a Markov chain using an appropriate updating scheme, but how do we do these updates? There are several standard approaches.

5.3.1 Metropolis Hastings

A general way to construct MCMC samplers is as a form of generalised rejection sampling, where values are drawn from distributions that approximate a target distribution and are "corrected" in order that, asymptotically, they behave as random observations from the target distribution. This is what happened in the rejection method of Example 5.1, and is the motivation for methods such as the *Metropolis Hastings algorithm*, which sequentially draws observations from a candidate distribution, conditional only upon the last observation, thus inducing a Markov chain. The most important aspect of such algorithms is not the Markov property, but the fact that the approximating candidate distributions are improved at each step in the simulation.

Most MCMC methods are special cases of the general Metropolis Hastings (MH) scheme (Chib and Greenberg 1995), which has already been illustrated in Example 4.2. This method is based upon the observation that given a Markov chain with transition density $\mathcal{K}(\boldsymbol{\theta}, \boldsymbol{\phi})$ and exhibiting what is called detailed balance for π i.e.

$$\pi(\boldsymbol{\theta})\mathcal{K}(\boldsymbol{\theta}, \boldsymbol{\phi}) = \pi(\boldsymbol{\phi})\mathcal{K}(\boldsymbol{\phi}, \boldsymbol{\theta}), \tag{5.2}$$

the chain has stationary density, $\pi(\cdot)$.

The candidate generating distribution (or proposal distribution) typically depends upon the current state of the chain, $\boldsymbol{\theta}^k$, with this distribution denoted

MARKOV CHAIN MONTE CARLO

by $q(\phi|\theta^k)$. The choice of proposal density is essentially arbitrary. However, in general, the induced chain will not satisfy the reversibility condition of Equation (5.2), so we introduce an acceptance function $\alpha(\theta^k, \phi)$. We then accept the candidate observation, and set $\theta^{k+1} = \phi$, with probability $\alpha(\theta^k, \phi)$; otherwise if the candidate observation is rejected, the chain remains at θ^k, so that $\theta^{k+1} = \theta^k$.

It can be shown that the optimal form for the acceptance function (Peskun 1973), in the sense that suitable candidates are rejected least often and computational efficiency is maximised, is given by

$$\alpha(\theta^k, \phi) = \min\left(1, \frac{\pi(\phi|x)q(\theta^k|\phi)}{\pi(\theta^k|x)q(\phi|\theta^k)}\right).$$

In order to specify the corresponding transition kernel, suppose that the parameters θ can take values in the set Θ. (For example, suppose that $\theta = \{\theta_1, \ldots, \theta_p\}$, then if each parameter is continuous and able to takes all possible real values, $\Theta = \mathcal{R}^p$; alternatively if the parameters are all probabilities, then $\Theta = [0, 1]^p$). For some set A, the transition kernel is given by

$$\mathcal{P}_H(\theta, A) = \int_A \mathcal{K}_H(\theta, \phi)d\phi + r(\theta)I_A(\theta),$$

where

$$\mathcal{K}_H(\theta, \phi) = q(\phi|\theta)\alpha(\theta, \phi),$$

and

$$r(\theta) = 1 - \int_\Theta \mathcal{K}_H(\theta, \phi)d\phi. \quad (5.3)$$

Thus $r(\theta)$ corresponds to the probability that the proposed parameter values, ϕ, are rejected. The term $I_A(\theta)$ is simply the identity function, so that,

$$I_A(\theta) = \begin{cases} 1 & \theta \in A \\ 0 & \theta \notin A. \end{cases}$$

The term $\mathcal{K}_H$ satisfies the reversibility condition of Equation (5.2), implying that the kernel, $\mathcal{P}_H$ also preserves detailed balance for π. For further discussion see Chib and Greenberg (1995).

Thus, the MH method can be written algorithmically by generating θ^{k+1} from θ^k in a two-step procedure, as follows.

Step 1. At iteration t, sample $\phi \sim q(\phi|\theta^k)$. Here ϕ is known as the *candidate* point and q is known as the *proposal* distribution.

Step 2. With probability

$$\alpha(\theta^k, \phi) = \min\left(1, \frac{\pi(\phi|x)q(\theta^k|\phi)}{\pi(\theta^k|x)q(\phi|\theta^k)}\right)$$

set $\theta^{k+1} = \phi$ (i.e. accept the candidate point) otherwise, set $\theta^{k+1} = \theta^k$ (i.e. reject the candidate point and the state remains unchanged).

Steps 1 and 2 are then repeated.

The algorithm is completed with the specification of the initial parameter values $\boldsymbol{\theta}^0$, which are chosen arbitrarily.

Example 5.2. Simulating Beta Random Variables Using MH

Suppose that we are interested in sampling from a $Beta(\alpha, \beta)$ distribution; we have already done this using rejection in Example 5.1. We shall now show how to perform the simulation using the MH algorithm. Suppose that the current parameter is θ. We simulate a candidate value ϕ from the proposal distribution of the form,

$$\phi|\theta \sim N(\theta, \sigma^2),$$

where σ^2 is to be specified. The corresponding acceptance probability is given by, $\alpha(\theta, \phi) = \min(1, A)$, where,

$$\begin{aligned}
A &= \frac{\pi(\phi|\boldsymbol{x})q(\theta|\phi)}{\pi(\theta|\boldsymbol{x})q(\phi|\theta)} \\
&= \frac{\pi(\phi|\boldsymbol{x})}{\pi(\theta|\boldsymbol{x})}
\end{aligned}$$

since $q(\theta|\phi) = q(\phi|\theta)$ as q is a symmetric distribution

$$= \begin{cases} \frac{\phi^{\alpha-1}(1-\phi)^{\beta-1}}{\theta^{\alpha-1}(1-\theta)^{\beta-1}} & \phi \in [0,1] \\ 0 & \phi \notin [0,1] \end{cases}$$

$$= \begin{cases} \left(\frac{\phi}{\theta}\right)^{\alpha-1} \left(\frac{1-\phi}{1-\theta}\right)^{\beta-1} & \phi \in [0,1] \\ 0 & \phi \notin [0,1]. \end{cases}$$

Note that if the candidate value ϕ is simulated outside the interval [0,1] (i.e. $\phi \notin [0,1]$), the corresponding acceptance probability for the move is simply zero. This is easily seen by considering the form of the numerator of the acceptance probability. The numerator is the evaluation of the posterior distribution at the candidate value. This value is equal to zero with both the likelihood function and the prior density at the candidate value simply equal to zero. However, for clarity, we explicitly state that the acceptance probability is equal to zero for $\phi \notin [0,1]$.

As an illustration, we take $\alpha = 163$ and $\beta = 15$, corresponding to the posterior distribution for the survival probability in the radio-marking of canvasback ducks, considered in Example 4.1. Figure 5.2 plots the resulting output for $\sigma = 0.05$. Notice that the acceptance function is independent of σ, but that the value of σ has a significant impact upon the acceptance probability of the chain. See Section 5.4.4 for further discussion and for improving the efficiency of the MH algorithm via pilot tuning.

We note that one possible method for reducing the proportion of proposed moves that are rejected is to ensure that the proposed parameter values always lie in the interval [0,1]. This can be done by specifying a (slightly) more

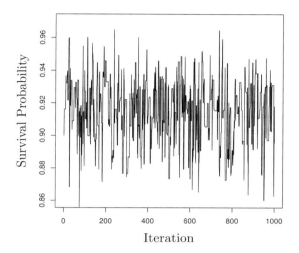

Figure 5.2 Sample path for a MH algorithm for the survival probability for radio-tagging data. The values are treated as a sample from the $Beta(163, 15)$ distribution, with mean 0.916.

complex proposal distribution. For example, we could specify,

$$\phi|\theta \sim U[-\epsilon_\theta, \epsilon_\theta],$$

where,

$$\epsilon_\theta = \min(\epsilon, \theta, 1 - \theta).$$

This ensures that $\phi \in [0, 1]$. Alternatively, we could specify,

$$\phi|\theta \sim TN(\theta, \sigma^2, 0, 1),$$

where $TN(\theta, \sigma^2, 0, 1)$ denotes the (truncated) normal distribution with mean μ and variance σ^2, constrained to the interval [0,1].

The corresponding acceptance probability is given by $\alpha(\theta, \phi) = \min(1, A)$, where,

$$\begin{aligned} A &= \frac{\pi(\phi|\boldsymbol{x})q(\theta|\phi)}{\pi(\theta|\boldsymbol{x})q(\phi|\theta)} \\ &= \left(\frac{\phi}{\theta}\right)^{\alpha-1} \left(\frac{1-\phi}{1-\theta}\right)^{\beta-1} \frac{q(\theta|\phi)}{q(\phi|\theta)}. \end{aligned}$$

Note that the proposal densities no longer cancel in the acceptance probability, in general.

Typically, for parameters whose posterior distributions are near boundaries, such constrained proposal distributions may increase the number of proposed values that are accepted. However, there is the trade-off of the additional complexity in the acceptance probability (and possible computation time). Note that, for simplicity, throughout this book we will generally consider the simpler (unconstrained) proposal distributions.

□

Within the MH scheme there are a number of special cases depending upon the (essentially arbitrary) choice of proposal distribution. We briefly discuss some of these here.

5.3.2 Single-Update Metropolis Hastings

Typically, the posterior distribution will be high-dimensional. The above algorithm has been specified in general, for the multi-dimensional case. However, the MH algorithm need not update all variables in the sample space simultaneously. It can be used in stages, updating variables one at a time. This is commonly called the *single-update MH algorithm*. Suppose that there are p parameters. In the single-update MH algorithm, we break each iteration of the algorithm into p individual updates, each using the 2-step algorithm described above. Let q_j denote a proposal distribution for candidates in the jth coordinate direction. Typically the proposal distribution q_j will not be dependent on the other coordinate directions (i.e. the parameter values $\theta_1, \ldots, \theta_{j-1}, \theta_{j+1}, \ldots, \theta_p$). Thus, for simplicity, we assume the proposal distribution is only dependent on the jth parameter (though the extension to all possible parameters is straightforward). For simulating the set of parameter values at iteration $k+1$, $\boldsymbol{\theta}^{k+1}$, given the values at iteration k, $\boldsymbol{\theta}^k$, the single update MH algorithm can be described as follows:

Step 1(1). Propose candidate value $\phi_1 \sim q_1(\phi_1|\theta_1^k)$.
Define $\boldsymbol{\phi}_1 = \{\phi_1, \theta_2^k, \ldots, \theta_p^k\}$, and $\boldsymbol{\theta}_1^k = \{\theta_1^k, \ldots, \theta_p^k\}$.

Step 2(1). Calculate the acceptance probability:

$$\alpha(\theta_1^k, \phi_1) = \min\left(1, \frac{\pi(\boldsymbol{\phi}_1|\boldsymbol{x})q_1(\theta_1^k|\phi_1)}{\pi(\boldsymbol{\theta}_1^k|\boldsymbol{x})q_1(\phi_1|\theta_1^k)}\right).$$

With probability $\alpha(\theta_1^k, \phi_1)$ set $\theta_1^{k+1} = \phi_1$; otherwise set $\theta_1^{k+1} = \theta_1^k$.

$$\vdots$$

Step 1(j). For $1 < j < p$, propose candidate value $\phi_j \sim q_j(\phi_j|\theta_j^k)$.
Define $\boldsymbol{\phi}_j = \{\theta_1^{k+1}, \ldots, \theta_{j-1}^{k+1}, \phi_j, \theta_{j+1}^k, \ldots, \theta_p^k\}$ and
$\boldsymbol{\theta}_j^k = \{\theta_1^{k+1}, \ldots, \theta_{j-1}^{k+1}, \theta_j^k, \theta_{j+1}^k, \ldots, \theta_p^k\}$.

Step 2(j). Calculate the acceptance probability:

$$\alpha(\theta_j^k, \phi_j) = \min\left(1, \frac{\pi(\boldsymbol{\phi}_j|\boldsymbol{x})q_j(\theta_j^k|\phi_j)}{\pi(\boldsymbol{\theta}_j^k|\boldsymbol{x})q_j(\phi_j|\theta_j^k)}\right).$$

With probability $\alpha(\theta_j^k, \phi_j)$ set $\theta_j^{k+1} = \phi_j$; otherwise set $\theta_j^{k+1} = \theta_j^k$.

⋮

Step 1(p). Propose candidate value $\phi_p \sim q_p(\phi_p|\theta_p^k)$.

Define $\boldsymbol{\phi}_p = \{\theta_1^{k+1}, \ldots, \theta_{p-1}^{k+1}, \phi_p\}$ and $\boldsymbol{\theta}_j^k = \{\boldsymbol{\theta}_p^k, \ldots, \theta_{p-1}^{k+1}, \theta_p^k\}$

Step 2(p). Calculate the acceptance probability:

$$\alpha(\theta_p^k, \phi_p) = \min\left(1, \frac{\pi(\boldsymbol{\phi}_p|\boldsymbol{x})q_p(\theta_p^k|\phi_p)}{\pi(\boldsymbol{\theta}_p^k|\boldsymbol{x})q_p(\phi_p|\theta_p^k)}\right).$$

With probability $\alpha(\theta_p^k, \phi_p)$ set $\theta_p^{k+1} = \phi_p$; otherwise set $\theta_p^{k+1} = \theta_p^k$.

This completes a transition from $\boldsymbol{\theta}^k$ to $\boldsymbol{\theta}^{k+1}$. Note that when we propose to update the jth component of the parameter vector $\boldsymbol{\theta}$, we use the *updated* values of the parameters $\theta_1, \ldots, \theta_{j-1}$.

Note that we can simplify the acceptance probability for the single-update algorithm, which may help to improve efficiency in terms of computational expense. Let $\boldsymbol{\theta}_{(i)}^k = \{\theta_1^{k+1}, \ldots, \theta_{i-1}^{k+1}, \theta_{i+1}^k, \ldots, \theta_p^k\}$. Hence $\boldsymbol{\theta}_{(i)}^k$ denotes the set of parameter values, excluding θ_i, such that the parameters $\theta_1, \ldots, \theta_i$ have been updated within the iteration of the Markov chain, whereas parameters $\theta_{i+1}, \ldots, \theta_p$ are still to be updated within the iteration of the Markov chain. We can then replace step 2(j) in the updating algorithm with:

Calculate the acceptance probability:

$$\alpha(\theta_j^k, \phi_j) = \min\left(1, \frac{\pi(\phi_j|\boldsymbol{x}, \boldsymbol{\theta}_{(j)}^k)q_j(\theta_j^k|\phi_j)}{\pi(\theta_j^k|\boldsymbol{x}, \boldsymbol{\theta}_{(j)}^k)q_j(\phi_j|\theta_j^k)}\right).$$

With probability $\alpha(\theta_j^k, \phi_j)$ set $\theta_j^{k+1} = \phi_j$; otherwise set $\theta_j^{k+1} = \theta_j^k$.

In other words we can simplify the acceptance probability and simply calculate the ratio of the full posterior *conditional* distribution of the parameter evaluated at the proposed value and current value, instead of the full posterior distribution.

Example 5.3. Radio-Tagging Data – Two Parameters

We extend Example 4.1 to consider a two-parameter version of the model. In particular, we produce two distinct survival probabilities: one for the probability of surviving to the first capture time, ϕ_1, and another for all following times, ϕ_2. Thus, this model essentially allows for there to be an initial change in the survival probability following the application of the radio-tags (for example, we might expect this to be a decrease in survival). Without any prior expert information, we specify independent $U[0,1]$ priors on ϕ_1 and ϕ_2. The corresponding multinomial cell probabilities are now given by,

$$p_j = \begin{cases} 1 - \phi_1 & j = 1 \\ \phi_1(1 - \phi_2) & j = 2 \\ \phi_1 \phi_2^{j-2}(1 - \phi_2) & j = 3, \ldots, 6 \\ \phi_1 \phi_2^{j-2} & j = 7. \end{cases}$$

The corresponding joint posterior distribution for both ϕ_1 and ϕ_2 is given by,

$$\begin{aligned}
\pi(\phi_1, \phi_2 | \boldsymbol{x}) &\propto f(\boldsymbol{x}|\phi_1, \phi_2) p(\phi_1, \phi_2) \\
&= \frac{(\sum_{i=1}^{7} x_i)!}{\prod_{i=1}^{7} x_i!} \prod_{i=1}^{7} p_i^{x_i} \times 1 \\
&\propto (1-\phi_1)^{x_1} (\phi_1[1-\phi_2])^{x_2} (\phi_1\phi_2[1-\phi_2])^{x_3} (\phi_1\phi_2^2[1-\phi_2])^{x_4} \\
&\quad \times (\phi_1\phi_2^3[1-\phi_2])^{x_5} (\phi_1\phi_2^4[1-\phi_2])^{x_6} (\phi_1\phi_2^5)^{x_7} \\
&= (1-\phi_1)^{x_1} \phi_1^{\sum_{i=2}^{7} x_i} (1-\phi_2)^{\sum_{i=2}^{6} x_i} \phi_2^{\sum_{i=3}^{7} (i-2) x_i} \\
&= (1-\phi_1)^3 \phi_1^{22} (1-\phi_2)^{11} \phi_2^{129}.
\end{aligned}$$

It is clear that ϕ_1 and ϕ_2 are independent, *a posteriori*, since we can write,

$$\pi(\phi_1, \phi_2 | \boldsymbol{x}) = \pi(\phi_1|\boldsymbol{x}) \pi(\phi_2|\boldsymbol{x}).$$

Thus we could easily (and directly) obtain the marginal posterior distribution for ϕ_1 and ϕ_2 (or their joint distribution). However, suppose that we did not notice this, and we wish to use the single-update MH algorithm. At iteration k of the Markov chain, the parameters values are denoted by (ϕ_1^k, ϕ_2^k). We initially propose to update ϕ_1 and then ϕ_2 (though we could do this in the reverse order if we wished – the order of updating is unimportant). We use what is called a *random walk MH step* for each parameter (see Section 5.3.3), with normal proposal distribution (see Example 5.2). We begin with proposing a new parameter value for ϕ_1. We simulate,

$$\omega_1 | \phi_1^k \sim N(\phi_1^k, \sigma_1^2),$$

for some proposal variance σ_1^2. The candidate value is accepted with probability, $\alpha(\phi_1^k, \omega_1) = \min(1, A_1)$ where,

$$\begin{aligned}
A_1 &= \frac{\pi(\omega_1, \phi_2^k|\boldsymbol{x}) q(\phi_1^k|\omega_1)}{\pi(\phi_1^k, \phi_2^k|\boldsymbol{x}) q(\omega_1|\phi_1^k)} \\
&= \begin{cases} \left(\frac{1-\omega_1}{1-\phi_1^k}\right)^3 \left(\frac{\omega_1}{\phi_1^k}\right)^{22} & \omega_1 \in [0,1] \\ 0 & \omega_1 \notin [0,1], \end{cases}
\end{aligned}$$

with the proposal densities cancelling as the proposal distribution is symmetric (see Section 5.3.3). Once again we note that if the candidate value ω_1 lies outside the interval [0,1], the acceptance probability is simply equal to zero since the posterior density is equal to zero for this value. If the candidate value is accepted, we set $\phi_1^{k+1} = \omega_1$; otherwise we set $\phi_1^{k+1} = \phi_1^k$.

We then update the parameter value for ϕ_2, conditional on the value ϕ_1^{k+1}. We propose a candidate value for ϕ_2, by simulating,

$$\omega_2 | \phi_2^k \sim N(\phi_2^k, \sigma_2^2),$$

for some proposal variance σ_2^2. The corresponding acceptance probability is

$\alpha(\phi_2^k, \omega_2) = \min(1, A_2)$, where

$$A_2 = \frac{\pi(\phi_1^{k+1}, \omega_2 | \boldsymbol{x}) q(\phi_2^k | \omega_2)}{\pi(\phi_1^{k+1}, \phi_2^k | \boldsymbol{x}) q(\omega_2 | \phi_2^k)}$$

$$= \begin{cases} \left(\frac{1-\omega_2}{1-\phi_2^k}\right)^{11} \left(\frac{\omega_2}{\phi_2^k}\right)^{129} & \omega_2 \in [0,1] \\ 0 & \omega_2 \notin [0,1]. \end{cases}$$

If the candidate value is accepted, we set $\phi_2^{k+1} = \omega_2$; otherwise we set $\phi_2^{k+1} = \phi_2^k$.

Note that in general the acceptance probabilities will be dependent on the other parameters in the model. Having the parameters *a posteriori* independent (under independent priors) is a special case. This is not true in the following example. □

Example 5.4. Example 2.3.1 Revisited – Ring-Recovery Data

We consider a simple ring-recovery example where there is a single cohort of animals, with a total of T recovery events. There are a total of R distinct individuals observed within the study, with m_t individuals recovered at time $t = 1, \ldots, T$. For notational convenience we let m_{T+1} denote the number of individuals that are not recovered within the study, i.e. $m_{T+1} = R - \sum_{i=1}^{T} m_t$. We assume a constant recovery and survival probability, so that there are only two model parameters, λ and ϕ. The corresponding likelihood is,

$$f(\boldsymbol{m} | \phi, \lambda) \propto \prod_{i=1}^{T} (\phi^{i-1} \lambda (1-\phi))^{m_i} \chi_i^{m_{T+1}},$$

where $\chi_i = 1 - \sum_{j=1}^{T} \phi^{j-1} \lambda (1-\phi)$, (see Equation 2.3).

We assume independent uniform priors on λ and ϕ over the interval $[0, 1]$. The joint posterior distribution over λ and ϕ is then proportional to the likelihood given above. We implement a single-update MH algorithm. Suppose that at iteration k of the Markov chain, the parameter values are denoted by λ^k and ϕ^k. We begin by updating the recovery probability, followed by the survival probability within each iteration of the Markov chain. For the recovery probability we propose the candidate value, where,

$$\omega = \lambda^k + u,$$

for,

$$u \sim U[-0.1, 0.1].$$

Thus, $q(\omega | \lambda^k) = 1/0.2$, for $\lambda^k - 0.1 \leq \omega \leq \lambda^k + 0.1$. Similarly, we have that $\lambda^k = \omega - u$, so that $q(\lambda^k | \omega) = 1/0.2$, for $\omega - 0.1 \leq \lambda^k \leq \omega + 0.1$. We accept the candidate value with probability $\alpha(\lambda^k, \omega) = \min(1, A_\lambda)$,

$$
\begin{aligned}
A_\lambda &= \frac{\pi(\phi^k, \omega|\boldsymbol{m})q(\lambda^k|\omega)}{\pi(\phi^k, \lambda^k|\boldsymbol{m})q(\omega|\lambda^k)} \\
&= \frac{f(\boldsymbol{m}|\phi^k, \omega)p(\phi^k)p(\omega)q(\lambda^k|\omega)}{f(\boldsymbol{m}|\phi^k, \lambda^k)p(\phi^k)p(\lambda^k)q(\omega|\lambda^k)} \\
&= \begin{cases} \dfrac{\prod_{i=1}^T ((\phi^k)^{i-1}\omega(1-\phi^k))^{m_i}\left(1-\sum_{i=1}^T(\phi^k)^{i-1}\omega(1-\phi^k)\right)^{m_{T+1}}}{\prod_{i=1}^T ((\phi^k)^{i-1}\lambda^k(1-\phi^k))^{m_i}\left(1-\sum_{i=1}^T(\phi^k)^{i-1}\lambda^k(1-\phi^k)\right)^{m_{T+1}}} & \omega \in [0,1] \\ 0 & \omega \notin [0,1] \end{cases}
\end{aligned}
$$

since $p(\omega) = 1 = p(\lambda^k)$; and $q(\lambda^k|\omega) = 5 = q(\omega|\lambda^k)$

$$
= \begin{cases} \dfrac{\prod_{i=1}^T (\omega)^{m_i}\left(1-\sum_{i=1}^T(\phi^k)^{i-1}\omega(1-\phi^k)\right)^{m_{T+1}}}{\prod_{i=1}^T (\lambda^k)^{m_i}\left(1-\sum_{i=1}^T(\phi^k)^{i-1}\lambda^k(1-\phi^k)\right)^{m_{T+1}}} & \omega \in [0,1] \\ 0 & \omega \notin [0,1]. \end{cases}
$$

If we accept the candidate value, we set $\lambda^{k+1} = \omega$; otherwise we set $\lambda^{k+1} = \lambda^k$.

Given the updated value for the recovery probability, we now consider the survival probability, and implement the analogous MH updating step. In particular, we propose the candidate value,

$$\nu = \phi^k + v,$$

where $v \sim U[-0.05, 0.05]$, so that ν has proposal density $q(\nu|\phi^k) = 1/0.1$ for $\phi^k - 0.05 \leq \nu \leq \phi^k + 0.05$. Similarly, $q(\phi^k|\nu) = 1/0.1$ for $\nu - 0.05 \leq \phi^k \leq \nu + 0.05$.

We accept the candidate value with the standard acceptance probability $\alpha(\phi^k, \nu) = \min(1, A_\phi)$, where,

$$
\begin{aligned}
A_\phi &= \frac{\pi(\nu, \lambda^{k+1}|\boldsymbol{m})q(\phi^k|\nu)}{\pi(\phi^k, \lambda^{k+1}|\boldsymbol{m})q(\nu|\phi^k)} \\
&= \frac{f(\boldsymbol{m}|\nu, \lambda^{k+1})p(\nu)p(\lambda^{k+1})q(\phi^k|\nu)}{f(\boldsymbol{m}|\phi^k, \lambda^{k+1})p(\phi^k)p(\lambda^{k+1})q(\nu|\phi^k)} \\
&= \begin{cases} \dfrac{\prod_{i=1}^T (\nu^{i-1}(1-\nu))^{m_i}\left(1-\sum_{i=1}^T \nu^{i-1}\lambda^{k+1}(1-\nu)\right)^{m_{T+1}}}{\prod_{i=1}^T ((\phi^k)^{i-1}\lambda^{k+1}(1-\phi^k))^{m_i}\left(1-\sum_{i=1}^T(\phi^k)^{i-1}\lambda^{k+1}(1-\phi^k)\right)^{m_{T+1}}} & \nu \in [0,1] \\ 0 & \nu \notin [0,1], \end{cases}
\end{aligned}
$$

since $p(\nu) = 1 = p(\phi^k)$; and $q(\phi^k|\nu) = 5 = q(\nu|\phi^k)$. If we accept the candidate value we set $\phi^{k+1} = \nu$; otherwise we set $\phi^{k+1} = \phi^t$.

This completes a single iteration of the Markov chain. □

Example 5.5. Example 1.1 Revisited – Dipper Data – Two-Parameter Model

We return to the capture-recapture data relating to the dippers, where we have a model with just two parameters – recapture probability p and survival probability ϕ. We have already encountered the likelihood when just a single cohort of marked birds is released, in Equation (2.4). When there are T such cohorts, the likelihood becomes

$$f(\boldsymbol{m}|p,\phi) \propto \prod_{i=1}^{T}\prod_{j=i}^{T}(p(1-p)^{j-i}\phi^{j-i+1})^{m_{i,j}}\chi_i^{m_{i,T+1}},$$

where,

$$\chi_i = 1 - \sum_{j=i}^{T}(p(1-p)^{j-i}\phi^{j-i+1}),$$

and $m_{i,j}$ denotes the number of individuals released at time i, subsequently caught for the first time at time $j+1$; and $m_{i,T+1}$ denotes the number of individuals released at time i that are not observed again (recall the likelihood expression given in Section 2.3.2).

For simplicity we once more assume independent $U[0,1]$ priors on each of the parameters. We begin by initially setting the parameter values to be $p^0 = 0.9$ and $\phi^0 = 0.5$. We then propose to update each parameter in turn, using a normal proposal density. The order of updating the parameters at each iteration is arbitrary, so we begin by initially updating p and then ϕ.

We consider the updating step at iteration k. To update the recapture probability, p^k, we begin by proposing a new candidate value, ω, such that,

$$\omega|p^k \sim N(p^k, 0.01).$$

The proposal density is then given by,

$$q(\omega|p^k) = \frac{1}{\sqrt{0.02\pi}}\exp\left(-\frac{(\omega-p^k)^2}{0.02}\right).$$

The proposed move is then accepted with the standard probability $\alpha(p^k,\omega) = \min(1, A)$, where

$$A = \frac{f(\boldsymbol{m}|\omega,\phi^k)p(\omega)p(\phi)q(p^k|\omega)}{f(\boldsymbol{m}|p^k,\phi^k)p(p^k)p(\phi^k)q(\omega|p^k)}.$$

Clearly, we have some simplifications that can be made to this expression. In particular, since we have independent priors specified on the parameters, $p(\phi^k)$ cancels in the numerator and denominator (irrespective of the prior distribution specified on ϕ). In addition, we have independent uniform priors on the parameters, so that $p(p^k) = 1 = p(\omega)$ (for $p^k, \omega \in [0,1]$). We also have that,

$$\begin{aligned}q(\omega|p^k) &= \frac{1}{\sqrt{0.02\pi}}\exp\left(-\frac{(\omega-p^k)^2}{0.02}\right)\\ &= \frac{1}{\sqrt{0.02\pi}}\exp\left(-\frac{(p^k-\omega)^2}{0.02}\right) = q(p^k|\omega).\end{aligned}$$

Thus, the proposal distributions also cancel in the acceptance probability. This leaves the acceptance probability equal to the ratio of the likelihood evaluated at the proposed and current value of the parameter. Mathematically, we have the acceptance probability, $\min(1, A)$, where,

$$A = \begin{cases} \frac{f(\boldsymbol{m}|\omega,\phi^k)}{f(\boldsymbol{m}|p^k,\phi^k)} & \omega \in [0,1] \\ 0 & \omega \notin [0,1]. \end{cases}$$

$$= \begin{cases} \frac{\prod_{i=1}^{T} \prod_{j=i}^{T} (\omega(1-\omega)^{j-i}(\phi^t)^{j-i+1})^{m_{i,j}} \chi_i(\omega,\phi^k)^{m_{i,T+1}}}{\prod_{i=1}^{T} \prod_{j=i}^{T} (p^t(1-p^k)^{j-i}(\phi^k)^{j-i+1})^{m_{i,j}} \chi_i(p^k,\phi^k)^{m_{i,T+1}}} & \omega \in [0,1] \\ 0 & \omega \notin [0,1], \end{cases}$$

where,

$$\chi_i(p,\phi) = 1 - \sum_{j=i}^{T}(p(1-p)^{j-i}\phi^{j-i+1}).$$

We then accept the proposed value, ω, with probability $\alpha(p^k,\omega)$ and set $p^{k+1} = \omega$; otherwise if we reject the move, we set $p^{k+1} = p^k$.

We repeat the updating procedure, for the survival probability, given the updated value of the recapture probability. In particular, we propose the candidate value,

$$\nu|\phi^k \sim N(\phi^k, 0.01).$$

Thus, as for the recapture probability above, we have that,

$$q(\nu|\phi^k) = q(\phi^k|\nu).$$

The acceptance probability of the proposed value is given by $\alpha(\phi^k, \nu) = \min(1, A)$, where,

$$A = \frac{f(\boldsymbol{m}|p^{k+1},\nu)p(p^{k+1})p(\nu)q(\phi^k|\nu)}{f(\boldsymbol{m}|p^{k+1},\phi^k)p(p^{k+1})p(\phi^k)q(\nu|\phi^k)}$$

$$= \begin{cases} \frac{\prod_{i=1}^{T} \prod_{j=i}^{T} (p^{k+1}(1-p^{k+1})^{j-i}\nu^{j-i+1})^{m_{i,j}} \chi(p^{k+1},nu)_{i}^{m_{i,T+1}}}{\prod_{i=1}^{T} \prod_{j=i}^{T} (p^{k+1}(1-p^{k+1})^{j-i}(\phi^k)^{j-i+1})^{m_{i,j}} \chi_i(p^{k+1},\phi^k)^{m_{i,T+1}}} & \nu \in [0,1] \\ 0 & \nu \notin [0,1]. \end{cases}$$

We then accept the proposed value, ν, with probability $\alpha(\phi^k,\nu)$ and set $\phi^{k+1} = \nu$; otherwise if we reject the move, we set $\phi^{k+1} = \phi^k$.

For example, using the above algorithm we run 10 iterations and provide the current and proposed values of the parameters and corresponding acceptance probability in Table 5.1. Note that any values that are proposed outside the range $[0,1]$ automatically have an acceptance probability equal to 0. Other zero acceptance values are only equal to 0.000 to (at least) three decimal places.

The two-dimensional trace plot for these initial 10 iterations of the Markov chain is given in Figure 5.3. Figure 5.3(a) gives the raw trace plot and (b) the values of the parameter values identified at each iteration number. Note that horizontal or vertical movements correspond to only one of the proposal parameter values being accepted in the iteration (vertical for accepted proposal values for ϕ [iterations 1, 2, 5, 6 and 9] and horizontal for p [iteration 4]); diagonal movements correspond to both proposal parameters being accepted within a single iteration [iteration 3]; and no movement corresponds to both

MARKOV CHAIN MONTE CARLO

Table 5.1 Sample Values from the MH Algorithm Above for the Parameters p and ϕ for the Dipper Data

	p			ϕ		
Iteration t	p^t	Proposed	Acceptance probability	ϕ^t	Proposed	Acceptance probability
1	0.900	1.017	0.000	0.500	0.605*	1.000
2	0.900	1.066	0.000	0.605	0.560*	1.000
3	0.900	0.946*	0.196	0.560	0.526*	0.654
4	0.946	0.866*	0.872	0.526	0.477	0.004
5	0.866	1.031	0.000	0.526	0.544*	1.000
6	0.866	0.971	0.014	0.544	0.606*	0.496
7	0.866	0.801	0.656	0.606	0.624	0.254
8	0.866	0.996	0.000	0.606	0.690	0.000
9	0.866	0.746	0.001	0.606	0.581*	1.000
10	0.866	0.859	0.802	0.581	0.497	0.170

Note: The proposed values indicated by a * denote values that were accepted in the Markov chain.

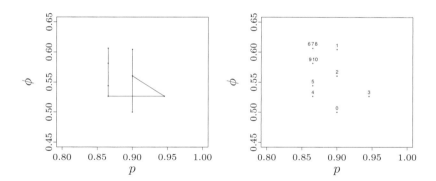

Figure 5.3 Sample paths for the single-update MH algorithm for the two-parameter dipper example; (a) raw trace plot and (b) iteration values.

proposal parameters being rejected within a single iteration [iterations 7, 8 and 10].

Running the simulations for longer, we can obtain a much larger sample of parameter values for the joint posterior distribution. Figure 5.4(a) provides the two-dimensional trace plot for p and ϕ simultaneously for 1000 iterations; and Figures 5.4(b) and (c) the corresponding trace plots of p and ϕ.

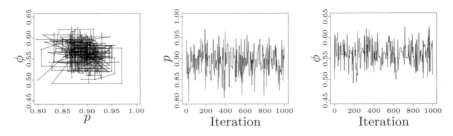

Figure 5.4 Sample paths for the single-update MH algorithm for the two-parameter dipper example; (a) trace plot for p and ϕ simultaneously; (b) marginal trace plot for p; and (c) marginal trace plot for ϕ.

□

Example 5.6. Covariate Regression Analysis – Lapwings

We revisit Example 4.11 where we have (m-array) ring-recovery data, $\boldsymbol{m}$, corresponding to the UK lapwing population. In addition, for this analysis we logistically regress the (first-year and adult) survival probabilities on the covariate $fdays$, and the recovery probability on time. In other words,

$$\begin{aligned} \text{logit } \phi_{1,t} &= \alpha_1 + \beta_1 f_t; \\ \text{logit } \phi_{a,t} &= \alpha_a + \beta_a f_t; \\ \text{logit } \lambda_t &= \alpha_\lambda + \beta_\lambda u_t, \end{aligned}$$

where f_t and u_t denote the normalised values for the covariates $fdays$ and time, respectively.

The parameters in the model are the regression coefficients $\alpha_1, \beta_1, \alpha_a, \beta_a, \alpha_\lambda$ and β_λ. To implement the single-update MH algorithm, we cycle through each of these regression coefficients, proposing to update each in turn. As usual, the update order is arbitrary and so we propose to order them in the order presented above.

Suppose that at iteration k of the Markov chain, the regression parameters are denoted by $\alpha_1^k, \beta_1^k, \alpha_a^k, \beta_a^k, \alpha_\lambda^k$ and β_λ^k. Similarly, the corresponding demographic parameters are denoted by $\boldsymbol{\phi}_1^k, \boldsymbol{\phi}_a^k, \boldsymbol{\lambda}^k$, where the vector notation $\boldsymbol{\phi}_1^k$ denotes the set of all first-year survival probabilities for all times, and similarly for $\boldsymbol{\phi}_a^k$ and $\boldsymbol{\lambda}^k$ at iteration i. We begin by proposing to update parameter α_1. We note that $\boldsymbol{\phi}_1$ is the only demographic parameter that is a function of this parameter. We propose a new candidate value, α_1', say, such that,

$$\alpha_1' \sim N(\alpha_1^k, \sigma_\alpha^2).$$

Thus, the proposal density $q(\alpha_1'|\alpha_1)$ is simply the normal density of the given distribution above.

MARKOV CHAIN MONTE CARLO

Since ϕ_1 is a deterministic function of α_1, this also induces the new values,

$$\text{logit } \phi'_{1,t} = \alpha'_1 + \beta_1^k f_t,$$

or alternatively (taking the inverse of the logit function),

$$\phi'_{1,t} = \frac{\exp(\alpha'_1 + \beta_1^k f_t)}{\exp(\alpha'_1 + \beta_1^k f_t) + 1} = \frac{1}{1 + \exp(-\alpha'_1 - \beta_1^k f_t)}.$$

Finally, we use ϕ'_1 to denote the full set of candidate values proposed for the first-year survival probabilities. The corresponding acceptance probability is given by $\alpha(\alpha_1^k, \alpha'_1) = \min(1, A_\alpha)$, where,

$$\begin{aligned} A_\alpha &= \frac{\pi(\phi'_1, \phi_a^k, \boldsymbol{\lambda}^k | \boldsymbol{m}) q(\alpha_1 | \alpha'_1)}{\pi(\phi_1^k, \phi_a^k, \boldsymbol{\lambda}^k | \boldsymbol{m}) q(\alpha'_1 | \alpha_1)} \\ &= \frac{f(\boldsymbol{m} | \phi'_1, \phi_a^k, \boldsymbol{\lambda}^k) p(\alpha'_1)}{f(\boldsymbol{m} | \phi_1^k, \phi_a^k, \boldsymbol{\lambda}^k) p(\alpha_1^k)}. \end{aligned}$$

Note that the proposal density functions cancel in the acceptance probability, since the normal proposal distribution is symmetrical. In addition, the priors are specified on the regression parameters (rather than explicitly on the demographic parameters), and so the priors for the regression parameters appear in the acceptance probability when the posterior distribution is decomposed into the product of the likelihood and prior. Note that since all the other regression parameters remain the same (except for α_1), the prior terms cancel in the acceptance probability, leaving only the prior on α_1. As usual, if we accept the proposed parameter value, we set $\alpha_1^{k+1} = \alpha'_1$; otherwise we set $\alpha_1^{k+1} = \alpha_1^k$. Let the corresponding first-year survival probabilities be denoted by ϕ_1^*.

We then turn to the next parameter to be updated. In this example, we propose to update the parameter β_1, associated again with the first-year survival probability. We propose a candidate value,

$$\beta'_1 \sim N(\beta_1^k, \sigma_\beta^2).$$

We calculate the corresponding proposed first-year survival probabilities to be,

$$\phi'_{1,t} = \frac{1}{1 + \exp(-\alpha_1^{k+1} - \beta'_1 f_t)}.$$

Recall that we need to use the updated value of α_1. The corresponding (standard) acceptance probability is of the analogous form as above, so omitting the same steps, we have $\alpha(\beta_1^k, \beta'_1) = \min(1, A_\beta)$, where,

$$A_\beta = \frac{f(\boldsymbol{m} | \phi'_1, \phi_a^k, \boldsymbol{\lambda}^k) p(\beta'_1)}{f(\boldsymbol{m} | \phi_1^*, \phi_a^k, \boldsymbol{\lambda}^k) p(\beta_1^k)}.$$

If we accept the proposed parameter value, we set $\beta_1^{k+1} = \beta'_1$; otherwise we set $\beta_1^{k+1} = \beta_1^k$. The corresponding first-year survival probabilities are,

$$\phi_{1,t}^{k+1} = \frac{1}{1 + \exp(-\alpha_1^{k+1} - \beta_1^{k+1} f_t)}.$$

We repeat the above steps for $\alpha_a, \beta_a, \alpha_\lambda$ and β_λ, remembering always to use the updated parameter values when moving on to the next parameter to be updated.

□

Example 5.7. Covariate Regression Analysis – White Storks
We revisit the capture-recapture data, m, for the white stork data presented in Exercise 3.5. We assume a constant recapture probability p, and logistically regress the survival probability, ϕ_t, on a single environmental covariate, corresponding to the amount of rainfall at a meteorological weather station in the Sahel (in particular, the Kita weather station). Thus, we set,

$$\text{logit } \phi_t = \beta_0 + \beta_1 w_t,$$

where w_t denotes the (normalised) covariate value at time t. In this example we have a combination of an arbitrary (but constant) recapture probability and a survival probability dependent on the given environmental covariate. We specify the priors,

$$\begin{aligned} p &\sim Beta(\alpha_p, \beta_p); \\ \beta_i &\sim N(0, \sigma_i^2) \qquad \text{for } i = 0, 1. \end{aligned}$$

We consider a single-update MH algorithm. Suppose that at iteration k of the Markov chain, the parameters are p^k, β_0^k and β_1^k, with corresponding survival probabilities, ϕ^k. We propose to update the recapture probabilities, and then the regression coefficients. For the recapture probability, we use a random walk MH update with a uniform proposal, so that we propose the candidate value,

$$p' = p^k + \epsilon,$$

where $\epsilon \sim U[-\delta_p, \delta_p]$, and δ_p is typically chosen via pilot tuning. The proposal density is $q(p'|p^k) = \frac{1}{2\delta_p}$ for $p^k - \delta_p \leq p' \leq p^k + \delta_p$. Here we set $\delta = 0.05$. Alternatively, we can write the proposal distribution for p' in the form,

$$p' \sim U[p^k - \delta_p, p^k + \delta_p].$$

The parameter being updated is the recapture probability, and so lies within the interval [0,1]. Thus, if $p' \notin [0,1]$, the corresponding acceptance probability is equal to zero, i.e. we automatically reject the proposed value. Otherwise, if $p' \in [0,1]$, we accept the proposed value with probability, $\alpha(p^t, p') = \min(1, A_p)$, where,

$$\begin{aligned} A_p &= \frac{\pi(p', \phi^k|m) q(p^k|p')}{\pi(p^k, \phi^k|m) q(p'|p^k)} \\ &= \frac{f(m|p', \phi^k) p(p')}{f(m|p^k, \phi^k) p(p^k)}, \end{aligned}$$

MARKOV CHAIN MONTE CARLO

and $p(\cdot)$ denotes the beta prior density function. Note that the prior densities for β_0 and β_1, and proposal densities, q, cancel in the acceptance probability. If the proposed value is accepted, we set $p^{k+1} = p'$; otherwise we set $p^{k+1} = p^k$.

Cycling to the next parameter, we consider the intercept term, β_0, of the logistic regression expression for the survival probabilities. We once more use a uniform proposal distribution, and propose the candidate value,

$$\beta_0' \sim U[\beta_0^k - \delta_0, \beta_0^k + \delta_0],$$

where we set $\delta_0 = 0.1$. We let the corresponding survival probabilities associated with the proposed candidate value be denoted by ϕ', where,

$$\text{logit } \phi_t' = \beta_0' + \beta_1^k w_t,$$

or, equivalently,

$$\phi_t' = \frac{1}{1 + \exp(-\beta_0' - \beta_1^k w_t)}.$$

Unlike the recapture probability, the regression coefficients are not constrained by any boundaries, and can take any value on the real line. Thus, we accept the proposed value with probability $\alpha(\beta_0^k, \beta_0') = \min(1, A_0)$, where,

$$A_0 = \frac{f(\boldsymbol{m}|p^{k+1}, \boldsymbol{\phi}')p(\beta_0')}{f(\boldsymbol{m}|p^{k+1}, \boldsymbol{\phi})p(\beta_0)},$$

where $p(\cdot)$ denotes the normal proposal density, and once more cancelling the uniform proposal densities and priors specified on all other parameters not being updated. Note that we use the updated value of the recapture probability, p^{k+1}, as usual. If we accept the move, we set $\beta_0^{k+1} = \beta_0'$; otherwise we set $\beta_0^{k+1} = \beta_0$. In addition, we update the corresponding survival probabilities (which are a function of the $\boldsymbol{\beta}$ parameters), setting,

$$\phi_t^* = \frac{1}{1 + \exp(-\beta_0^{k+1} - \beta_1^k w_t)},$$

with standard vector notation $\boldsymbol{\phi}^*$.

Finally, there is the slope coefficient to be updated within the MCMC iteration (i.e. β_1). This is analogous to the updating of the intercept term. We once more use a uniform proposal density to propose the value,

$$\beta_1' \sim U[\beta_1^k - \delta_1, \beta_1^k + \delta_1],$$

where we set $\delta_0 = 0.1$. The corresponding proposal survival probabilities are given by ϕ', such that,

$$\phi_t' = \frac{1}{1 + \exp(-\beta_0^{k+1} - \beta_1' w_t)},$$

recalling that we once again use the updated value of β_0. The acceptance probability is analogous to that obtained when updating the intercept term, and given by $\alpha(\beta_1^k, \beta_1') = \min(1, A_1)$, where,

$$A_1 = \frac{f(\boldsymbol{m}|p^{k+1}, \boldsymbol{\phi}')p(\beta_1')}{f(\boldsymbol{m}|p^{k+1}, \boldsymbol{\phi}^*)p(\beta_1^k)},$$

where $p(\cdot)$ denotes the normal prior density. If we accept the move we set $\beta_1^{k+1} = \beta_1'$; otherwise we set $\beta_1^{k+1} = \beta_1^k$.

This completes a single iteration of the MCMC algorithm. The algorithm is repeated to obtain a sample from the posterior distribution, following convergence of the Markov chain to its stationary distribution.

□

Typically, it is the single-update MH algorithm that is implemented in practice, and it generally performs well. Heuristically, the reason for this is that any autocorrelation should be less for the individual parameters, since we only need to propose a "good" move for one parameter for it to be accepted, without regard to the other parameters. In contrast, when we consider multiple parameter updates at the same time, a move may be rejected through being "poor" for some parameters, but "good" for other parameters. Thus, for multiple parameter moves, we generally need to propose relatively small changes compared with the current values, as we are forced to take a smaller proposal variance than for single-parameter updates. However, this is not always the case, for example, where parameters are highly correlated. In this case, using a single-update algorithm, only small moves are accepted, as the updated parameter is highly dependent on the other parameters; thus we can move around the parameter space faster if we update parameters together, rather than only managing very small updates when individual steps are taken. Thus, sometimes it is convenient to do something that is intermediate between MH for all parameters simultaneously and single-update MH, and update parameters in blocks (Roberts and Sahu 1997), as we shall discuss later in Section 5.4.4.

5.3.3 Random Walk Updates

One basic and common choice is to base the proposal around the current point θ^k, for example, $\phi \sim N(\theta^k, \sigma^2)$. This is known as a *random walk* (RW) algorithm (Brooks et al. 2000a). More generally, since at iteration t of the Markov chain the parameters in the model are denoted by θ^k, we then propose the set of new parameters ϕ. If $q(\phi|\theta) = f(|\phi - \theta|)$ for some arbitrary density f, we have a random walk MH step, since the candidate observation is of the form $\theta^{k+1} = \theta^k + z$, where $z \sim f$. There are many common choices for f, including the uniform, normal, or t distribution. Note that when the candidate generating function is symmetric i.e. $q(\phi|\theta^k)$
$= q(\theta^k|\phi)$ (as for the uniform, normal or t-distribution) the acceptance probability reduces to:

$$\alpha(\theta^k, \phi) = \min\left(1, \frac{\pi(\phi|\boldsymbol{x})}{\pi(\theta^k|\boldsymbol{x}))}\right).$$

MARKOV CHAIN MONTE CARLO

This special case is the original Metropolis algorithm (Brooks et al. 2000a). Examples 5.2 through 5.7 are all examples of random walk MH updates with symmetric proposal densities.

5.3.4 The Independence Sampler

If $q(\phi|\theta^k) = f(\phi)$, then the candidate observation is drawn independently of the current state of the chain. In this case, the acceptance probability can be written as:

$$\alpha(\theta^k, \phi) = \min\left(1, \frac{w(\phi)}{w(\theta^k)}\right),$$

where $w(\phi) = \pi(\phi|\boldsymbol{x})/f(\phi)$ is the importance weight function that would be used in importance sampling given observations generated from f. The independence sampler is not often implemented within a MH algorithm, since the acceptance probabilities are typically much lower than for other methods, so that the chain is slow to explore the parameter space.

5.3.5 The Gibbs Sampler

A special case of the single-update MH algorithm arises if we set the proposal distribution for any parameter to be the conditional posterior distribution of that parameter given the current value of the others. In this case, the acceptance probability is always exactly 1. This is known as a Gibbs sampler (or simply Gibbs) update (Casella and George 1992). It is attractive as it does not require rejection.

We now give the full detail of this algorithm, due to its high usage within Bayesian analyses. Mathematically, suppose that we have the parameters $\boldsymbol{\theta} = (\theta_1, \ldots, \theta_p) \in \mathbb{R}^p$, with distribution $\pi(\boldsymbol{\theta})$. The Gibbs sampler uses the set of full conditional distributions of π to sample indirectly from the marginal distributions. Specifically, let $\pi(\theta_i|\boldsymbol{\theta}_{(i)})$ denote the induced full conditional distribution of θ_i, given the values of the other components $\boldsymbol{\theta}_{(i)} = (\theta_1, \ldots, \theta_{i-1}, \theta_{i+1}, \ldots, \theta_r)$, $i = 1, \ldots, r$, $1 < r \leq p$. Then, given the state of the Markov chain at iteration k, $\boldsymbol{\theta}^k = (\theta_1^k, \ldots, \theta_r^k)$, the Gibbs sampler successively makes random drawings from the full posterior conditional distributions as follows:

$$\begin{aligned}
\theta_1^{k+1} &\quad \text{is sampled from} \quad \pi(\theta_1|\boldsymbol{\theta}_{(1)}^k) \\
\theta_2^{k+1} &\quad \text{is sampled from} \quad \pi(\theta_2|\theta_1^{k+1}, \theta_3^k, \ldots, \theta_r^k) \\
&\quad \vdots \\
\theta_i^{k+1} &\quad \text{is sampled from} \quad \pi(\theta_i|\theta_1^{k+1}, \ldots, \theta_{i-1}^{k+1}, \theta_{i+1}^k, \ldots, \theta_r^k) \\
&\quad \vdots \\
\theta_r^{k+1} &\quad \text{is sampled from} \quad \pi(\theta_r|\boldsymbol{\theta}_{(r)}^{k+1}).
\end{aligned}$$

This completes a transition from $\boldsymbol{\theta}^k$ to $\boldsymbol{\theta}^{k+1}$. The transition probability for going from $\boldsymbol{\theta}^k$ to $\boldsymbol{\theta}^{k+1}$ is given by

$$\mathcal{K}_G(\boldsymbol{\theta}^k, \boldsymbol{\theta}^{k+1}) = \prod_{l=1}^{r} \pi(\theta_l^{k+1}|\theta_j^{k+1}, \ j < l \text{ and } \theta_j^k, \ j > l), \tag{5.4}$$

and has stationary distribution π.

Thus, in two dimensions a typical trajectory of the Gibbs sampler may look something like that given in Figure 5.5.

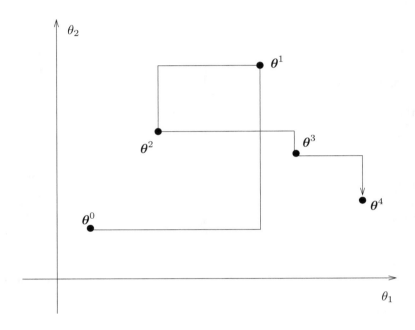

Figure 5.5 Typical Gibbs sampler path in two dimensions.

Conceptually, the Gibbs sampler appears to be a rather straightforward algorithmic procedure. Ideally, each of the conditional distributions will be of the form of a standard distribution and suitable prior specification often ensures that this is the case.

Example 5.8. Example 5.3 Revisited – Radio-Tagging with Two Parameters
We return to the radio-tagging data and consider the two-parameter model introduced in Example 5.3. The joint posterior distribution (assuming independent $U[0,1]$ priors) for the survival probabilities is given in the form,

$$\pi(\phi_1, \phi_2|\boldsymbol{x}) \propto (1-\phi_1)^{x_1} \phi_1^{\sum_{i=2}^{7} x_i} (1-\phi_2)^{\sum_{i=2}^{6} x_i} \phi_2^{\sum_{i=3}^{7}(i-2)x_i}.$$

By fixing the terms that we condition upon, we can see that the conditional posterior distributions are given by,

$$\phi_1|\phi_2, \boldsymbol{x} \sim Beta\left(1 + \sum_{i=2}^{7} x_i, 1 + x_1\right) \equiv Beta(23, 4)$$

$$\phi_2|\phi_1, \boldsymbol{x} \sim Beta\left(1 + \sum_{i=3}^{7}(i-2)x_i, 1 + \sum_{i=2}^{6} x_i\right) \equiv Beta(130, 12).$$

From the form of the posterior conditional distributions of ϕ_1 and ϕ_2 it is once more clear that the parameters are independent *a posteriori* (since ϕ_2 does not appear in the conditional distribution for ϕ_1 and vice versa). Implementing the Gibbs sampler algorithm, with parameters ϕ_1^t and ϕ_2^t at iteration t of the Markov chain, we begin by simulating,

$$\phi_1^{t+1}|\phi_2^t, \boldsymbol{x} \sim Beta\left(23, 4\right).$$

Given this updated parameter value, we update the next parameter, which in this case is the second survival probability, ϕ_2. We simulate,

$$\phi_2^{t+1}|\phi_1^{t+1}, \boldsymbol{x} \sim Beta\left(130, 12\right).$$

This completes a single iteration of the Gibbs sampler. Note that in this example, the algorithm simulates directly from the posterior distribution of ϕ_1 and ϕ_2, since the parameters are independent (*a posteriori*).

□

5.3.6 Comparison of Metropolis Hastings and Gibbs Sampler

One useful feature of the Gibbs update (beyond no longer having the accept-reject step) is that it requires no pilot tuning of the proposal parameters, compared to the MH algorithm. In addition, the proposal distribution is adaptive in that it depends upon the current state of the other parameters and so changes from iteration to iteration rather than having at least some components (such as the proposal variance) fixed throughout the simulation.

Gibbs updates tend to be highly efficient in terms of MCMC convergence, but can be computationally more burdensome if sampling from the relevant posterior conditional distributions cannot be achieved efficiently. In practice we often use a mixture of MH and Gibbs updates. For parameters with standard conditional distributions we use Gibbs updates and, for the rest we might use random walk MH updates. Such an algorithm is often referred to as a *hybrid* algorithm or *MH within Gibbs* (see Example 5.9 for an example). Since Gibbs updates are so easy to implement (and often very efficient), we typically like to use them as often as possible. Thus, conjugate priors are often adopted so that the posterior conditional distributions are of standard form, and the Gibbs sampler can be easily implemented. Note that it is not essential to have a standard posterior conditional distribution in order to use the Gibbs sampler. For example Vounatsou and Smith (1995) use a generalised version of the ratio-of-uniforms method in order to sample from the posterior conditional distributions of the parameters from a ring-recovery model. However, in practice with an "off-the-shelf" and easy-to-use MH algorithm, this is typically the default updating algorithm for non-standard conditional distributions.

Example 5.9. Metropolis Hastings with Gibbs – Capture-Recapture Dipper Data

We consider capture-recapture data (such as the dippers), denoted by $\boldsymbol{m}$, where the survival probabilities are time dependent and recapture probabilities are constant. This is denoted by C/T in the model notation described in Section 3.3.1. We assume that there are T cohorts (i.e. T years of releases) and also a total of T years of recaptures. We specify the following priors on the parameters. For the recapture probability, we specify,

$$p \sim Beta(\alpha_p, \beta_p).$$

Alternatively, for the survival probabilities, we use the reparameterisation,

$$\text{logit } \phi_t = \gamma + \epsilon_t,$$

where,

$$\gamma \sim N(\mu, \sigma^2);$$
$$\epsilon_t \sim N(0, \tau).$$

Finally, we assume a hierarchical prior by specifying a prior on the variance term, τ, with,

$$\tau \sim \Gamma^{-1}(\alpha_\tau, \beta_\tau).$$

Note that this hierarchical prior can be regarded as a random effects model (see Section 8.5). The corresponding DAG is given in Figure 5.6.

We consider the posterior conditional distribution for each of the parameters. It is easy to show that the posterior conditional distributions of p, γ and $\boldsymbol{\epsilon} = \{\epsilon_1, \ldots, \epsilon_T\}$ are all of non-standard form. For example, consider the posterior conditional distribution of p. Note that the likelihood for the general model T/T is provided in Equation (2.5). Simplifying this to consider a constant recapture probability is trivial. In particular, we have the likelihood function,

$$f(\boldsymbol{m}|p, \boldsymbol{\epsilon}, \gamma) p(p) \propto \prod_{i=1}^{T} \prod_{j=i}^{T} \left(p(1-p)^{j-i} \prod_{k=i}^{j} \phi_k \right)^{m_{i,j}} \chi_i^{m_{i,T+1}},$$

where $m_{i,j}$ as usual denotes the number of individuals released at time i that are next recaptured at time $j+1$; $m_{i,T+1}$, the number of individual released at time i that are never seen again; and χ_i, the probability that an individual

MARKOV CHAIN MONTE CARLO

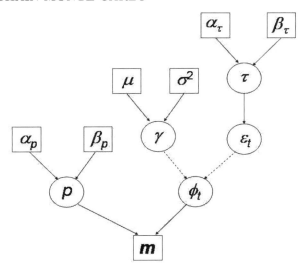

Figure 5.6 The DAG for Example 5.9 for capture-recapture data and model C/T.

is not observed again, given that it was released at time i given by,

$$\chi_i = 1 - \sum_{j=i}^{T} \left(p(1-p)^{j-i} \prod_{k=i}^{j} \phi_k \right).$$

Combining the above likelihood function with the prior for p, we obtain the posterior conditional distribution for p to be of the form,

$$\begin{aligned}
\pi(p|\boldsymbol{m},\boldsymbol{\epsilon},\boldsymbol{\gamma},\boldsymbol{\tau}) &\propto f(\boldsymbol{m}|p,\boldsymbol{\epsilon},\boldsymbol{\gamma})p(p) \\
&\propto \prod_{i=1}^{T}\prod_{j=i}^{T} \left(p(1-p)^{j-i} \prod_{k=i}^{j} \phi_k \right)^{m_{i,j}} \chi_i^{m_{i,T+1}} p^{\alpha-1}(1-p)^{\beta-1} \\
&\propto p^{\alpha-1+\sum_{i=1}^{T}\sum_{j=i}^{T} m_{i,j}}(1-p)^{\beta-1+\sum_{i=1}^{T}\sum_{j=i}^{T}(j-i)m_{i,j}} \chi_i^{m_{i,T+1}}.
\end{aligned}$$

The first two terms of this expression appear to be of the form of a beta distribution, however, the χ_i terms also contain the parameter p. Thus, this distribution is of non-standard form. Similar results follow for $\boldsymbol{\epsilon}$ and $\boldsymbol{\gamma}$, although their posterior conditional distributions could be regarded as even more complex since they are also on the logistic link scale.

However, we now consider the posterior conditional distribution for τ. We have,

$$\pi(\tau|\boldsymbol{m},p,\boldsymbol{\epsilon},\gamma) \propto f(\boldsymbol{m}|p,\boldsymbol{\epsilon},\gamma)p(\boldsymbol{\epsilon}|\tau)p(\tau)$$

$$\propto \prod_{i=1}^{T} \frac{1}{\sqrt{\tau}} \exp\left(-\frac{(\epsilon_t - \mu)^2}{2\tau}\right) \tau^{-(\alpha_\tau+1)} \exp\left(-\frac{\beta_\tau}{\tau}\right)$$

since the likelihood does not explicitly depend on τ

$$\propto \tau^{-(\alpha_\tau+T/2+1)} \exp\left(-\left(\frac{\frac{1}{2}\sum_{t=1}^{T}(\epsilon_t-\mu)^2 + \alpha_\tau}{\tau}\right)\right).$$

This is the form of an inverse gamma distribution (so this is an example of a conjugate prior). In particular, we have,

$$\tau|p,\boldsymbol{\epsilon},\gamma,\boldsymbol{m} \sim \Gamma^{-1}\left(\alpha_\tau + \frac{T}{2}, \beta_\tau + \frac{1}{2}\sum_{t=1}^{T}(\epsilon_t-\mu)^2\right).$$

Thus, for this example, we might update the parameters p, γ and $\boldsymbol{\epsilon}$ using the MH algorithm and τ using the Gibbs sampler.

□

5.4 Implementing MCMC

As with most numerical techniques, there are various MCMC algorithms that can be implemented for a given problem. In addition, there are numerous posterior summary statistics that can be used to summarise the posterior distribution of interest. We now discuss the most common issues that arise when implementing an MCMC algorithm, together with practical advice.

5.4.1 Posterior Summaries

We have already seen in Section 4.4 that posterior distributions are typically summarised via a number of statistics such as the mean, standard deviation, credible intervals, correlations, etc. Such summary statistics are easily obtained from the MCMC sample as their standard empirical estimates. For example, in order to estimate the $100(1-\alpha)\%$ symmetric credible interval for parameter θ, we first order our MCMC-based sample $\{\theta^k : k = 1, \ldots, n\}$, and take $[\theta^{(j)}, \theta^{(l)}]$ where $j = \alpha n/2$, $l = (1-\alpha)n/2$ and $\theta^{(j)}$ denotes the jth largest observation of θ in the sample.

There are any number of additional summary statistics that may be of interest and can be estimated from the posterior sample obtained via Monte Carlo integration. To provide a flavour of the type of statistics we can calculate, suppose that we wish to estimate the posterior probability that the parameter θ is positive (or negative), i.e. $\mathbb{P}(\theta > 0|\boldsymbol{x})$. This is easily estimated by taking the proportion of realisations sampled from the Markov chain that are positive. Marginal density estimates are also often of interest and basic histograms or standard kernel-smoothed empirical estimates can be used. Alternatively,

IMPLEMENTING MCMC

Rao-Blackwell estimates can be used to obtain estimates of marginal densities by taking

$$\widehat{\pi}(\theta_i) = \frac{1}{n} \sum_{t=1}^{n} \pi(\theta_i^k | \theta_{(i)}^k).$$

If the posterior conditional distribution is standard, then the density estimate, $\widehat{\pi}$, is easily normalised; see for example Casella and Robert (1996) and (Robert 2007) for further discussion.

Note that the ideas presented for the single-parameter case are all directly generalisable to the multi-parameter case. For example, point estimates can be obtained for correlations between parameters; posterior credible intervals generalise to posterior credible regions with dimension equal to the number of parameters. For example, in the two-parameter case, where $\boldsymbol{\theta} = \{\theta_1, \theta_2\}$, the $100(1-\alpha)\%$ posterior credible interval is now a two-dimensional region, R, such that,

$$\mathbb{P}((\theta_1, \theta_2) \in R | \boldsymbol{x}) = 1 - \alpha.$$

Clearly, a computer is needed in order to plot these more complex regions. Often, in practice, the marginal posterior density intervals may be calculated for each single parameter, rather than the more complex higher-dimensional density regions for the full set of parameters.

5.4.2 Run Lengths

There are two elements to be considered when determining the simulation length:

- the time required for convergence, and
- the post-convergence sample size required for suitably small Monte Carlo errors.

We deal with convergence assessment here: we want to determine how long it takes for the Markov chain to converge to the target distribution. In practice, we discard observations from the start of the chain, during the burn-in period, and just use observations from the chain once it has converged.

The simplest method to determine the length of the burn-in period is to look at trace plots. You can often see the individual parameters converging from their starting position to values based around a constant mean. For example, consider the dipper example (Example 1.1), where there are the two parameters – survival probability, ϕ, and recapture probability, p. We consider two different chains starting at very different starting values. The first chain begins with values, $p = 0.01$ and $\phi = 0.01$; and the second chain starts with values $p = 0.99$ and $\phi = 0.99$. The same updating algorithm is used from both starting values, namely a single-update random walk MH algorithm. In particular, we use a normal proposal distribution with standard deviation equal to 0.05 for each parameter. Figure 5.7 provides the corresponding trace plots for the two simulations. Clearly it appears that convergence is achieved very quickly for both of the simulations, both by approximately 200 iterations. However, we might also suggest that the posterior distribution is reached in simulation 2 slightly faster, resulting from the initial values being closer to the area of high posterior mass than in simulation 1. When specifying the burn-in being used, it is best to be conservative and err on the side of overestimation to ensure that convergence has been achieved when obtaining the sample to be used to form Monte Carlo estimates of the parameters of interest. Thus, for this example, we might use a burn-in of 500, or even 1000, say.

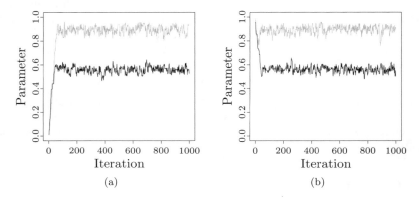

Figure 5.7 Trace plots for ϕ (in black) and p (in grey) for (a) simulation 1, with starting value of $p = 0.01$ and $\phi = 0.01$ and (b) simulation 2, with starting value $p = 0.99$ and $\phi = 0.99$.

The use of trace plots is a fairly efficient method, but it is not robust. For example, an ad hoc interpretation of the first trace plot in Figure 5.8 might suggest that the chain had converged after around 500 iterations. However, when the chain is run for longer, it is clear that the chain has not converged within these first 500 iterations.

Running several replications from different over-dispersed starting points provides additional reassurance when one is trying to check that convergence has been achieved. Basically, if you run the chain several times from different starting points and they all give you the same posterior estimates then this suggests that no major modes have been missed in any one simulation and that each has probably converged.

This approach is formalised in the so-called Brooks-Gelman-Rubin (BGR) diagnostic (Brooks and Gelman 1998). There are various implementations of this diagnostic procedure, all based upon the idea of using an analysis of variance to determine whether or not there are any differences in estimates from different replications. The simplest implementation is to compare the width of the empirical 80% credible interval obtained from all chains combined, with the corresponding mean within-chain interval width. Convergence is assumed

IMPLEMENTING MCMC

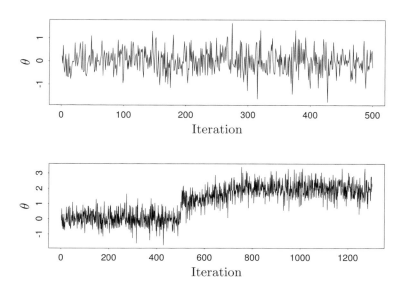

Figure 5.8 MCMC sample paths.

when these are roughly equal, implying that all chains have roughly equal variability, i.e. similar amounts of information. More sophisticated techniques compare within- and between-chain variances rather than interval widths, but the principle remains the same. For further discussion, alternative convergence diagnostic tools and reviews of convergence diagnostics; see for example, Gelman and Rubin (1992), Geweke (1992), Cowles and Carlin (1996) and Gelman (1996).

Example 5.10. Dippers – Two-Parameter Model
We implement the BGR statistic, using the width of the 80% symmetric credible interval for the dippers example for model C/C with two parameters p and ϕ for the trace plots in Figure 5.7, discarding the first n iterations as burn-in, for simulations of length $2n$. For this example there are two chains started from over-dispersed starting points, as described above. We then calculate the statistic $\hat{R}$ (following the assumed burn-in of the first n iterations), defined to be,

$$\hat{R} = \frac{\text{width of 80\% credible interval of pooled chains}}{\text{mean of width of 80\% credible interval of individual chains}}.$$

The corresponding BGR statistic, $\hat{R}$, is plotted in Figure 5.9 for both the survival probability, ϕ, and recapture probability, p.

Alternatively, the BGR statistic can be reported in tabular format, for spec-

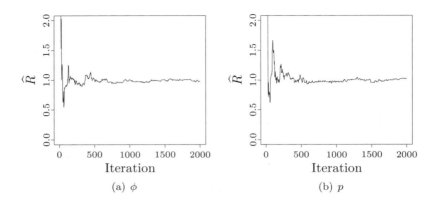

Figure 5.9 BGR statistic for the model C/C fitted to the dipper data for (a) survival probability, ϕ, and (b) recapture probability, p.

ified values of $2n$ (the total number of iterations, including the first n discarded as burn-in). These are provided in Table 5.2. Using either the plotted BGR statistic or the tabular form we would assume that convergence has been achieved by iteration 1000 (this is still somewhat conservative). However, it is better to use a conservative burn-in, as discussed above.

Table 5.2 The BGR Statistic for ϕ and p for the Two-Parameter Dippers Example for a Total of $2n$ Simulations with the First n Discarded as Burn-In

Total Number of Iterations	$\hat{R}$ ϕ	p	Total Number of Iterations	$\hat{R}$ ϕ	p
100	0.900	0.149	1100	0.981	0.995
200	1.008	1.159	1200	0.965	1.005
300	0.921	1.061	1300	0.984	1.043
400	1.051	0.960	1400	0.989	1.005
500	0.984	1.018	1500	1.004	0.962
600	1.004	0.932	1600	1.017	0.995
700	0.962	0.966	1700	0.992	0.995
800	0.946	0.969	1800	1.010	1.004
900	0.965	0.980	1900	0.980	1.018
1000	1.008	0.981	2000	0.984	1.022

□

IMPLEMENTING MCMC

We emphasise that all of these convergence diagnostics are not foolproof, in that they can only provide evidence of *lack of convergence*, rather than proof of convergence – these are very different things!

5.4.3 Monte Carlo Errors

MCMC integration is a method of estimating statistics of interest. It is a simulation-based estimation technique and, as such, is subject to what we call Monte Carlo (MC) error, which decreases with increasing sample size. Monte Carlo error essentially measures the variation you would expect to see in your estimates if you ran multiple replications of your MCMC chain. It is related to the autocorrelation of your chain, but if the sample size increases by a factor n, then the Monte Carlo error decreases by a factor $n^{1/2}$ (Ripley 1987).

Formally, it can be shown that ergodic averages satisfy the central limit theorem. So, for example, suppose that we wish to estimate the posterior mean of the parameter, θ. We obtain a sample $\theta^1, \ldots, \theta^n$ from the posterior distribution, and estimate the posterior mean $\mathbb{E}_\pi(\theta)$ by the sample mean, $\bar{\theta} = \frac{1}{n}\sum_{i=1}^{n} \theta^i$ (i.e. use Monte Carlo integration). Then, using the central limit theorem, we have the result that,

$$\bar{\theta} \sim N\left(\mathbb{E}_\pi(\theta), \frac{\sigma^2}{n}\right).$$

The term σ^2/n is the Monte Carlo variance, or more commonly, $\sqrt{\sigma^2/n}$ is the Monte Carlo error.

In order to estimate the Monte Carlo error, we essentially need to estimate $\sigma/\sqrt{n}$. The most common method for estimating the Monte Carlo error is called *batching* (Roberts 1996). Suppose we wish to estimate the posterior mean of θ from a sample $\theta^1, \ldots, \theta^n$, as above, where $n = mT$. We divide the sample into m batches, each of size T, where T is assumed to be sufficiently large to provide reasonably reliable estimates of $\mathbb{E}_\pi(\theta)$. We calculate the sample mean of each θ for each of the m batches, which we denote by $\bar{\theta}_1, \ldots, \bar{\theta}_m$; and the corresponding sample variance of these mean values using,

$$\frac{1}{m-1}\sum_{i=1}^{m}(\bar{\theta}_i - \bar{\theta})^2.$$

This gives us the Monte Carlo variance associated with samples of size T. To obtain the Monte Carlo variance for the full sample (i.e. estimate σ^2/n), we divide the sample variance above by m. Thus, to estimate the Monte Carlo error, we simply take the square root of this expression, i.e. use the formula:

$$\sqrt{\frac{1}{m(m-1)}\sum_{i=1}^{m}(\bar{\theta}_i - \bar{\theta})^2}.$$

Example 5.11. Dippers – Two-Parameter Model

We apply this method to the C/C model for the dippers data with parameters ϕ and p. From the above convergence diagnostics, we use a conservative burn-in of 1000 iterations and simulation 2 (starting values of 0.99). We then take the batch sizes to be of length $T = 100$ and calculate the corresponding Monte Carlo error for the estimated posterior means for increasing numbers of parameters. These are provided in Table 5.3. Clearly for the dippers example, the Monte Carlo errors can be considered to be relatively very small compared to the parameter values.

Table 5.3 The Estimated Posterior Mean and Monte Carlo Error for (a) p and (b) ϕ for the Model C/C Fitted to the Dipper Data for a Total of $n = 200, \ldots, 1000$ Iterations, Using a Batch Size of $T = 100$

	(a) p		(b) ϕ	
n	Posterior Mean	Monte Carlo Error	Posterior Mean	Monte Carlo Error
200	0.897	0.0065	0.550	0.0356
300	0.894	0.0080	0.558	0.0242
400	0.893	0.0080	0.559	0.0176
500	0.896	0.0059	0.558	0.0136
600	0.893	0.0045	0.561	0.0117
700	0.895	0.0031	0.561	0.0099
800	0.895	0.0027	0.562	0.0086
900	0.896	0.0025	0.562	0.0076
1000	0.895	0.0024	0.563	0.0069

□

5.4.4 Improving Performance

The performance of an MCMC algorithm depends jointly on the posterior of interest and the proposal distributions used. Since the posterior is determined by the combination of prior and data, the performance of our algorithm can be changed only through the proposal distributions. We consider two different approaches for improving convergence using pilot tuning and block updates.

Pilot Tuning

With the exception of the Gibbs update, most MCMC updates require a degree of so-called pilot tuning in order to ensure adequate convergence and acceptable Monte Carlo errors. Firstly, the proposal distributions for the parameters in the model are specified. For example, the specification of a normal random

IMPLEMENTING MCMC

walk proposal, a uniform random walk proposal, or even an independent sampler. Alternatively, for parameters which have boundaries, constrained proposal distributions could be used (such as a truncated normal distribution), to ensure that the proposed parameter values always lie within the given boundary values (see Example 5.2). Once the proposal distribution is defined, pilot tuning typically involves adjusting the relevant proposal variances so as to obtain MH acceptance probabilities of between 20 and 40% (Gelman et al. 1996). In many cases this can be automated by running the algorithm for, say, 100 or 1000 iterations, calculating the average acceptance ratio for each parameter during that time and then increasing the corresponding proposal variance if the acceptance probabilities are too high and decreasing the variances if the probabilities are too low. This is what is meant by *pilot tuning*.

Example 5.12. Pilot Tuning – Example 4.1 Revisited – Radio-Tagging Data
We consider the radio-tagging data (assuming a uniform prior) and the MH algorithm described in Example 5.2. A random walk MH algorithm is adopted, using a normal proposal distribution with variance parameter σ^2. However, we need to specify a value for σ^2. We consider three possible values, $\sigma = 0.005, 0.05$ and 0.5 (or equivalently, $\sigma^2 = 0.000025, 0.0025, 0.25$). Typical trace plots of 1000 iterations for each of these proposal parameter values are provided in Figure 5.10. The corresponding mean acceptance probabilities for the MH updates are $0.921, 0.421$ and 0.053, for $\sigma = 0.005, 0.05, 0.5$, respectively. Note that none of these values of σ appear to provide a mean acceptance probability between 20 and 40%, although $\sigma = 0.05$ is very close. Thus, a further step would be to consider additional values of σ and repeat the process. For example, consider $\sigma = 0.1$.

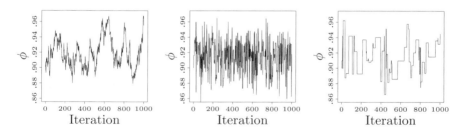

Figure 5.10 Sample paths for MH algorithms with (a) $\sigma = 0.005$, (b) $\sigma = 0.05$ and (c) $\sigma = 0.5$.

Clearly, the specification of the value of σ^2 has a significant impact upon the acceptance probability of the chain (although recall that the acceptance probability itself is independent of σ^2). This is because the proposal with $\sigma^2 = 0.05$ is much closer to the target distribution, so that more sensible candidate values are generated and subsequently accepted. However, the proposal with $\sigma^2 = 0.5$, generates candidate observations too far out in the tail to come from the target distribution (and often outside the interval $[0, 1]$) and these are subsequently rejected. Conversely, the proposal with $\sigma = 0.005$ generates candidate values very similar to the current value that are typically accepted, but this means that it takes a long time to move over the parameter space. The movement around the parameter space is often referred to as *mixing*.

□

Note that there is typically a trade-off between the time spent on the pilot tuning process and the computational time of the required simulations. For very fast computational simulations, extensive pilot tuning is often unnecessary, whereas for lengthy simulations, pilot tuning can be very important in reducing the computational expense.

Block Updates*

The use of multi-parameter updates can be used to avoid especially slowly-converging chains in the presence of high posterior correlations between parameters. If highly-correlated parameters are updated one at a time, then the fact that one is updated whilst the others remain fixed constrains the range of acceptable new values for the parameter being updated. This can cause extremely slow convergence and is best overcome by updating parameters in blocks so that highly correlated parameters are updated together (Roberts and Sahu 1997, Rue 2001). This will usually overcome the problem, though the specification of suitable multi-dimensional proposal distributions can be difficult. In practice, (for continuous parameters) pilot runs can be used to gain rough estimates of posterior means and covariances and then multivariate normal approximations can be used as proposal distributions for a much longer simulation that forms the basis for inference.

Example 5.13. Block Updates – Highly Correlated Variables
Suppose that we wish to sample from a bivariate normal distribution using the MH algorithm. We consider two different approaches – the single-update algorithm and the block updating algorithm (i.e. simultaneously updating both parameters). For notational purposes, we assume the parameters are denoted by $\boldsymbol{\theta} = (\theta_1, \theta_2)^T$, and that,

$$\boldsymbol{\theta} \sim \mathcal{N}_2(\boldsymbol{\mu}, \boldsymbol{\Sigma}),$$

where,

$$\boldsymbol{\mu} = \begin{pmatrix} 0 \\ 0 \end{pmatrix}; \quad \text{and} \quad \boldsymbol{\Sigma} = \begin{pmatrix} 1 & 0.7 \\ 0.7 & 1 \end{pmatrix}.$$

In other words the variances of θ_1 and θ_2 are both equal to 1, and the covariance (and correlation) of θ_1 and θ_2 is equal to 0.7. Thus, the parameters are fairly strongly correlated.

We begin by considering the single-update MH algorithm. In particular we

IMPLEMENTING MCMC

consider a random walk updating step, using a normal proposal distribution with variance 1 for each parameter individually. Figure 5.11 gives a sample path of the MH algorithm for (a) the joint trace plot, (b) the marginal trace plot for θ_1 and (c) marginal trace plot for θ_2. The mean acceptance probabilities for individually updating θ_1 and θ_2 are both approximately 0.46.

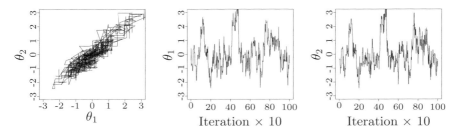

Figure 5.11 Sample paths for a single-update random walk MH algorithm for a bivariate normal distribution; (a) joint trace plot; (b) marginal trace plot for θ_1, and (c) marginal trace plot for θ_2.

We consider an alternative block updating algorithm, where we propose to update both θ_1 and θ_2 simultaneously using a random walk bivariate normal proposal distribution. For comparability, we assume a proposal variance of 1 for both θ_1 and θ_2, and covariance of 0.7. Figure 5.12 gives a sample path of the parameters using the MH algorithm for (a) joint trace plot, (b) marginal trace plot for θ_1 and (c) marginal trace plot for θ_2. The mean acceptance probability for simultaneously θ_1 and θ_2 is equal to 0.44 (so comparable to the above mean acceptance probability for the single update MH algorithm for updating θ_1 and θ_2). Comparing Figures 5.11 and 5.12 we can see that the block updates appear to have better mixing properties, traversing more of the parameter space (this is easily seen within the marginal trace plots for θ_1 and θ_2).

□

Example 5.14. Block Updates – Example 1.1 Revisited – C/C Model for the Dipper Data

We return to the C/C model for the dipper data with survival probability ϕ and recapture probability p. In Example 5.5 we considered a single MH updating algorithm, where we updated each parameter in turn. This algorithm appeared to perform well, with a mean acceptance probability of approximately 30% for each parameter. However, suppose that the single update MH algorithm did not appear to perform well, as a result of highly correlated parameters. We might consider the following block updating algorithm where we simultaneously propose to update both parameters at each iteration of the Markov chain.

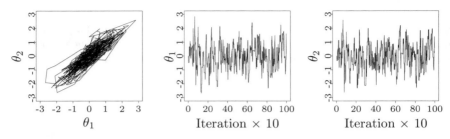

Figure 5.12 Sample paths for block update random walk MH algorithm for a bivariate normal distribution; (a) joint trace plot; (b) marginal trace plot for θ_1, and (c) marginal trace plot for θ_2.

Suppose that at iteration k of the Markov chain, the parameter values are p^k and ϕ^k. We then simulate a set of candidate values for these parameter values denoted by $\boldsymbol{\omega} = (\omega_1, \omega_2)^T$. In particular we consider a bivariate normal distribution given by,

$$\boldsymbol{\omega}|p^k, \phi^k \sim \mathcal{N}_2 \left(\begin{pmatrix} p^k \\ \phi^k \end{pmatrix}, \begin{pmatrix} \sigma_p^2 & \sigma_{p\phi} \\ \sigma_{p\phi} & \sigma_\phi^2 \end{pmatrix} \right).$$

In other words, we specify a bivariate normal distribution, where the mean vector is equal to the current parameter values with covariance matrix such that the proposal variance of ω_1 is σ_p^2 and the proposal variance of ω_2 is σ_ϕ^2, with covariance $\sigma_{p\phi}$. We specify $\sigma_p^2 = \sigma_\phi^2 = 0.01$, for comparability with Example 5.5. In order to specify a value for $\sigma_{p\phi}$ we use an initial pilot tuning algorithm, using the previous single-update MH algorithm given in Example 5.5. In particular, we obtain the posterior correlation of p and ϕ to be equal to -0.16. Thus, we consider a prior correlation between the two parameters of -0.2. This corresponds to a prior covariance of $\sigma_{p\phi} = -0.002$. Note we consider a slightly larger magnitude for the correlation term in the proposal distribution in an attempt to ensure that the parameter space is well explored.

The acceptance probability is given by $\alpha((p^k, \phi^k)^T, \boldsymbol{\omega}) = \min(1, A)$, where,

$$\begin{aligned} A &= \frac{\pi(\omega_1, \omega_2 | \boldsymbol{m}) q(p^k, \phi^k | \boldsymbol{\omega})}{\pi(p^k, \phi^k | \boldsymbol{m}) q(\boldsymbol{\omega} | p^k, \phi^k)} \\ &= \begin{cases} \frac{f(\boldsymbol{m}|\omega_1, \omega_2)}{f(\boldsymbol{m}|p^k, \phi^k)} & \omega_1, \omega_2 \in [0,1] \\ 0 & \omega_1 \notin [0,1] \text{ or } \omega_2 \notin [0,1], \end{cases} \end{aligned}$$

since the proposal distribution q is symmetric and cancels in the acceptance probability; and we assume independent $U[0,1]$ priors for p and ϕ.

Figure 5.13 gives the trace plot for the updates of p and ϕ for 1000 iterations. This can be compared to Figure 5.3(a) for the single-update MH algorithm. One of the differences between these two plots is that there are vertical and horizontal lines in Figure 5.3 for the single update algorithm corresponding to

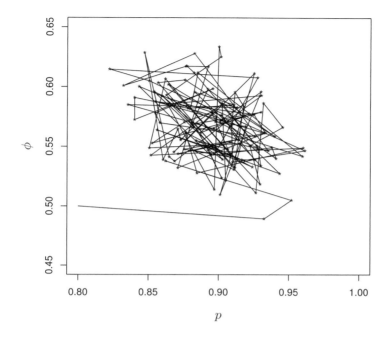

Figure 5.13 Sample paths for the MH block updating algorithm for the two-parameter model for the dipper data.

updating ϕ and not p, or vice versa, within a single iteration of the Markov chain. Clearly, this does not occur for the block updating algorithm (i.e. using the bivariate normal), since either both parameters are simultaneously proposed and accepted in an single iteration, or both remain at their current values. As a result, in the block updating algorithm, we always obtain diagonal lines. Note that for the block updating algorithm, we obtain a mean acceptance probability of 0.13. This is only marginally higher than using a MH algorithm for simultaneously updating p and ϕ using independent normal proposal distributions (i.e. setting $\sigma_{p\phi} = \rho = 0$), with a mean acceptance probability of 0.11. In addition, for the single updating algorithm, we obtain a mean acceptance probability of approximately 0.3 individually for p and ϕ. Thus, in this case there are more moves accepted using the single-update MH algorithm. The reason for this is that the correlation between the parameters is relatively small (a posterior correlation of -0.16). In this instance, due to the relatively small correlation between parameters, a single update MH algorithm, as described in Example 5.5, would appear to perform better.

Note that for comparability, we considered proposal variances of 0.01. However, within the pilot tuning algorithm that was performed to obtain a proposal covariance value for the parameters, we could equally obtain (approximate) estimates for the posterior variances for each parameter. In particular, we obtain posterior standard deviations of (approximately) 0.03 for each parameter. Thus, we could also use these values within the proposal distributions. For example, setting $\sigma_p^2 = \sigma_\phi^2 = 0.03^2$ and $\sigma_{p\phi} = -0.00018$ (once more giving a prior correlation of -0.2), we obtain a mean acceptance probability of 0.51. This is significantly higher and also higher than the 20 to 40% acceptance probability generally accepted.

□

5.4.5 Autocorrelation Functions and Thinning

The performance of the chain, in terms of mixing and exploring the parameter space, is often initially performed by eye from trace plots, or from consideration of the mean acceptance rate, if implementing a MH algorithm. However, another useful tool is the *autocorrelation function* (ACF). This is simply defined to be the correlation between the given parameter value in the Markov chain separated by l iterations. The term $l > 1$ is usually referred to as *lag*. Mathematically, suppose that we are interested in parameter θ, that takes value θ^k at iteration k of the Markov chain. The autocorrelation of the parameter at lag l is simply defined to be $cor(\theta^k, \theta^{k+l})$. This is typically calculated for values $l = 1, \ldots, l_{max}$ and plotted on a graph. Note that the autocorrelation function is always equal to 1 for the value $l = 0$, since $cor(\theta^k, \theta^k) = 1$.

Ideally, for good mixing Markov chains, there should be a fast decrease in the value of the autocorrelation function as the lag increases. In other words, in the ACF plot, this would be represented by a sharp gradient at low values of l. This would imply that there is little relationship between values of the Markov chain within a small number of iterations. Conversely, poorly mixing chains will typically have a very shallow gradient in the ACF plot, with high autocorrelation values for even relatively large values of l (say, $l \geq 20$). For MH algorithms, pilot tuning the parameter values or using block updates may improve the mixing (and hence ACF plot). However, when using the Gibbs sampler, there is no tuning involved in its implementation, so that it is typically more difficult to adjust, however, block updating may still be implemented.

One suggested method for reducing the autocorrelation within a Markov chain is to simply keep every i^{th} iteration of the sampled values from the Markov chain. For example, every 10th observation. This is typically referred to as thinning. The thinned values have reduced autocorrelation, but discards a (very) large number of sampled values. The discarded values (although possibly very highly autocorrelated) still provides information concerning the posterior distribution. Thus thinning should only be used if there are issues

IMPLEMENTING MCMC

relating to the storage and/or memory allocation of the large number of sampled values.

Example 5.15. Example 4.1 Revisited – Radio-Tagging Data
We once more consider the radio-tagging data, and in particular consider the different MH updating algorithms described in Example 5.12. Recall that a random walk MH algorithm is used with three different proposal parameter values for the normal distribution standard deviation: (a) $\sigma = 0.005$, (b) $\sigma = 0.05$ and (c) $\sigma = 0.5$. Corresponding sample trace plots are given in Figure 5.10. From visual inspection of the trace plots, it is clear that setting $\sigma = 0.05$ results in the best mixing within the Markov chain, in terms of a reasonably high acceptance rate and exploration of the parameters space. This is also clearly demonstrated when we consider the ACF plots of each of these trace plots given in Figure 5.14.

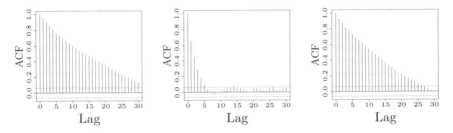

Figure 5.14 ACF plots of the sample paths of the MH algorithm with (a) $\sigma = 0.005$, (b) $\sigma = 0.05$ and (c) $\sigma = 0.5$ for the radio-tagging data.

Clearly, from Figure 5.14 we can see that the lowest levels of autocorrelation (at all lags $l \geq 1$), is for $\sigma = 0.05$ (plot (b)). This was the algorithm that appeared to have the best mixing properties in Example 5.12. Alternatively, we can see that both Figures 5.14(a) and (c) both have very high levels of autocorrelation, and are very similar to each other. However, the reasons for the relatively high autocorrelations in plots (a) and (c) are different. For Figure 5.14(a) the proposal distribution is very tight, so that only small moves from the current value are proposed (and generally accepted with high probability). This means that although the probability of accepting a proposed value is high in the MH step, the distance travelled is small. Alternatively, the high levels of autocorrelation seem in Figure 5.14(c) are a result of a small acceptance probability in the MH step. This means that the parameter values often remain at the same value without being updated to a new value for a number of iterations in the Markov chain, resulting in only a small decrease in the autocorrelation function for increasing lag.

□

5.4.6 Model Checking – Bayesian p-values

Typically when fitting models to data, we not only wish to estimate the parameters within the model, but also assess whether the model is a good fit to the data. The most commonly used Bayesian model-checking procedure is the Bayesian p-value (Gelman and Meng 1996). Essentially, the underlying principle of Bayesian p-values is to assess the predictive ability of the model by matching predicted or imputed data against that observed. In other words, assessing whether simulated data from the posterior distribution looks like the data observed.

In order to calculate a Bayesian p-value, the first step involves generating a new data set $\boldsymbol{x}^k$ from the model at each MCMC iteration, for iterations, $k = 1, \ldots, n$. In general, a discrepancy statistic $D(\boldsymbol{x}^k, e^k)$ is used to measure the distance between the generated data and the corresponding expected values, e^k, at that iteration. This is compared to the discrepancy function evaluated at the observed data value. In particular, the proportion of times that $D(\boldsymbol{x}, e^k) < D(\boldsymbol{x}^k, e^k)$ is recorded. The Bayesian p-value is simply this calculated value. If the model is a good fit to the data, then we would expect that the observed data should be similar to the simulated data. Thus, we would expect the Bayesian p-value to be close to 0.5. The most commonly used discrepancy function is probably simply the negative of the log-likelihood function. However, alternative functions also include the Pearson chi-squared statistic or the Freeman-Tukey statistic (see Brooks et al. 2000a for further discussion in relation to capture-recapture data). We note that scatter plots of the discrepancy functions $D(\boldsymbol{x}, e^k)$ and $D(\boldsymbol{x}^k, e^k)$ may also be useful. See for example Bayarri and Berger (1998) and Brooks et al. (2000a) for further details. Further, we note that, the Bayesian p-values are implicitly dependent on the priors specified on the parameters. This is clearly seen since the Bayesian p-value is calculated using a sample from the posterior distribution, which is a function of the prior distribution (via Bayes' Theorem). See Brooks et al. (2000a) for further discussion.

Example 5.16. Example 1.1 Revisited – Dipper Data
To illustrate the ideas associated with Bayesian p-values, we once more consider the dipper data. In particular, we consider the goodness-of-fit of the two models, C/C and $C/C2$ (i.e. the models with constant recapture and survival probability; and constant recapture probability and separate survival probabilities for flood and non-flood years). We calculate the Bayesian p-values for these models (assuming independent uniform priors on each parameter), using the deviance as the discrepancy function, so that $D(\boldsymbol{x}, e^k) = -2 \log f(\boldsymbol{x}|\boldsymbol{\theta})$, where $\boldsymbol{\theta}$ denotes the parameter values. The corresponding Bayesian p-values obtained are 0.04 for model C/C and 0.13 for model $C/C2$. The Bayesian p-value for model C/C may be regarded as small, and so we may suspect the goodness-of-fit of this model. The corresponding Bayesian p-value scatter plots of the discrepancy functions are provided in Figure 5.15 for models C/C

IMPLEMENTING MCMC

and $C/C2$. The Bayesian p-values are simply the proportion of points that lie above the line $x = y$.

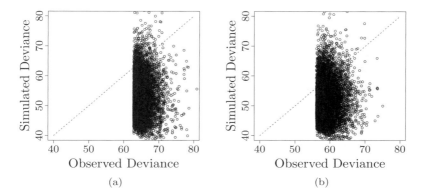

Figure 5.15 The Bayesian p-value scatter plot of the deviance of the observed data versus the deviance of the simulated data for the dipper data for (a) model C/C and (b) model $C2/C$. The dashed line corresponds to the line $x = y$.

Figure 5.15 has some interesting characteristics. In particular, the spread of the deviance values of the simulated data (i.e. the discrepancy function of the simulated data) is greater than that for the deviance of the observed data (i.e. discrepancy function of the observed data). This is usually the case, since there is the additional uncertainty relating to the simulation of data, in addition to parameter uncertainty. Further, there appears to be a lower limit for the deviance evaluated for the observed data. In this instance (due to the uniform priors on the parameters), this is essentially the deviance evaluated at the MLEs of the parameters. Note that a "limit" is not always as marked as in the above examples and will be dependent, for example, on the choice of the discrepancy function, and shape of the posterior distribution, including the gradient of the distribution within the area of highest posterior mass.

□

Example 5.17. Example 4.1 Revisited – Radio-Tagging Data
We reconsider the radio-tagging data in Example 4.1. In particular we consider two different discrepancy functions: the deviance and Freeman-Tukey statistic. The Freeman-Tukey statistic is defined to be,

$$\sum_i (\sqrt{x_i} - \sqrt{e_i})^2,$$

where e_i is the expected data value. In this case, e_i is the expected cell entry in the multinomial distribution, so that $e_i = 36 p_i$, where p_i is the corresponding multinomial cell probabilities. Once more we assume a $U[0,1]$ prior on the survival parameter. The corresponding Bayesian p-value plots are given in Figure 5.16.

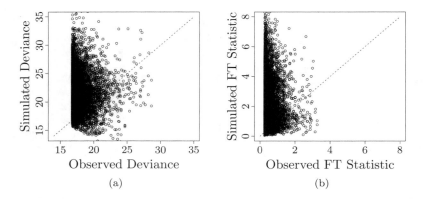

Figure 5.16 Bayesian p-value scatter plots for the radio-tagging canvasback data using the discrepancy functions (a) deviance with corresponding Bayesian p-value of 0.88, and (b) Freeman-Tukey (FT) statistic with Bayesian p-value of 0.91. The dashed line corresponds to the line $x = y$.

Once more we have interesting patterns emerging in the Bayesian p-value plots. The discrepancy functions of the observed data once again have a clear limit, with the discrepancy function for the simulated data much more overdispersed. The Bayesian p-values are very similar for both the discrepancy functions, 0.88 for the deviance function, compared to 0.91 for the Freeman-Tukey statistic. This need not always be the case – in some cases, Bayesian p-values can differ substantially, depending on the discrepancy function. This in itself is an interesting point, and often motivates further investigation as to the underlying reasons for these discrepancies. Gelman et al. (2003, p. 172) suggest that an advantage of the Bayesian p-values is the ability to consider a number of different discrepancy functions.

□

Finally, we briefly mention here an alternative model-checking algorithm of cross-validation. This algorithm essentially involves removing one or more observations, x_i, from the data and then imputing it within the model-fitting process to get $\widehat{x}_i$. Comparisons are then made between the imputed and corresponding true value and averaged over data points to get a cross-validation statistic,

$$CV = \frac{1}{I} \sum_{i=1}^{I} x_i / \widehat{x}_i.$$

If the model fits the data well, we would expect a cross-validation value of around 1. For further discussion, see for example or Gelman et al. (2003) and Carlin and Louis (2001).

5.5 Summary

The posterior distribution is often high-dimensional and complex, so that little inference can be obtained directly from it. However, simulation techniques can be used to obtain a sample from the posterior distribution, which can then be used to obtain Monte Carlo estimates of the parameters of interest. The simulation technique most commonly used is Markov chain Monte Carlo. Most MCMC methods are special cases of the MH scheme, a notable such special case being Gibbs sampling. A useful approach when possible is to include Gibbs updates within MH, an approach, which is facilitated by the use of conjugate priors. Without fast computers, these methods would not be computationally feasible.

Care needs to be taken when implementing the computationally intensive techniques. Many results rely on asymptotic arguments relating to the stationary distribution of the constructed Markov chain. It is difficult to prove conclusively that a constructed Markov chain has reached its stationary distribution, although there are a number of graphical and analytic tools that will provide evidence of (or an indication of) lack of convergence. Although the MH algorithm is an off-the-shelf procedure, typically some (minor) pilot tuning is required. This can aid the convergence of the Markov chain to the stationary distribution by having an efficient updating algorithm, shortening the burn-in and optimising computational effort. Multiple replications of the simulations, starting from over-dispersed starting points, is also advised to explore a larger parameter space and to check post-burn-in convergence of the posterior estimates obtained. Finally, Bayesian p-values permit a formal model checking approach by essentially comparing whether the observed data resembles data simulated from the posterior distribution.

5.6 Further Reading

There are several books that focus on the MCMC algorithm in detail, and typically any Bayesian reference text (such as those discussed in Section 4.7) contains some discussion of the ideas to some degree. A particularly clear and excellent book describing MCMC and associated issues is Gamerman (1997). Additional books worth note include Gilks et al. (1996), Robert and Casella (2004), Givens and Hoeting (2005) and Robert (2007). These provide an excellent general introduction to the MCMC algorithm, extending to more advanced techniques using a variety of different applied problems. For example, Robert and Casella (2004) and Givens and Hoeting (2005) both present the slice sampler (Neal 2003), an alternative updating algorithm that utilises the local conditional distribution in proposing a candidate value. This technique is less generic than the MH algorithm, but is implemented, for example, in the computer package WinBUGS (see Appendix C). Alternatively, Gilks and Roberts (1996) discuss a number of methods for improving the efficiency of MCMC updates, such as reparameterisation and direction sampling. With particular reference to ecological data, Link and Barker (2008) consider efficient MH updates for capture-recapture data. Alternatively, for ecological data, Newman et al. (2009) improve the MCMC mixing by integrating out a number of parameters, thereby reducing the dimension of the posterior distribution being sampled from, before implementing a MH algorithm.

Detailed implementation of MCMC algorithms for ecological data (including the posterior conditional distributions) are also provided by Vounatsou and Smith (1995), Brooks et al. (2000a) and Brooks et al. (2000b). For example, Vounatsou and Smith (1995) provide a detailed description of both the use of the Gibbs sampler (sampling from a non-standard posterior conditional distribution) and MH algorithm for a Bayesian analysis of ring-recovery data. Brooks et al. (2000a) present the posterior conditional distributions for both ring-recovery data and capture-recapture data and discuss the merits of the different updating procedures.

MCMC is the most popular approach for obtaining a sample from the posterior distribution of interest; however, there are alternative techniques. These include sequential importance sampling (Liu 2001) and perfect sampling (Propp and Wilson 1996). These techniques are less well developed than MCMC, and typically involve a greater amount of preparatory work before they can be implemented. In particular, perfect sampling is typically of more interest to theoreticians, and has not yet been developed to become a standard Bayesian tool. Any stochastic algorithm (including MCMC) relies on the simulation of random variables from a range of statistical distributions. A valuable resource for simulating random variables is Morgan (2008).

5.7 Exercises

5.1 Show that the Gibbs sampler is a special case of the single-update MH algorithm, where the proposal distribution is the posterior conditional distribution of the parameter.

5.2 Classical and Bayesian analyses of integrated recapture and recovery data on shags, ringed both as young and as adults, result in the estimates of annual survival shown in Table 5.4. Here ϕ_{1i} is the annual survival probability of first-year birds in the ith year of study, (and $\overline{\phi}_1$ the mean first-year survival probability, averaged over years), ϕ_{imm} is the annual survival probability of immature birds, and ϕ_a is the annual survival probability of adult birds.

For the Bayesian analysis, the prior distribution was the product of independent $U[0,1]$ distributions for each of the parameters in the model, and the mean and standard deviation (SD) follow from MCMC. The classical analysis results in the MLEs.

EXERCISES

Table 5.4 Parameter Estimates for the Shag Data; (a) Provides the Classical MLEs and Standard Errors (se) and (b) Provides the Posterior Means and Standard Deviations (SD)

	(a)			(b)	
	Estimate	se		Mean	SD
ϕ_{11}	0.325	0.037	ϕ_{11}	0.328	0.038
ϕ_{12}	0.439	0.035	ϕ_{12}	0.439	0.035
ϕ_{13}	0.193	0.035	ϕ_{13}	0.198	0.035
ϕ_{14}	0.732	0.064	ϕ_{14}	0.726	0.061
ϕ_{15}	0.441	0.051	ϕ_{15}	0.441	0.051
ϕ_{16}	0.613	0.088	ϕ_{16}	0.610	0.083
ϕ_{17}	0.397	0.076	ϕ_{17}	0.403	0.076
ϕ_{18}	0.231	0.114	ϕ_{18}	0.259	0.102
ϕ_{19}	0.767	0.083	ϕ_{19}	0.732	0.097
$\overline{\phi}_1$	0.460	0.023	$\overline{\phi}_1$	0.460	0.023
ϕ_{imm}	0.698	0.021	ϕ_{imm}	0.696	0.021
ϕ_a	0.866	0.012	ϕ_a	0.864	0.013

Source: Table (a) is reproduced with permission from Catchpole et al. (1998, p. 40) published by the International Biometric Society.

Comment on the similarities and dissimilarities between the two sets of results.

5.3 Consider capture-recapture data with constant recapture and survival probabilities, p and ϕ (i.e. model C/C).

   (a) Write down an explicit expression for the posterior conditional distribution of p and ϕ.

   (b) Suggest how the Gibbs sampler may be implemented given the posterior conditional distributions.

   (c) Suggest an MH updating algorithm and discuss how pilot tuning may be performed to determine the parameters in the proposal distribution.

5.4 Consider model C/C for the dippers data.

   (a) The trace plot in Figure 5.17 was produced using an independent MH algorithm with $U[0,1]$ proposal distribution. Describe the main properties we would expect for the corresponding ACF for the parameter ϕ.

   (b) Suggest alternative proposal distributions for p and ϕ, assuming an independent MH updating algorithm.

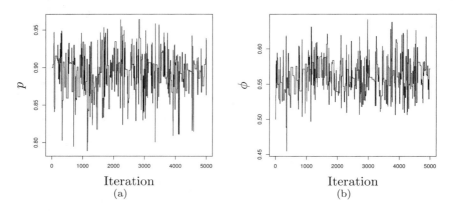

Figure 5.17 Trace plots for (a) p and (b) ϕ for model C/C for the dippers data using an independent MH updating algorithm with $U[0,1]$ proposal distribution.

(c) Consider the ACF plot given in Figure 5.18, for parameter p using a random walk MH algorithm with normal proposal distribution and proposal variance, σ^2.

What can we say regarding the proposal variance of the MH algorithm that resulted in the ACF plot for the MCMC output? What additional information is useful in order to determine how we should change the proposal variance σ^2 in an attempt to decrease the autocorrelation?

5.5 Consider the ring-recovery data corresponding to the lapwing data described in Section 2.3.1, where the first-year survival probability and adult survival probabilities are logistically regressed on $fdays$. So that,

$$\phi_{1,t} = \alpha_1 + \beta_1 f_t;$$
$$\phi_{a,t} = \alpha_a + \beta_a f_t,$$

where f_t denotes the covariate value for $fdays$ in year t. For simplicity we consider a constant recovery probability λ. We specify independent $N(0,10)$ priors on each logistic regression parameter and a $U[0,1]$ prior on λ.

(a) Calculate the corresponding prior on the first-year survival probability $\phi_{1,t}$, say, in an "average" year, where the covariate value is equal to the mean observed covariate value, under the following situations:

   (i) where the covariate values are normalised (so that the mean covariate value is equal to 0); and
   (ii) where the covariate values are unnormalised, such that the mean observed covariate value is equal to 10.

EXERCISES

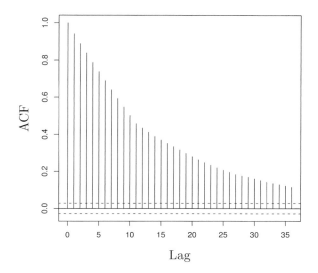

Figure 5.18 ACF plot for p for the dippers C/C model using a random walk MH updating algorithm with normal proposal distribution.

(b) Using a single-update random walk MH algorithm with normal proposal density with variance parameter σ^2, derive the corresponding acceptance probability for each parameter α_1, β_1, α_a, β_a and λ.

5.6 Consider the system where,

$$N_t \sim Po(N_{t-1}\phi\rho),$$

where N_t denotes the number of individuals in a given area at time $t = 1, \ldots, T$, ϕ, the survival probability and ρ, the productivity rate for the population under study. The true population sizes are not observed, only an estimate, denoted by y_t, where we assume,

$$y_t \sim N(N_t, \sigma^2),$$

for $t = 1, \ldots, T$, such that σ^2 is known. For notational purposes, we let $\boldsymbol{y} = \{y_1, \ldots, y_T\}$ and are the observed data. Finally, we assume that N_0 is known.

(a) Write down the joint posterior distribution, $\pi(\phi, \rho, \boldsymbol{N}|\boldsymbol{y})$, where $\boldsymbol{N} = \{N_1, \ldots, N_T\}$.

(b) Reparameterise the model to remove any parameter redundancy in terms of the model parameters.

(c) Suppose that we use a single-update random walk MH algorithm. Suggest a proposal distribution and calculate the corresponding acceptance probabilities for each individual parameter.

(d) Suppose that we use a normal approximation to the Poisson distribution. Describe a hybrid Metropolis within Gibbs algorithm where we update demographic parameters using a random walk MH algorithm, and update the parameters $N_1, \ldots, N_T$ using a Gibbs sampler. (Explicitly state the posterior conditional distribution of the parameters used in the Gibbs update).

(e) Extend the algorithm to the case where σ^2 is also unknown, specifying an appropriate prior and describe the MCMC updating algorithm.

5.7 Discuss the advantages and disadvantages of using the following summary statistics for describing the posterior distribution of the parameter of interest: mean, median, variance, 95% symmetric credible interval, 95% highest posterior density interval and correlation between pairs of parameters. In addition, discuss the computational implications of calculating these posterior summary statistics and methods for reducing memory allocation within their calculation, if we do not wish to retain the full set of sampled values from an MCMC algorithm.

CHAPTER 6

Model Discrimination

6.1 Introduction

In the last chapter we formed probability models for ecological data, and obtained estimates of the posterior distribution of the parameters. The different models fitted will typically be based on biological knowledge of the system. However, in many cases there may be several competing biological hypotheses represented by different underlying models. Typically in such circumstances, discriminating between the competing models is of particular scientific interest. For example, consider ring-recovery data, collected from animals that were marked as adults of unknown age, where there is simply a survival probability (common to all animals) and a recovery probability; this is the model C/C of Section 3.3.1. There are potentially four different models, dependent on whether the survival and/or recovery probability is time dependent or constant over time. Each of these may be biologically plausible, a priori, but describe different biological hypotheses, corresponding to constant survival/recovery probabilities versus time dependent probabilities. We have already encountered this situation in Examples 3.4 through 3.6.

In Example 2.5 we saw an example of different models for large mammals with age dependence of the survival probabilities (for the different age groups) dependent on a number of different covariates. The subscript notation used there indicates age classes, and terms in brackets describe the covariate or time dependence for each given age group. Typically, there are a number of possible covariates that could be fitted to the data and the corresponding posterior distributions of the parameters estimated. For example, consider the following model for Soay sheep:

$$\phi_1(BW, P), \phi_{2:7}(W), \phi_{8+}(C, N)/p_1, p_{2+}/\lambda_1, \lambda_{2+}(t).$$

This denotes that the first-year survival probability is dependent on birth weight (BW) and population size (P); animals in years of life 2 to 7 have a common survival probability dependent on current weight (W); all older individual (animals in year 8 and older) have a survival probability dependent on coat type (C) and NAO index (N). The recapture and recovery probabilities are different for lambs compared to all older individuals, all of which are constant over time except the recovery probability of sheep in years 2 of life and above which have an arbitrary time dependence.

However, we could have fitted the alternative model:

$$\phi_1(BW, P), \phi_{2:7}(W), \phi_{8+}(C, P)/p_1, p_{2+}/\lambda_1, \lambda_{2+}(t),$$

i.e. changing the dependence of the senior sheep survival probability from the NAO index (N) to the population size (P). In other words the second model would correspond to the biological hypothesis that it is not the harshness of the winter that affects their survival probability, but population size, which could be regarded as a surrogate for the competition of the available resources. The choice of covariates used to explain the underlying temporal and/or individual heterogeneity can have a direct impact on the posterior distributions of the parameters of interest and hence the corresponding interpretation of the results. The different possible models, in terms of the covariate dependence of the demographic parameters essentially represent competing biological hypotheses concerning the underlying dynamics of the system. Hence discriminating between the different models allows us to discriminate between the competing biological hypotheses. We note that discriminating between the competing models can be regarded as an hypothesis test. For example, for the above models, testing the dependence of the senior survival probability on NAO index versus population size, given the observed data (and possible prior information in the Bayesian setting). Clearly this kind of procedure is often of particular biological interest. We note here that although the dependence of the recapture and recovery probabilities on age and/or time may not be of interest biologically, the underlying model specified for these nuisance parameters may influence the corresponding results and interpretation for the parameters of primary interest (i.e. estimates of regression parameters and model selection procedure).

Typically a number of models may be deemed to be biologically plausible. We wish to discriminate between these models that represent these biological hypotheses in order to identify the underlying dependence structure of the parameters, and hence increase our understanding of the system. Within the classical framework, each individual model may be fitted to the data, and the models compared via some criterion, such as an information criterion. We have seen this done in Section 3.3. We shall now discuss the alternative Bayesian ideas behind model discrimination before describing the computational tools necessary for this approach.

6.2 Bayesian Model Discrimination

In the Bayesian framework the information criteria described in Section 3.3 no longer have a fixed value, but a distribution because the parameter vector, $\boldsymbol{\theta}$, has a distribution. An information criterion (the Deviance Information Criterion) has been suggested that takes into account the fact that $\boldsymbol{\theta}$ has a distribution, rather than a fixed value. This approach once again considers a trade-off between the complexity of the model and the corresponding fit of the model to the data. A very different approach, and as we shall see a natural extension to the construction of the posterior distribution of interest using Bayes' Theorem, is to consider the model itself to be a parameter within the analysis. This approach allows us to calculate posterior model probabili-

BAYESIAN MODEL DISCRIMINATION

ties which quantitatively discriminate between competing models in an easily interpretable manner. When appropriate, it also permits model-averaged estimates of parameters of interest, taking into account model uncertainty. We shall now discuss each of these methods in turn.

6.2.1 Bayesian Information Criteria

The AIC and BIC given in Section 3.3 are defined "classically," where the likelihood is evaluated at the MLE of the parameters. These information criteria could also be considered in the Bayesian case, in terms of calculating their expected value, where expectation is taken with respect to the posterior distribution. However, an alternative information criterion, suggested by Spiegelhalter et al. (2002), is the Deviance Information Criterion (DIC). This is similar to the previous information criteria, in that it discriminates between models by considering both the fit and complexity of the models, but was developed specifically for use within the Bayesian framework. Once more the fit of model m is evaluated in terms of the deviance, while the complexity of the model is defined in terms of the *effective number of parameters*, denoted by $p_D(m)$. We can express this effective number of parameters for model m in the form,

$$p_D(m) = -2\mathbb{E}_\pi(\log f_m(\boldsymbol{x}|\boldsymbol{\theta})) + 2\log f_m(\boldsymbol{x}|\widetilde{\boldsymbol{\theta}}),$$

where the expectation is with respect to the posterior distribution, and $\widetilde{\boldsymbol{\theta}}$ is a posterior estimate of $\boldsymbol{\theta}$. A typical choice is $\widetilde{\boldsymbol{\theta}} = \overline{\boldsymbol{\theta}} = \mathbb{E}_\pi(\boldsymbol{\theta}|\boldsymbol{x})$, so that $\overline{\boldsymbol{\theta}}$ denotes the posterior mean of the parameters $\boldsymbol{\theta}$. Thus, the effective number of parameters is the difference between the posterior expectation of the deviance and the deviance evaluated at the posterior mean of the parameters.

The DIC for model m can then be expressed in the form,

$$DIC_m = -2\mathbb{E}_\pi(\log f_m(\boldsymbol{x}|\boldsymbol{\theta})) + p_D(m),$$

where, once more, the expectation is with respect to the posterior distribution. The DIC can generally be easily calculated within any MCMC procedure with practically no additional computational expense. However, we note that the effective number of parameters, p_D is dependent on the choice of $\widetilde{\boldsymbol{\theta}}$. The posterior mean is the most commonly used value for $\widetilde{\boldsymbol{\theta}}$. By definition, the effective number of parameters uses a posterior point estimate of the parameter values (i.e. $\widetilde{\boldsymbol{\theta}}$). This can be problematic in cases where, for example, the posterior distribution is bimodal (which is often the case for mixture models, for example), resulting in the mean being a very poor point estimate with very low posterior support. In addition, in some applications, a negative effective number of parameters can be obtained, (again this is often the case when fitting mixture models). Finally we note that the DIC cannot be used (at least in the form given above, using the posterior mean) where the parameters are discrete-valued, since $\overline{\theta}$ will typically not be equal to one of these discrete values. For further discussion of these and other related issues, see Spiegelhalter et al. (2002) and Celeux et al. (2006). We note that the DIC is provided in both MARK and WinBUGS for standard models. However, for the reasons discussed above, and others, the DIC cannot be calculated in WinBUGS for mixture models or in the presence of missing information. As a consequence, we suggest that the DIC is used very cautiously. For example, within a preliminary or exploratory analysis of the data.

When applying the DIC, we need to fit each possible model of interest to the data, which may be infeasible if there are a large number of possible models to be considered. In addition, the DIC does not provide a quantitative comparison between competing models which is readily interpretable. A more natural approach to Bayesian model discrimination is via posterior model probabilities. These have the advantage that they are a quantitative comparison of competing models and permit additional model averaging ideas. We discuss posterior model probabilities next.

6.2.2 Posterior Model Probabilities

A more formal and natural approach to discriminating between competing models within the Bayesian framework is via posterior model probabilities. These are obtained by a simple extension to Bayes' Theorem to take into account model uncertainty. Essentially, this alternative approach involves considering the model itself to be parameter. It allows us to compare competing models quantitatively, and to incorporate any information concerning the relative a priori probabilities of the different models. We concentrate on this latter approach throughout the rest of the book for discriminating among competing models.

Within the Bayesian framework, we can consider the model itself to be a discrete parameter to be estimated. The set of possible values that this parameter can take is simply the set of models deemed to be biologically plausible. By applying Bayes' Theorem, we can form the joint posterior distribution over both the model and the parameters. Formally, we have,

$$\pi(\boldsymbol{\theta}_m, m | \boldsymbol{x}) \propto f_m(\boldsymbol{x} | \boldsymbol{\theta}_m) p(\boldsymbol{\theta}_m | m) p(m),$$

where we use a subscript on the parameters $\boldsymbol{\theta}_m$ to denote the set of parameters in model m; $f_m(\boldsymbol{x}|\boldsymbol{\theta}_m)$ denotes the likelihood given model m, and the corresponding parameter values; $p(\boldsymbol{\theta}_m|m)$ denotes the prior distribution for the parameters in model m; and $p(m)$ denotes the prior probability for model m. Thus, since we are treating the model as a parameter to be estimated, we need to specify the corresponding prior probability function for each possible model. Typically a parameter may be present in more than one model, for example, a regression coefficient corresponding to a given covariate. In general, we can specify a different set of prior distributions for the parameters in each of the different models. However, it is common to specify (independent) priors on the parameters such that the same prior is specified on the parameter for each model containing that parameter. This is particularly the case when we specify non-informative priors.

BAYESIAN MODEL DISCRIMINATION   151

As an example, suppose that we have ring-recovery data and we have two different models that we want to discriminate between: C/C and T/C. In other words we want to test whether the survival probability was constant or time dependent, assuming that the recovery probability is constant. For each of these two models we need to specify the corresponding prior distributions. We begin with the priors on the parameters for each model. Consider the model C/C. Without any prior information, we may set $\phi \sim U[0,1]$ and $\lambda \sim U[0,1]$. Alternatively, consider model T/C. Once again we might specify the priors such that $\phi_t \sim U[0,1]$ for $t = 1, \ldots, T-1$, and $\lambda \sim U[0,1]$. Finally, we need to specify a prior for the two models C/C and T/C. In the absence of any strong prior information, an obvious choice is to define the prior probability of each model to simply be $1/2$, i.e. they have the same prior support, before observing any data. As we noted above, in practice, it is often the case that parameters may be common to more than one model. Typically, this parameter is specified to have the same prior distribution for each model for which it is present, as was the case for λ in the above example. However, this is not a necessary condition, in general.

Of particular interest within this model uncertainty framework is the posterior marginal distribution of the different possible models. In other words, this is the updated support for each possible model, following the data being observed. Suppose that set of biologically plausible models is denoted by $\boldsymbol{m} = (m_1, \ldots, m_K)$. So, returning to our ring-recovery example above where the survival and recovery probabilities are either constant over time or time dependent, the full set of possible models is $\boldsymbol{m} = \{C/C; C/T; T/C; T/T\}$. The corresponding posterior model probability for model m_i, given observed ring-recovery data $\boldsymbol{x}$, can be written as,

$$\pi(m_i|\boldsymbol{x}) = \frac{f(\boldsymbol{x}|m_i)p(m_i)}{\sum_{i=1}^{K} f(\boldsymbol{x}|m_i)p(m_i)}, \tag{6.1}$$

where,

$$f(\boldsymbol{x}|m_i) = \int f_{m_i}(\boldsymbol{x}|\boldsymbol{\theta}_{m_i})p(\boldsymbol{\theta}_{m_i}|m_i)d\boldsymbol{\theta}_{m_i}, \tag{6.2}$$

and $f_{m_i}(\boldsymbol{x}|\boldsymbol{\theta}_{m_i})$ denotes the likelihood of the data, $\boldsymbol{x}$, given model m_i and corresponding parameters $\boldsymbol{\theta}_{m_i}$. The marginal posterior distribution of the models that is given in Equation (6.1) quantitatively discriminates between different models by calculating the corresponding posterior model probability of each model. Related (and equivalent) statistics relating to the posterior model probabilities of different models are their corresponding Bayes factors.

6.2.3 Bayes Factors

Bayes factors can be used to compare two competing models, given the data observed, $\boldsymbol{x}$. Suppose that we are comparing two models, labeled m_1 and m_2.

The Bayes factor of model m_1 against model m_2 can be expressed as,

$$B_{12} = \frac{\pi(m_1|\boldsymbol{x})/\pi(m_2|\boldsymbol{x})}{p(m_1)/p(m_2)},$$

where $p(\cdot)$ denotes the prior for the given model; and $\pi(\cdot|\boldsymbol{x})$ the posterior probability of the model, given the data. Thus the Bayes factor can be interpreted as the ratio of the posterior odds to its prior odds of model m_1 to model m_2. Using Bayes' Theorem, as given in Equation (6.1), we can also express the Bayes factor in the form,

$$B_{12} = \frac{f(\boldsymbol{x}|m_1)}{f(\boldsymbol{x}|m_2)},$$

where $f(\boldsymbol{x}|m)$ denotes the likelihood under model m (see Equation (6.2)).

Kass and Raftery (1995) give the following rule of thumb guide for interpreting Bayes factors:

Bayes factor	Interpretation
< 3	Not worth mentioning
3–20	Positive evidence for model m_1 compared to m_2
20–150	Strong evidence for model m_1 compared to m_2
> 150	Very strong evidence for model m_1 compared to m_2

We shall show in Section 6.5 how we are able to use posterior model probabilities in order to estimate and average parameters across all the different plausible models, and thereby include model uncertainty into our parameter estimation procedure. However, before we consider how to use these posterior model probabilities (or Bayes factors) in estimating parameters of interest, we first explain how we may calculate them.

6.3 Estimating Posterior Model Probabilities

We can write down an expression for the posterior model probabilities in the form of an integral using Equations (6.1) and (6.2). However the integration over the parameter space in Equation (6.1) is analytically intractable in general. There are a number of different methods that have been proposed for estimating posterior model probabilities. Here we focus on the two most widely used approaches: (i) Monte Carlo estimates; and (ii) MCMC-type estimates. We consider each of these in turn and discuss a few of the most popular methods for implementing the approaches.

6.3.1 Monte Carlo Estimates

The Monte Carlo approach involves obtaining an estimate of Equation (6.2), for each of the different possible models. These can then be used to combine

ESTIMATING POSTERIOR MODEL PROBABILITIES

with the prior model probabilities in order to obtain an estimate of the posterior model probabilities, given in Equation (6.1). However, obtaining an estimate of the expression $f(\boldsymbol{x}|m_i)$ in Equation (6.2) is non-trivial. We consider a number of different approaches next, essentially all based on the following principle. We can express Equation (6.2) in the form,

$$f(\boldsymbol{x}|m_i) = \mathbb{E}_p[f_{m_i}(\boldsymbol{x}|\boldsymbol{\theta})],$$

i.e. the expectation of the likelihood, $f_{m_i}(\boldsymbol{x}|\boldsymbol{\theta})$, with respect to the prior distribution $p(\boldsymbol{\theta})$. (Note that we drop the subscript m_i notation for the set of parameters $\boldsymbol{\theta}$ for notational convenience since it is clear from the context that the set of parameters refers to those for model m_i). We can estimate this expectation by drawing n observations from the prior distribution, denoted by $\boldsymbol{\theta}^1, \cdots, \boldsymbol{\theta}^n$ and then use the Monte Carlo estimate,

$$\widehat{f}_1(\boldsymbol{x}|m_i) = \frac{1}{n} \sum_{j=1}^{n} f_{m_i}(\boldsymbol{x}|\boldsymbol{\theta}^j).$$

In other words we estimate the mean of the likelihood with respect to the prior distribution by the sample mean where the parameter values are drawn from the prior distribution (see Section 5.1).

We then estimate the posterior probability of model m_i as,

$$\pi(m_i|\boldsymbol{x}) \propto p(m_i)\widehat{f}_1(\boldsymbol{x}|m_i).$$

Thus, the estimate converges to the posterior model probability (up to proportionality). We repeat this process for each model $m = m_1, \ldots, m_K$ and renormalise the estimates obtained for each model to obtain an estimate of the corresponding posterior model probabilities.

However, the above estimate of the likelihood is generally inefficient and very unstable, as a result of the parameters being drawn from the prior distribution. The corresponding likelihood values are often very small, as the prior, in general, is far more dispersed over the parameter space than the likelihood. Thus, the expectation is heavily dominated by only a few sampled values, even for large values of n, resulting in a very large variance, and the instability of the estimate.

An alternative Monte Carlo estimate for the likelihood can be derived using importance sampling (see Morgan 1984, p. 168). The idea here is to sample from a distribution for which the values are in high-likelihood areas, and then re-weight the estimate. Formally, if $\boldsymbol{\theta}^1, \ldots, \boldsymbol{\theta}^n$ are drawn from an importance sampling distribution $q(\boldsymbol{\theta})$, a new estimate of the expression of Equation (6.2) is given by,

$$\widehat{f}_2(\boldsymbol{x}|m_i) = \frac{\sum_{j=1}^{n} f_{m_i}(\boldsymbol{x}|\boldsymbol{\theta}^j)p(\boldsymbol{\theta}^j)/q(\boldsymbol{\theta}^j)}{\sum_{j=1}^{n} p(\boldsymbol{\theta}^j)/q(\boldsymbol{\theta}^j)}.$$

In order to avoid the instability problems, as for estimate $\widehat{f}_1(\boldsymbol{x}|m_i)$, with samples drawn from the prior distribution, we need to specify an importance sampling distribution, where draws from this distribution lie in high-likelihood areas. An obvious choice for the importance sampling distribution is the posterior distribution, which we can sample from using the MCMC algorithm, in order to obtain the parameters $\boldsymbol{\theta}^1, \ldots, \boldsymbol{\theta}^n$ from the posterior distribution. Setting the importance sampling distribution above to be the posterior distribution, the estimate of the likelihood in Equation (6.2) simplifies to the harmonic mean, i.e.

$$\widehat{f_3}(\boldsymbol{x}|m_i) = \left[\frac{1}{n} \sum_{j=1}^n \frac{1}{f_{m_i}(\boldsymbol{x}|\boldsymbol{\theta}^j)} \right]^{-1}.$$

However, this estimate is very sensitive to small likelihood values and can once again be very unstable. This estimate actually has an infinite asymptotic variance. There are various other (typically more complex) Markov chain estimation procedures and for further details of these estimation procedures see, for example, Kass and Raftery (1995) and Gamerman (1997, Chapter 7.2).

Example 6.1. Dippers Continued

We consider the capture-recapture data relating to the dipper data set described in Example 1.1 and given in Table 1.1. There are a number of competing biological models that we wish to discriminate between, and these are denoted by T/T, T/C, C/C and $C/C2$. Without any prior information, we specify independent $U[0,1]$ priors on each of the recapture and survival probabilities in each model. Finally, we specify an equal prior probability on each of the four models.

A sample of size 100,000 is drawn from the prior distribution and used to obtain estimates of the posterior probabilities (using $\hat{f}_1$) for each individual model, and presented in Table 6.1. Clearly, there appears to be very strong posterior support for a constant recapture probability over time. In addition, there is positive evidence to support the $C2$ model of survival (flood dependent survival probability) compared to a constant survival probability over time, with a Bayes factor of 3.88. This is the same conclusion drawn from classical analysis in Example 3.2 though of course in that case it was limited to just the single best model. See Brooks et al. (2000a) for further discussion of this data set and related issues.

□

The Monte Carlo estimates are the easiest to program and conceptualise. However, they are often very inefficient and do not always converge within a feasible number of iterations. In addition, each individual model needs to be considered in turn, which may be infeasible when there are a large number of models. In such instances search algorithms can be implemented to reduce the number of possible models considered. We describe the most common algorithm of this type, Occam's window, before discussing more general approaches, which allows us to calculate feasibly the posterior model probabilities of all possible models.

ESTIMATING POSTERIOR MODEL PROBABILITIES 155

Table 6.1 Posterior Model Probabilities for Four Models for the Dipper Data Set, Assuming Independent $U[0,1]$ Priors on Each Recapture and Survival Probability

Model	Posterior Probability
T/T	0.000
T/C	0.000
C/C	0.205
$C/C2$	0.795

Source: Reproduced with permission from Brooks et al. (2000a, p. 368) published by the Institute of Mathematical Statistics.

Occam's Window*

Even if we can obtain estimates of the posterior model probabilities for each model, using the Monte Carlo estimates discussed previously, the number of possible models may be prohibitively large. An approach to overcome this is suggested by Madigan and Raftery (1994), using Occam's window as a selection algorithm to reduce the number of possible models to be considered.

The search algorithm involves initially obtaining a subset of models $\mathcal{A}'$, say, that reasonably predict the data. The set is defined to be,

$$\mathcal{A}' = \left\{ m_i : \frac{\max_j(\pi(m_j|\boldsymbol{x}))}{\pi(m_i|\boldsymbol{x})} \leq c \right\},$$

for some predefined constant c. The constant c is generally chosen subjectively, with values between 10 and 100 in most cases considered adequate. Thus the models with very small posterior model probabilities, given the data, are excluded. The set is then refined by removing further models for which a nested model has larger support. So that if we define the set,

$$\mathcal{B} = \left\{ m_i : \text{ there exists } m_j \in \mathcal{A}'; m_j \subseteq m_i; \frac{\pi(m_j|\boldsymbol{x})}{\pi(m_i|\boldsymbol{x})} > 1 \right\},$$

then the overall set of models to be considered within the analysis is defined to be $\mathcal{A} = \mathcal{A}' \backslash \mathcal{B}$ (i.e. the models that are in $\mathcal{A}'$, but not $\mathcal{B}$).

The algorithm suggested by Madigan and Raftery (1994) for obtaining the set $\mathcal{A}$ involves pairwise comparisons of nested models; the search algorithm involves both adding and removing terms from the given model. The main principle is that if in a comparison a larger model is proposed and accepted, then the smaller model, and all nested models, are rejected. For comparing two models, the posterior odds ratios are compared. If the posterior model probability is larger for the smaller model, then the larger model is rejected. However, for the smaller model to be rejected in favour of the larger model, there needs to be more substantial support. If the ratio is inconclusive, in that it falls between the values needed to reject either of the models, i.e. it falls in Occam's window, neither model is rejected.

Once the set of possible models has been obtained, the posterior model probabilities can be estimated, using the approaches outlined above. Thus, this approach assumes that we can feasibly estimate the posterior model probabilities, which in practice may not always be the case. In addition we limit ourselves to a smaller class of models.

We now consider an alternative method for estimating the posterior model probabilities, which does not restrict the class of models. This involves constructing a Markov chain over model space, such that the stationary distribution provides a posterior distribution for the models. Thus a Markov chain is constructed with stationary distribution $\pi(m, \boldsymbol{\theta}_m | \boldsymbol{x})$, i.e. the joint posterior of the models and parameters. The parameters $\boldsymbol{\theta}_m$ can then be integrated out in the Markov chain to estimate the corresponding posterior model probabilities, and the parameters within each model are estimated from the single chain. We now discuss how we may construct such a Markov chain in more detail.

6.3.2 Reversible Jump MCMC

We have a posterior distribution (over parameter and model space) defined up to proportionality:

$$\pi(\boldsymbol{\theta}_m, m | \boldsymbol{x}) \propto f(\boldsymbol{x} | \boldsymbol{\theta}_m, m) p(\boldsymbol{\theta}_m | m) p(m).$$

If we can sample from this posterior distribution, then we are able to obtain posterior estimates of summary statistics of interest. However, the distribution is too complex to sample from directly, and so we consider an MCMC-type approach. Recall that the basic idea is simply to construct a Markov chain with stationary distribution equal to π. If we run the Markov chain long enough, then realisations of the Markov chain can be regarded as a (dependent) sample from the posterior distribution and used to estimate posterior summary statistics of interest. In particular, the posterior model probabilities can be estimated as the proportion of time that the chain is in the given model. However, there is an additional problem here in terms of how we construct a Markov chain. We need the Markov chain to move over both parameter and model space. To move over the parameter space (given the model) we can use the MH algorithm described in Section 5.3. However, different models generally have a different number of parameters, and so moving between models can involve a change in the dimension of the Markov chain. As a result we cannot use the MH algorithm, since this is only defined for moves which do not alter the dimension of the Markov chain, i.e. for moves that do not alter the number of parameters. Thus we need to consider an alternative algorithm.

The reversible jump (RJ) MCMC algorithm (Green 1995) allows us to construct a Markov chain with stationary distribution equal to the joint posterior distribution of both models and parameters, $\pi(m, \boldsymbol{\theta}_m | \boldsymbol{x})$. This algorithm can be seen as an extension of the MH algorithm. Thus, the algorithm is ideally

ESTIMATING POSTERIOR MODEL PROBABILITIES   157

suited to model discrimination problems where we wish to explore both parameter and model space simultaneously. Note that we are able to traverse the posterior model space, in terms of the models and corresponding parameters, in a single Markov chain. Within each iteration of the Markov chain, the algorithm involves two steps:

Step 1. Update the parameters, $\boldsymbol{\theta}_m$, conditional on the model using the MH algorithm, and

Step 2. Update the model, m, conditional on the current parameter values using the RJMCMC algorithm described below.

The posterior model probabilities can be simply estimated as the proportion of time that the constructed Markov chain is in any given model, effectively integrating out over the parameters in the model, $\boldsymbol{\theta}_m$ (as required for Equation (6.2)). Note that the standard MCMC issues also apply here, such as using an appropriate burn-in for the Markov chain, considering prior sensitivity analyses, etc. We now describe the reversible jump step for updating the model within the Markov chain in more detail.

6.3.3 RJ Updating Step

The RJ step of the algorithm involves two steps:

Step 1. Proposing to move to a different model with some given parameter values, and

Step 2. Accepting this proposed move with some probability.

Thus, the general structure of the RJ step is the same as for the MH algorithm.

We initially describe the reversible jump step for a simple example, before describing the more general structure of the reversible jump step for proposing to move between nested models, and finally presenting the general algorithm.

Example 6.2. Testing for a Covariate in Logistic Regression

To motivate the general idea we initially consider a simple example (which we will revisit for more complex RJMCMC move types throughout this section). Suppose that the underlying model for a time dependent survival probability ϕ_t is unknown and that there are two possible models:

$$\text{Model } m_1: \quad \text{logit } \phi_t = \beta_0;$$
$$\text{Model } m_2: \quad \text{logit } \phi_t = \beta_0 + \beta_1 y_t.$$

The parameters β_0, and β_1 (for model m_2), are to be estimated and y_t denotes some time varying covariate. For each step of the MCMC algorithm we first update each parameter, conditional on the model, using the MH algorithm (i.e. update β_0 if the chain is in model m_1, and update β_0 and β_1 if the chain is in model m_2); then we update the model, conditional on the current parameter values of the Markov chain, using the RJ algorithm. We only describe the updating of the model here and consider each step of the RJ algorithm in turn.

Step 1: Propose New Model

Suppose that at iteration k, the Markov chain is in state $(\boldsymbol{\theta}, m)_k$, where $\boldsymbol{\theta} = \{\beta_0\}$ and $m = m_1$. Since there are only two possible models, we always propose to move to the alternative model. We propose to move to model $m' = m_2$, with parameter values $\boldsymbol{\theta}' = \{\beta_0', \beta_1'\}$. We set,

$$\beta_0' = \beta_0$$
$$\beta_1' = u,$$

where $u \sim q(u)$, i.e. we simulate a value of u from some (arbitrary) proposal distribution q which has the same (or larger) support as β_1.

Step 2: Accept/Reject Step

We accept the proposed model move to state $(\boldsymbol{\theta}', m')$ and set $(\boldsymbol{\theta}, m)_{k+1} = (\boldsymbol{\theta}', m')$ with probability $\min(1, A)$, where,

$$A = \frac{\pi(\boldsymbol{\theta}', m'|\boldsymbol{x})\mathbb{P}(m|m')}{\pi(\boldsymbol{\theta}, m|\boldsymbol{x})\mathbb{P}(m'|m)q(u)} \left| \frac{\partial(\beta_0', \beta_1')}{\partial(\beta_0, u)} \right|,$$

where $\mathbb{P}(m|m')$ denotes the probability of proposing to move to model m, given the current state of the chain is m', and vice versa (which in this case is always equal to one), and the final expression denotes a Jacobian which is given by,

$$\left| \frac{\partial(\beta_0', \beta_1')}{\partial(\beta_0, u)} \right| = \begin{vmatrix} \frac{\partial \beta_0'}{\partial \beta_0} & \frac{\partial \beta_1'}{\partial \beta_0} \\ \frac{\partial \beta_0'}{\partial u} & \frac{\partial \beta_1'}{\partial u} \end{vmatrix}$$
$$= \begin{vmatrix} 1 & 0 \\ 0 & 1 \end{vmatrix} = 1.$$

(The Jacobian is often denoted by $|J|$ for simplicity). Otherwise, if we reject the proposed move, we set, $(\boldsymbol{\theta}, m)_{k+1} = (\boldsymbol{\theta}, m)_k$.

For the reverse move, where we propose to move from state $(\boldsymbol{\theta}', m')$, to state $(\boldsymbol{\theta}, m)$, (so effectively removing the parameter β_1' from the model), we set,

$$u = \beta_1'$$
$$\beta_0 = \beta_0'.$$

Note that this reverse move is completely determined given the move from model m_1 to model m_2, as described above. This move is then accepted with probability $\min(1, A^{-1})$, where A is given above. □

Example 6.3. Dippers Continued
We return to the capture-recapture data relating to dippers (see Example 1.1), and for illustration simply focus on the model for the survival probabilities (assuming a constant recapture probability). In particular we are interested

ESTIMATING POSTERIOR MODEL PROBABILITIES

in whether the survival probability of the dippers is different for flood years versus non-flood years. Thus, there are just two different models that we wish to consider:

$$\text{Model } m_1 \text{ (C):} \quad \phi;$$
$$\text{Model } m_2 \text{ (C2):} \quad \phi_f, \phi_n,$$

where, as before, ϕ denotes a constant survival probability over time (model m_1), and ϕ_f and ϕ_n denote the survival probability in flood and non-flood years (model m_2). We can regard these models as nested, with model m_1 as the special case where $\phi_f = \phi_n$. As noted above, we assume a constant recapture probability over time, and as this is not of interest, we omit this parameter in our description of the reversible jump update. As above, suppose that at iteration k of the Markov chain, the state is denoted by $(\boldsymbol{\theta}, m)_k$. We begin by considering the move from the smaller to the larger model. Thus, we assume that $m = m_1$ and $\boldsymbol{\theta} = \{\phi\}$. As there are only two models, as above, we always propose to move to the alternative model, where $m' = m_2$, and $\mathbb{P}(m|m') = \mathbb{P}(m'|m) = 1$. The parameters in the proposed model are $\boldsymbol{\theta}' = \{\phi'_f, \phi'_n\}$. We need to define the bijective function g. (A bijective function is simply a one-to-one and onto function so that every point is mapped to only a single value, and all values in the range of possible values are mapped to by the function). One possible function would be to set,

$$\phi'_f = \phi + u;$$
$$\phi'_n = \phi - u,$$

where u has some proposal density q. In other words, we set the new survival probabilities for flood and non-flood years to be a simple perturbation from the current value, where the perturbation is of the same magnitude but of opposite signs for the two new parameters. A typical choice for the proposal distribution of u might be,

$$u \sim U[-\epsilon, \epsilon],$$

where ϵ is chosen via pilot tuning (typically set to be relatively small); or,

$$u \sim N(0, \sigma^2),$$

where σ^2 is also chosen via appropriate pilot tuning. Note that both of these proposal distributions are symmetrical, but this need not be the case. In addition, they may result in proposed values for ϕ'_f and ϕ'_n that are outside the range $[0, 1]$, in which case the model move is accepted with probability zero, (i.e. is automatically rejected). Alternative proposal distributions can be defined such that new proposed values for survival probabilities are constrained to the interval $[0, 1]$, for example,

$$u \sim U[-\epsilon_\phi, \epsilon_\phi],$$

where,

$$\epsilon_\phi = \min(\epsilon, \phi, 1 - \phi),$$

and where ϵ is once again chosen via pilot tuning (see Example 5.2 for further discussion of such proposal distributions). An alternative approach is to parameterise models on the logistic scale. Clearly, constraining the proposed parameter values to be in the interval $[0,1]$ introduces more complex proposal distributions q. Typically, if the posterior distribution of the parameters values is sufficiently away from the parameter boundaries, then the simpler proposal distributions should work well, given sensible pilot tuning; however, if the posterior distribution is close to the boundaries, the acceptance probabilities of the model moves (and hence mixing within the Markov chains) may be improved by adapting the more complex proposal distributions.

Given, the proposal distribution for u, we finally need to calculate the Jacobian in order to obtain the acceptance probability of the model move. The Jacobian is calculated as:

$$\left| \frac{\partial(\phi'_f, \phi'_n)}{\partial(\phi, u)} \right| = \begin{vmatrix} \frac{\partial \phi'_f}{\partial \phi} & \frac{\partial \phi'_n}{\partial \phi} \\ \frac{\partial \phi'_f}{\partial u} & \frac{\partial \phi'_n}{\partial u} \end{vmatrix}$$

$$= \begin{vmatrix} 1 & 1 \\ -1 & 1 \end{vmatrix} = 2.$$

We accept the proposed model move with probability $\min(1, A)$, where,

$$A = \frac{\pi(\boldsymbol{\theta}', m'|\boldsymbol{x})\mathbb{P}(m|m')}{\pi(\boldsymbol{\theta}, m|\boldsymbol{x})\mathbb{P}(m'|m)q(u)} \left| \frac{\partial(\phi'_f, \phi'_n)}{\partial(\phi, u)} \right|$$

$$= \frac{2\pi(\boldsymbol{\theta}', m'|\boldsymbol{x})}{\pi(\boldsymbol{\theta}, m|\boldsymbol{x})q(u)},$$

after substituting in the Jacobian calculated above and the corresponding posterior probabilities for moving between the different models. If we accept the move, we set $(\boldsymbol{\theta}, m)_{k+1} = (\boldsymbol{\theta}', m')$; otherwise we set $(\boldsymbol{\theta}, m)_{k+1} = (\boldsymbol{\theta}, m)_k$. Note that, (as for the MH algorithm), if $\phi'_f \notin [0,1]$ or $\phi'_n \notin [0,1]$, the term A is simply equal to zero, since both the likelihood function and corresponding prior evaluated at the parameter values in the proposed move are both equal to zero (and hence the acceptance probability is also zero).

Finally, the reverse move is completely defined given the bijective function g above. So, suppose that the current values of the chain are $\boldsymbol{\theta}' = \{\phi'_f, \phi'_n\}$ and m', as given above. We propose to move to model m with parameters $\boldsymbol{\theta} = \{\phi\}$ and u. The inverse of the bijective function, g^{-1} is specified such that,

$$\phi = \frac{1}{2}(\phi'_f + \phi'_n);$$
$$u = \frac{1}{2}(\phi'_f - \phi'_n).$$

The move is then accepted with probability $\min(1, A^{-1})$, for A given above.

We implement the reversible jump algorithm above using a normal proposal distribution for the reversible jump step, with $\sigma^2 = 0.01$. This value was chosen via pilot tuning. The simulations are run for 20,000 iterations, with the first 10% discarded as burn-in. The trace plots for the parameters p, ϕ_n and ϕ_f and the model itself are given in Figure 6.1 (model m_1 denotes C/C and model m_2, $C/C2$) and suggest that the burn-in is very conservative. Clearly, there appears to be good mixing both within models and between models. The posterior probabilities for models C/C and $C/C2$ are 0.217 and 0.783, respectively. These are very similar to those obtained using the Monte Carlo approach in Example 6.1 (allowing for Monte Carlo error).

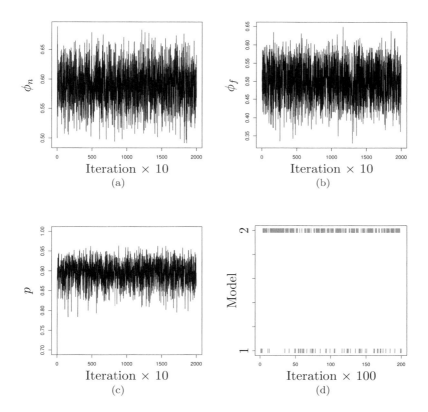

Figure 6.1 Trace plots for (a) recapture probability; (b) survival probability for non-flood years; (c) survival probability for flood years; and (d) model indicator, where model 1 is C/C and model 2 is $C/C2$.

□

MODEL DISCRIMINATION

Example 6.4. Covariate Analysis – White Storks

We extend the above ideas and once more consider capture-recapture data, $\boldsymbol{x}$, relating to the white stork data, initially described in Exercise 3.5. We assume a constant recapture probability p, and logistically regress the survival probabilities on a number of different covariates, but where there is uncertainty in the covariate dependence. There are a total of 10 covariates, corresponding to the amount of rainfall at 10 different locations in the Sahel. Notationally, we let $cov_{i,t}$ denote the covariate value for location $i = 1, \ldots, 10$ at time $t = 1, \ldots, T-1$. Note that we normalise the covariate values for each individual region. For the saturated model, the survival probability at time t is given by,

$$\text{logit } \phi_t = \beta_0 + \sum_{i=1}^{10} \beta_i cov_{i,t}.$$

Submodels are defined by setting β_i values equal to zero, so that the survival probability is not dependent on the corresponding i^{th} covariate. For convenience, we let the model be denoted by $m = \{i : \beta_i \neq 0\}$ and set $\boldsymbol{\beta}_m = \{\beta_i : i = 0, \ldots, 10\}$. For example, suppose that the survival probability is dependent on covariates 1, 2 and 9. The logistic regression parameters in this model are $\boldsymbol{\beta}_m = \{\beta_0, \beta_1, \beta_2, \beta_9\}$. Thus, the model is denoted by $m = \{0, 1, 2, 9\}$. Note that $\boldsymbol{\beta}_m$ necessarily defines the model m, however, it is useful to use the separate notation.

Within the RJMCMC algorithm, we initially update each parameter in the model (i.e. p and β_i, such that $i \in m$ or equivalently, the values of $\beta_i \neq 0$ for $i = 0, \ldots, 10$). We use the analogous updating algorithm described in Example 5.7. We then update the model using the RJ step. As usual we consider the two distinct steps: proposing the new model, and the accept/reject step.

Step 1: Propose New Model

At iteration k of the Markov chain, we let the current state of the chain be equal to $(\boldsymbol{\theta}, m)_k$, where $\boldsymbol{\theta} = \{p, \boldsymbol{\beta}\}$. We randomly select $r \in [1, 2, \ldots, 10]$. If $r \notin m$ (i.e. survival probability is independent of covariate r, so that $\beta_r = 0$) we propose to add the covariate dependence; else if $r \in m$ (i.e. survival probability is dependent on covariate r, so that $\beta_r \neq 0$), we propose to remove the covariate dependence. (Note that if we wanted to perform model selection on the intercept term as well, we simply randomly select $r \in [0, 1, \ldots, 10]$.) We initially consider the first case, where $r \in m$ (or equivalently, $\beta_r = 0$). The proposed model is given by $m' = m \cup r$, and the parameters in the proposed model are denoted by $\boldsymbol{\theta}' = \{p', \boldsymbol{\beta}'\}$, where,

$$\begin{aligned} p' &= p \\ \beta_i' &= \beta_i \quad \text{for } i \neq r \\ \beta_r' &= u, \end{aligned}$$

where $u \sim q(u)$, for proposal density q. In particular, we set,

$$u \sim N(0, \sigma_r^2).$$

ESTIMATING POSTERIOR MODEL PROBABILITIES

Note that for this example, we set $\sigma_r^2 = 0.5$ (for $r = 1, \ldots, 10$).

Alternatively, suppose that $\beta_r \neq 0$. Then, we have the proposed model $m' = m \backslash r$ with corresponding parameter values $\boldsymbol{\theta}' = \{p', \boldsymbol{\beta}'\}$, where,

$$p' = p$$
$$\beta'_i = \beta_i \quad \text{for } i \neq r$$
$$u = \beta_r$$
$$\beta'_r = 0.$$

Step 2: Accept/Reject Step

We begin with the first case above, for $r \notin m$, where the current value of the parameter $\beta_r = 0$, and we propose to add the covariate dependence to the survival probability. We accept this proposed model move with the standard probability $\min(1, A)$, where,

$$A = \frac{\pi(\boldsymbol{\theta}', m'|\boldsymbol{x})\mathbb{P}(m|m')}{\pi(\boldsymbol{\theta}, m|\boldsymbol{x})\mathbb{P}(m'|m)q(u)} \left|\frac{\partial \boldsymbol{\theta}'}{\partial(\boldsymbol{\theta}, u)}\right|,$$

where $\mathbb{P}(m'|m) = \mathbb{P}(m|m') = \frac{1}{10}$; and the Jacobian is equal to unity, since the function describing the relationship between the current and proposed parameters is equal to the identity function. Thus, we can simplify the acceptance probability to be equal to,

$$A = \frac{\pi(\boldsymbol{\theta}', m'|\boldsymbol{x})}{\pi(\boldsymbol{\theta}, m|\boldsymbol{x})q(u)}$$
$$= \frac{f(\boldsymbol{x}|\boldsymbol{\theta}', m')p(\beta'_r)}{f(\boldsymbol{x}|\boldsymbol{\theta}, m)q(u)},$$

where $u = \beta'_r$, $q(\cdot)$ is the corresponding normal density with proposal variance σ_r^2 and cancelling the prior densities for the parameters that are not updated within the RJ step.

Alternatively, in the second case, where $r \in m$, or equivalently, $\beta_r \neq 0$, the corresponding acceptance probability is equal to $\min(1, B)$, where,

$$B = \frac{f(\boldsymbol{x}|\boldsymbol{\theta}', m')q(u)}{f(\boldsymbol{x}|\boldsymbol{\theta}, m)p(\beta_r)}).$$

It can be seen that $B = 1/A$ (allowing for the notational differences within the model moves). □

6.3.4 General (Nested) RJ Updates

We again describe the RJ step, but using a more general notation for moving between nested models, to allow, for example, for moving between models that differ by more than a single parameter or for more general proposal schemes for the parameters. In order to illustrate the general method we will revisit

Example 6.2 above where there is essentially a variable selection problem. We extend the set of possible models by considering an additional covariate, denoted by z_t. Thus, we consider 4 possible models:

$$\begin{aligned}
\text{Model } m_1: \quad \text{logit } \phi_t &= \beta_0; \\
\text{Model } m_2: \quad \text{logit } \phi_t &= \beta_0 + \beta_1 y_t; \\
\text{Model } m_3: \quad \text{logit } \phi_t &= \beta_0 + \beta_2 z_t; \\
\text{Model } m_4: \quad \text{logit } \phi_t &= \beta_0 + \beta_1 y_t + \beta_2 z_t.
\end{aligned}$$

Thus, once again we have a general nested structure for the set of possible models, so that we can explore all possible models by simply moving between nested models. However, note that models m_2 and m_3 are not nested. We will return to the general RJMCMC algorithm for moving between such non-nested models in Section 6.3.5.

More generally, suppose that at iteration k of the Markov chain, the chain is in state $(\boldsymbol{\theta}, m)_k$. Without loss of generality we propose to move to state $(\boldsymbol{\theta}', m')$. Let the number of parameters in model m be denoted by n, and the number of parameters in model m' by n'. Generally, there may be several different move types that can be used to move between the different models. These are often in the form of adding or deleting a given number of parameters from the model and we give several examples of these in Sections 6.7, 9.3 and 11.4. Typically, there are a number of possible models that we may propose to move to, and we propose to move to each of these with some predefined probability. We let the probability of proposing to move to model m', given that the chain is in model m, by $\mathbb{P}(m'|m)$. For example, for the above covariate example, we have $\boldsymbol{m} = \{m_1, \ldots, m_4\}$. Suppose that $m = m_1$. We could consider three different possible moves: to models m_2, m_3, or m_4. We have, effectively, already described a RJMCMC algorithm for moving to models m_2 or m_3, since this only involves adding a single parameter to the model. However, we need to define the probability of moving to all of the models. For example, we may propose to move to each of these models with equal probability, i.e. $\mathbb{P}(m_i|m_1) = \frac{1}{3}$ for $i = 2, \ldots, 4$. We repeat this process, conditional on each model that the chain may be in, with the restriction that proposed moves are only between nested models. Thus, this would mean that $\mathbb{P}(m_i|m_4) = \frac{1}{3}$ for $i = 1, \ldots, 3$. However, models m_2 and m_3 are non-nested, so that under the restriction of only moving between nested models, we may specify, $\mathbb{P}(m_1|m_2) = \mathbb{P}(m_4|m_2) = \frac{1}{2}$ and $\mathbb{P}(m_1|m_3) = \mathbb{P}(m_4|m_3) = \frac{1}{2}$; but $\mathbb{P}(m_2|m_3) = \mathbb{P}(m_3|m_2) = 0$.

Now, suppose that the proposed move increases the dimension of the model, so that $n' > n$ (we shall consider the alternative case later). For example, suppose that the current state of the Markov chain is model m_1, and that we propose to move to model m_4. Then $n = 1$ and $n' = 3$. In general, we generate $p = n' - n$ random variables $\boldsymbol{u}$ from some given distribution $q(\boldsymbol{u})$. We then set the new vector parameter $\boldsymbol{\theta}'$ as some known bijective (i.e. one-to-one and onto) function of the current parameters, $\boldsymbol{\theta}$, and generated random variables

ESTIMATING POSTERIOR MODEL PROBABILITIES

u. Thus, $\theta' = g(\theta, u)$, for some known bijective function g. We accept this proposed model update with probability,

$$\alpha(\theta, \theta') = \min(1, A),$$

where

$$A = \frac{\pi(m', \theta'|x)\mathbb{P}(m|m')}{\pi(m, \theta|x)\mathbb{P}(m'|m)q(u)} \left|\frac{\partial \theta'}{\partial(\theta, u)}\right|.$$

The final term is simply the Jacobian defined to be,

$$\left|\frac{\partial \theta'}{\partial(\theta, u)}\right| = \begin{vmatrix} \frac{\partial \theta'_1}{\partial \theta_1} & \frac{\partial \theta'_2}{\partial \theta_1} & \cdots & \frac{\partial \theta'_{n'}}{\partial \theta_1} \\ \vdots & \vdots & & \vdots \\ \frac{\partial \theta'_1}{\partial \theta_n} & \frac{\partial \theta'_2}{\partial \theta_n} & \cdots & \frac{\partial \theta'_{n'}}{\partial \theta_n} \\ \frac{\partial \theta'_1}{\partial u_1} & \frac{\partial \theta'_2}{\partial u_1} & \cdots & \frac{\partial \theta'_{n'}}{\partial u_1} \\ \vdots & \vdots & & \vdots \\ \frac{\partial \theta'_1}{\partial u_p} & \frac{\partial \theta'_2}{\partial u_p} & \cdots & \frac{\partial \theta'_{n'}}{\partial u_p} \end{vmatrix}$$

The reverse move from model m' to model m is deterministically defined, given the bijective function g. In other words, $(\theta, u) = g^{-1}(\theta')$. The corresponding acceptance probability of the reverse move is simply

$$\alpha(\theta', \theta) = \min\left(1, \frac{1}{A}\right),$$

for A given above.

Example 6.5. Example 6.2 Extended
Returning to our covariate example above, we propose to move from model $m = m_1$ to model $m' = m_4$, such that $p = 2$. In the notation of the model move described, $\theta = \{\beta_0\}$, $u = \{u_1, u_2\}$, and the parameters in the proposed model m_4 are given by $\theta' = \{\beta'_0, \beta'_1, \beta'_2\}$. We define the bijective function g, such that,

$$\begin{aligned} \beta'_0 &= \beta_0 \\ \beta'_1 &= u_1 \\ \beta'_2 &= u_2. \end{aligned}$$

This is simply an identity function for g, and hence the corresponding Jacobian $|J| = 1$ (compare with the calculation of the Jacobian of Example 6.2). Finally, we need to define the proposal distribution for the parameters u. A typical choice would be to set independent proposal distributions $u_i \sim N(0, \sigma_i^2)$ for $i = 1, 2$. In other words each u_i is independently normally distributed about mean 0 (corresponding to no covariate dependence) with some variance σ_i^2, chosen via pilot tuning. This completely defines the reversible jump move given the current and proposed models (i.e. the bijective function g and proposal distribution q). The corresponding acceptance probability is $\min(1, A)$, where

$$A = \frac{\pi(m', \boldsymbol{\theta}'|\boldsymbol{x})\mathbb{P}(m|m')}{\pi(m, \boldsymbol{\theta}|\boldsymbol{x})\mathbb{P}(m'|m)q(u_1, u_2)},$$

where $q(u_1, u_2)$ denotes the (joint) proposal normal density function of u_1 and u_2. This is simply equal to the product of the normal density functions for u_1 and u_2, since we defined the proposal distributions to be independent. Clearly, this may not be the case in general. If we accept the proposed move we set $(\boldsymbol{\theta}, m)_{k+1} = (\boldsymbol{\theta}', m')$; otherwise we set $(\boldsymbol{\theta}, m)_{k+1} = (\boldsymbol{\theta}, m)_k$.

Conversely, the reverse move is completely defined, given the above description. So, let the current state of the chain be $(\boldsymbol{\theta}', m')$, where $\boldsymbol{\theta}' = \{\beta'_0, \beta'_1, \beta'_2\}$ and $m' = m_4$. Given that we propose to move to model $m = m_1$, we set,

$$\beta_0 = \beta'_0$$
$$u_1 = \beta'_1$$
$$u_2 = \beta'_2,$$

and accept the proposed move with probability, $\min(1, A^{-1})$, where A is given above.

□

For further implementational issues concerning the reversible jump procedure, see for example Richardson and Green (1997). We emphasise that once the model move has been described for increasing the number of parameters in the model, then the reverse move for moving to a model with a reduced number of parameters is completely determined to ensure the reversibility constraints. The RJ algorithm described is restricted to moves between nested models, for the given proposal distribution for the parameters. More general RJ updates are possible, and we briefly describe these next.

6.3.5 General (Non-Nested) RJ Updates*

The RJ algorithm described above is a special case of the general reversible jump algorithm that can be implemented for moves between any two different models (i.e. they need not be nested). For example, we may wish to propose model moves that simultaneously add and remove a number of covariates from the current model. Such a move does not fit into the nested framework described above. The structure of this general RJ update within the Markov chain is identical to before:

Step 1: propose a new model with given parameter values, and

Step 2: accept/reject the model move.

Thus we begin with the first step. Suppose that at iteration k the Markov chain is in model m with parameters $(\boldsymbol{\theta}, m)_k$ and that we propose to move to state $(\boldsymbol{\theta}', m')$. To do this, we define a (bijective) function g, such that, $(\boldsymbol{\theta}', \boldsymbol{u}') = g(\boldsymbol{\theta}, \boldsymbol{u})$, where $\boldsymbol{u}$ and $\boldsymbol{u}'$ are sets of random variables with density

ESTIMATING POSTERIOR MODEL PROBABILITIES   167

function $q(\boldsymbol{u})$ and $q'(\boldsymbol{u}')$, respectively. We accept this proposed move with probability $\min(1, A)$, where,

$$A = \frac{\pi(\boldsymbol{\theta}', m'|\boldsymbol{x})\mathbb{P}(m|m')q'(\boldsymbol{u}')}{\pi(\boldsymbol{\theta}, m|\boldsymbol{x})\mathbb{P}(m'|m)q(\boldsymbol{u})} \left| \frac{\partial(\boldsymbol{\theta}', \boldsymbol{u}')}{\partial(\boldsymbol{\theta}, \boldsymbol{u})} \right|,$$

where $\mathbb{P}(m|m')$ denotes the probability of proposing to move from model m to model m'. If the move is accepted, we set $(\boldsymbol{\theta}, m)_{k+1} = (\boldsymbol{\theta}', m')$; otherwise we set $(\boldsymbol{\theta}, m)_{k+1} = (\boldsymbol{\theta}, m)_k$. Note that the formula given is not as complicated as it may appear!

We discuss the term $\mathbb{P}(m'|m)$ in more detail for clarity. Recall that we treat the model as a discrete parameter, so that the term $\mathbb{P}(m'|m)$ is simply the proposal distribution for this discrete parameter. Thus within the RJMCMC algorithm, we need to specify the probability of moving between each pair of models (these probabilities need not be symmetrical). Typically, we often propose to only move to "neighbouring" models, and set the probability of moving to any other model as equal to zero. Then we may move to each neighbouring model with equal probability. Clearly, as the number of neighbouring models will typically differ for the different models, then the probability function is not symmetric in general (i.e. $\mathbb{P}(m'|m) \neq \mathbb{P}(m|m')$). Alternatively, we may propose to move to any possible model, but this is typically only used when the number of models is small. We emphasise that it is typically not the setting up of the probability of moving between the different models that is important in achieving efficient RJ algorithms, but instead the specification of the proposal distributions q and q'. As such, we discuss the construction of efficient proposal distributions in further detail in Section 6.3.6. However, we initially present a couple of examples of the general RJ algorithm.

Example 6.6. Covariate Analysis Continued
We revisit the covariate analysis, where there are 2 possible covariates, and hence a total of four possible models. We now consider moving between models, which simultaneously adds and removes a parameter within the same model move. So, in the notation of Example 6.5, we are moving between models m_2 and m_3. Thus, we now consider moving between each of the four possible models with equal probability, i.e. $\mathbb{P}(m_i|m_j) = \frac{1}{3}$ for $i \neq j$. To demonstrate the model move between non-nested models, suppose that at iteration k of the Markov chain the state of the chain is $(\boldsymbol{\theta}, m)_k$, where $m = m_2$ and $\boldsymbol{\theta} = \{\beta_0, \beta_1\}$. We propose to move to each other model with equal probability, and we randomly choose to move to model $m' = m_3$, with parameters $\boldsymbol{\theta}' = \{\beta_0', \beta_2'\}$. This is a non-nested model move.

We initially define $\boldsymbol{u} = \{u\}$ and $\boldsymbol{u}' = \{u'\}$. We then define the bijective function g, such that,

$$\begin{aligned} \beta_0' &= \beta_0 \\ \beta_2' &= u \\ u' &= \beta_1. \end{aligned}$$

Finally, we need to define the corresponding proposal distributions, denoted by q and q', for u and u'. In particular, we assume that these are both independent normal distributions, each with mean 0 and standard deviation σ and σ', respectively. This once more completely defines the reversible jump move. The bijective function g is simply the identity function, so that the corresponding Jacobian is equal to unity. The corresponding acceptance probability for moving between models $m = m_2$ and $m' = m_3$ is simply, $\min(1, A)$, where

$$A = \frac{\pi(\boldsymbol{\theta}', m'|\boldsymbol{x})q'(\boldsymbol{u}')}{\pi(\boldsymbol{\theta}, m|\boldsymbol{x})q(\boldsymbol{u})},$$

since $\mathbb{P}(m|m') = \frac{1}{3} = \mathbb{P}(m'|m)$, and the Jacobian is equal to unity. We note that in this example, the number of parameters in models m and m' are the same. Thus, the reversible jump algorithm reduces to the MH algorithm (using an independent proposal distribution).

□

Example 6.7. Ring-Recovery Data

To provide a further illustrative example we also describe the corresponding application to ring-recovery models, where there is uncertainty with respect to the time dependence of the survival and recovery probabilities, where we assume that there is only a single survival and recovery probability for all ages for simplicity. The set of possible models is $\boldsymbol{m} = \{C/C,\ C/T,\ T/C,\ T/T\}$. Suppose that $m = C/T$. We consider two possible move types which involves either adding or removing the time dependence for either the survival probability or the recovery probability. This could be done in a number of ways. One way to do this would be to propose to change the time dependence on the survival probability with probability p for $0 < p < 1$; otherwise we propose to change the time dependence of the recovery probabilities. In other words, we choose $m' = T/T$ with probability p; otherwise $m' = C/C$ with probability $1 - p$. The value of p is arbitrary and set by the statistician, with a typical choice being $p = 0.5$. Note that we only propose to move between nested models, however, due to the form of the function g, the example falls into the general RJ case. We let n and n' denote the number of parameters in model m and m', respectively.

Now, suppose that the proposed move increases the dimension of the model, so that $n' > n$. We shall consider the alternative case later. For our example, this means that $m' = T/T$. We let the set of parameters in model m be denoted by $\boldsymbol{\theta} = \{\phi, \lambda_1, \ldots, \lambda_{T-1}\}$, and the parameters in model m' be $\boldsymbol{\theta}' = \{\phi'_1, \ldots, \phi'_{T-1}, \lambda'_1, \ldots, \lambda'_{T-1}\}$. Thus, $n = 1 + (T - 1) = T$ and $n' = 2(T - 1)$. To define the bijective function g we also introduce the parameters $\boldsymbol{u} = \{u_1, \ldots, u_{T-1}\}$ and $\boldsymbol{u}' = \{u'\}$. We set,

$$\begin{aligned}
\lambda'_i &= \lambda_i & i &= 1, \ldots, T-1; \\
\phi'_i &= u_i & i &= 1, \ldots, T-1; \\
u' &= \phi.
\end{aligned}$$

Once again, this is essentially defining the bijective function g to be the identity function, so that the corresponding Jacobian is simply equal to unity. Finally, to complete the specification of the model move, we need to define the proposal distributions for $\boldsymbol{u}$ and $\boldsymbol{u}'$, denoted by $q(\boldsymbol{u})$ and $q'(\boldsymbol{u}')$, respectively. Now, since $\boldsymbol{u}$ and $\boldsymbol{u}'$ are set equal to probabilities, proposal values outside the range $[0, 1]$ will result in an acceptance probability of 0 for the model move. Thus natural proposal distributions for $\boldsymbol{u}$ and $\boldsymbol{u}'$ are simply independent beta distributions for each of the parameters. Specification of the corresponding proposal parameters for these independent beta distributions completely specifies the model move. Then the acceptance probability for the model move is the standard expression, $\min(1, A)$, where,

$$A = \frac{\pi(\boldsymbol{\theta}', m'|\boldsymbol{x})\mathbb{P}(m|m')q'(\boldsymbol{u}')}{\pi(\theta, m|\boldsymbol{x})\mathbb{P}(m'|m)q(\boldsymbol{u})},$$

where q and q' denote the independent beta proposal distributions, $\mathbb{P}(m|m') = p = \mathbb{P}(m'|m)$, and omitting the Jacobian term in this instance, since it is equal to unity.

The reverse move follows in the standard way. Suppose that the current model is m' with parameters $\boldsymbol{\theta}'$. We use the bijective function described above, so that we propose the values, $\boldsymbol{\theta}$ such that,

$$\begin{aligned}
\lambda_i &= \lambda_i' & i = 1, \ldots, T-1; \\
u_i &= \phi_i' & i = 1, \ldots, T-1; \\
\phi &= u'.
\end{aligned}$$

We accept the move with probability $\min(1, A^{-1})$ for A given above.

□

See Section 6.7 for a further example using this general updating algorithm. For further implementational details of the use of the RJMCMC algorithm in the context of ecological models see, for example, King and Brooks (2002a,b; 2003b) and King et al. (2006, 2008).

In practice several model updates may be performed within each iteration of the Markov chain. For example, for the covariate example, we may cycle through each individual covariate and propose to add/remove each covariate independently. For the ring-recovery data, within each iteration we could propose to update the time dependence on the survival probability; then following the completion of the model update (i.e. either accept or reject the move), propose to update the time dependence of the recovery probabilities. Further worked examples are given for the reversible jump step in Examples 6.11, 6.12, 8.3, 9.2, 9.3, 10.3 and Section 11.4 for a number of different types of model moves, and these examples discuss different aspects of the algorithm, including prior specification, interpretation of the results, and the description of posterior summary estimates, often of interest within statistical analyses.

The reversible jump algorithm in the presence of model uncertainty has the advantage that the acceptance probability is easy to calculate for a given proposed move. In addition, irrespective of the number of possible models within an analysis, only a single chain needs to be run in order to obtain both the estimates of the posterior model probabilities and of the parameters within those models. Thus, this approach can be implemented even when consideration of all possible models individually is not feasible. Essentially, time is spent exploring the models with reasonable posterior support given the observed data, and models not supported by the data (and priors) are not well explored. However, we note that in order to implement the RJMCMC algorithm it is not possible to specify improper priors on parameters on which there is model uncertainty (i.e. parameters that are not common to all models), and obtain estimates of the corresponding posterior model probabilities. In other words, for parameters on which there is model uncertainty, proper priors need to be specified. This can be clearly seen from the form of the acceptance probability for the RJ step, since this contains the prior density function, which is not defined for improper priors (i.e. has an infinite normalisation constant).

6.3.6 Constructing Proposal Distributions

Typically, the larger the number of parameters that two models differ by, the more difficult it is to construct sensible model moves, in terms of defining the proposal distribution q, such that the corresponding acceptance probability is "reasonable." This is most intuitive to understand if we consider the simple nested case where we propose to move to a model with a larger number of parameters. For each additional parameter in the proposed model, we need to simulate a value for the parameter, from some proposal distribution. However, each of these parameter values needs to be in an area of relatively high posterior mass in order for the acceptance probability to be "reasonable" (i.e. for the numerator to be relatively high). Alternatively, if we specify the proposal distribution of these parameters to be very highly concentrated on the posterior mode of the parameters, say, then the denominator of the acceptance probability can be very high due to the evaluation of the proposal densities, resulting in a (relatively) low acceptance rate. The same reasoning essentially applies for non-nested models. Thus, once more (as with the MH algorithm), there is typically a trade-off between the size of the model move, and the corresponding acceptance probability. As a result of the (often) complex proposal distributions that need to be constructed for model moves, we have described some ideas for "larger" model moves in the real examples that we will consider in Sections 6.7 and 8.4.

6.4 Prior Sensitivity

Care should be taken in specifying the priors on the parameters $p(\boldsymbol{\theta}|m)$ in the presence of model uncertainty, since these priors can have a significant impact on the corresponding posterior model probabilities, a feature known

PRIOR SENSITIVITY

as *Lindley's paradox*. A prior sensitivity analysis should always be performed for a number of reasonable (i.e. sensible) priors and compared.

Example 6.8. Dippers Revisited

Consider the previous dippers example. Brooks et al. (2000a) consider two alternative priors for the model parameters and compare these with the uniform priors used previously (denoted prior 1). In particular, they consider a $Beta(1,9)$ prior on each survival probability and $U[0,1]$ on the recapture probability (prior 2); and a $Beta(1,9)$ prior on all survival and recapture probabilities (prior 3). Note that the $Beta(1,9)$ prior has a mean of 0.1 and standard deviation of 0.09. Thus this beta prior could be regarded as quite extreme, and is used here for illustrative purposes. The posterior distributions of the parameters, conditional on any models C/C and $C/C2$ (i.e. the models with some posterior support), appear to be largely insensitive to the prior specification (see Example 4.5 for particular discussion of model $C/C2$). However, this is not the case for the posterior model probabilities given in Table 6.2.

Table 6.2 The Posterior Model Probabilities of the Different Models for the Dipper Dataset under the Three Different Priors

	Posterior Probability		
Model	Prior 1	Prior 2	Prior 3
T/T	0.000	0.003	0.000
T/C	0.000	0.003	0.000
C/C	0.205	0.951	1.000
$C/C2$	0.795	0.004	0.000

Source: Reproduced with permission from Brooks et al. (2000a, p. 368) published by the Institute of Mathematical Statistics.

Thus, there are significant changes to the posterior probabilities under priors 2 and 3, compared to prior 1. In this case, this is essentially a result of an inconsistency between the prior specified on the parameters and the posterior distribution of the parameters. For example, for model $C/C2$, (under uniform priors) the posterior means of ϕ_f and ϕ_n are 0.472 and 0.609, with posterior standard deviations 0.043 and 0.031, respectively. These are significantly different from the prior distributions specified on the parameters in priors 2 and 3, and leads to the prior sensitivity in the posterior model probabilities. Assuming equal prior model probabilities, the posterior model probabilities can be seen to be proportional to the integral of the likelihood with respect to the prior – see Equations (6.1) and (6.2). Alternatively, we can interpret this as taking the expected likelihood value, with respect to the prior – see Equation (6.1), and is the motivation behind Monte Carlo estimates. Thus, when there is a conflict between the prior distributions specified on the parameter and the information contained in the data via the likelihood, the integration (or averaging) is essentially performed over unlikely parameter values (given the data), and so naturally penalises models with larger numbers of "conflicting" parameters. Thus the model $C2$ is penalised more than model C. Conversely, for prior 1 ($U[0,1]$ on all parameters), there is reasonable (relative) prior support on all parameter values supported by the data via the likelihood, so that within the integration, this likelihood has a greater contribution.

Alternatively, and more heuristically, we can understand the change in posterior model probabilities as follows. The priors specified represent our initial beliefs concerning the parameters before observing any data. Given the data that we observe, we then update these prior beliefs. Under priors 2 and 3, we essentially have fairly strong prior beliefs. Prior 2 corresponds to the (prior) belief that the survival probabilities are low; whereas prior 3 corresponds to the (prior) belief that the recapture and survival probabilities are all low. Given the data that we observe, we update our beliefs on the parameters and these are now very different from our prior beliefs. Thus, we could think of a level of surprise of the expert in terms of his/her change in beliefs, given the data that are observed. Increasing the number of conflicting parameters in the model will increase the level of surprise of the expert, (acting in some cumulative manner). In this way the corresponding relative posterior probability for models with larger numbers of conflicting parameters will be reduced. Thus for the dipper example, since model C has only a single survival parameter and model $C2$ two survival parameters, when placing the conflicting $Beta(1,9)$ prior on the survival probabilities, model C has larger posterior support than $C2$. For prior 1 ($U[0,1]$), there is no such conflict between the initial beliefs on the parameters and the information contained in the data. Thus, we favour model $C2$ that appears to fit the data better (in terms of the likelihood) than model C.

□

Clearly, within any Bayesian analysis, the priors need to be specified in a sensible manner (see Section 4.2). In addition, a prior sensitivity analysis should always be conducted for both the parameters and models (when there is model uncertainty). Once again, there is a trade-off between the information contained in the data and the priors specified on the parameters. However, typically, for the reasons discussed above, the posterior model probabilities may often be more sensitive to the prior specification than the posterior distribution of the parameters themselves.

6.5 Model Averaging

Bayesian model averaging obtains an estimate of a parameter, based on all plausible models, taking into account both parameter and model uncertainty (see, for example, Hoeting et al. 1999, Brooks et al. 2000a, King et al. 2008b and King and Brooks 2008). Thus, model averaging can overcome the prob-

MODEL AVERAGING

lem associated with different plausible models giving different parameter estimates. Essentially, the model averaging approach obtains a single parameter estimate based on all plausible models by weighting each according to their corresponding posterior model probability, and so allows us to incorporate model uncertainty into the estimate of the parameter. Formally, the model-averaged posterior distribution of some parameter vector $\boldsymbol{\theta}$ common to all models m_i, $i = 1, \ldots, K$ is given by,

$$\pi(\boldsymbol{\theta}|\boldsymbol{x}) = \sum_{i=1}^{K} \pi(\boldsymbol{\theta}|\boldsymbol{x}, m_i)\pi(m_i|\boldsymbol{x}).$$

For further details see for example, Hoeting et al. (1999) and Madigan and Raftery (1994).

Example 6.9. Dippers Revisited

We once more reconsider the dipper data set, as in Example 6.1. We assume uniform priors over both parameter and model space, i.e. $U[0, 1]$ priors on each parameter in the model and a prior probability of $1/4$ for models T/T, T/C, C/C and $C/C2$. There are only two models with posterior support greater than 0.000 – model $C/C2$ and C/C with posterior probability 0.795 and 0.205, respectively (see Table 6.1). Consider the model-averaged survival probability for both flood years, ϕ_f and non-flood years, ϕ_n. The model-averaged posterior distribution for the survival probability in flood years (ϕ_f) and non-flood years is given in Figure 6.2.

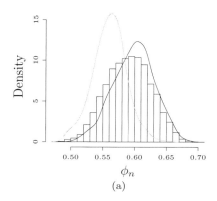
(a)

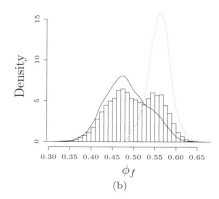
(b)

Figure 6.2 Histogram of the (model-averaged) sample from the posterior distribution for (a) survival probabilities in non-flood years, ϕ_n, and (b) survival probabilities in flood years, ϕ_f; and corresponding estimate of the posterior distribution for the parameter, conditional on each model (grey for model C/C and black for model $C/C2$).

Clearly, we can see that there is some (mild) bimodality for the flood years, with the left mode corresponding to the estimate under model $C/C2$ and the right mode for the estimate under model C/C, but still significant overlap between the posterior distributions, conditional on the given model. In addition, the posterior distribution for $C/C2$ has more weight on the model-averaged distribution of the parameter, reflecting the higher posterior probability associated with this model, compared to C/C. The posterior mean of ϕ_f is 0.472 and 0.561 under models $C/C2$ and C/C, respectively. The corresponding model-averaged mean for the survival probability in flood years is,

$$\begin{aligned}\mathbb{E}_\pi(\phi_f) &= (0.795 \times 0.472) + (0.205 \times 0.561) \\ &= 0.490.\end{aligned}$$

Similarly, the corresponding model-averaged variance for ϕ_f can be obtained, given the expected value of ϕ_f^2 under both models. Then, we have the model-averaged estimate of ϕ_f^2 is,

$$\mathbb{E}_\pi(\phi_f^2) = (0.795 \times 0.225) + (0.205 \times 0.315) = 0.243.$$

So that the posterior model-averaged variance of ϕ_f^2 is,

$$\begin{aligned}\mathrm{Var}_\pi(\phi_f) &= \mathbb{E}_\pi(\phi_f^2) - (\mathbb{E}_\pi(\phi_f^2))^2 \\ &= 0.003,\end{aligned}$$

corresponding to a standard deviation of 0.056. Thus, we note that the standard deviation for the model-averaged estimate for ϕ_f is greater than the corresponding standard deviation for ϕ_f under the individual models (0.025 and 0.031 for models C/C and $C/C2$, respectively). This reflects the additional uncertainty with respect to the model, in addition to the parameter.

□

Note that parameters should only be averaged over different models when the interpretation of the parameter in the different models is the same. For example, suppose that we specify the survival probability as a function of a number of covariates, where there is prior uncertainty relating to the covariates present in the model. Then the survival probability has the same interpretation within each possible model (although the set of covariates that it is a deterministic function of may vary), and hence a model-averaged estimate of the survival probability takes into account both the parameter uncertainty and model uncertainty, with respect to covariate dependence. However, now consider the regression coefficient of one of the possible covariates. The interpretation of this coefficient is different, depending on whether the covariate is present in the model or not. If the covariate is present in the model, then the regression coefficient describes how the survival probability is related to the given covariate; however, if the covariate is *not* present in the model, then there is no relationship between the survival probability and the covariate value (the regression coefficient is identically equal to zero). Thus, it does not make sense to provide a model-averaged estimate of the regression coefficient

MODEL AVERAGING

over all models; although it does make sense to provide a model-averaged estimate of the regression coefficient, conditional on the covariate being present in the model. In this instance, the averaging takes into account the uncertainty relating to the parameter and the dependence of the survival probabilities on all the other covariate values.

The reversible jump algorithm described above is able to calculate these model-averaged estimates of the parameters within the Markov chain. They can simply be obtained as the estimates of the parameters, irrespective of the model that the chain is in, i.e. we sum out over the different models. Thus, the reversible jump procedure obtains parameter estimates under individual models, together with the posterior model probabilities and finally model-averaged parameter estimates, all within a single chain.

Important Comments

Model averaging should always be undertaken with some thought (i.e. common sense). For example, consider the (extreme but highly demonstrative) case where there are only two plausible models, each having (approximately) equal posterior support. Suppose that the corresponding posterior distribution of the parameter θ, common to both models, is given by Figure 6.3. In other words, the posterior distribution (averaged over models) is distinctly bimodal, with no overlap between the different modes.

One mode of the distribution corresponds to the posterior distribution of

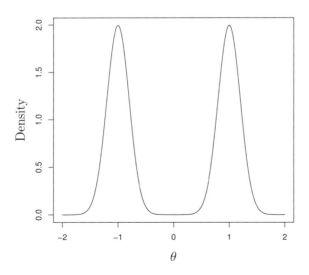

Figure 6.3 The bimodal model-averaged posterior distribution of a parameter.

the parameter given one of the possible models, and the other mode the posterior distribution from the alternative model. Using the posterior mean of the parameter, conditional on the model gives a good representation of the location of the parameter within the individual models. However, clearly taking the posterior model-averaged mean of θ is not a good summary description of the posterior distribution of θ. A better description would be the marginal density or at least the HPDI if a summary statistic is required. In this example, it is the use of the summary statistic that is at fault, rather than the idea of model averaging. This is nothing new, since the problem can occur within any Bayesian analysis when any parameter has a non-unimodal distribution. However, multimodal posterior distributions of parameters are more common within model averaging, since it is possible that the posterior distribution of the parameter may differ between the models with reasonable posterior support. Finally, we note that in some cases, identifying the reason for somewhat different parameter estimates for different models with reasonable posterior support can itself be of interest. Different models may make different predictions for good reasons.

6.6 Marginal Posterior Distributions

Typically in the presence of complex model structures and/or model discrimination on a number of different parameters, marginal posterior results are presented. For example, we may be interested in the posterior probability that the survival probability is dependent on a given covariate (irrespective of the presence/absence of other covariates). These marginal posterior results can help in understanding the complex posterior distribution over the model space. However, only considering the marginal posterior distributions can also obscure intricate details, in the presence of strong correlated results. For example, consider the perhaps conceptually easier case, where there are two parameters, θ_1 and θ_2, which are highly correlated. For illustration we consider a bivariate normal distribution with mean zero, variance one, and correlation of 0.9. In this example, the joint distribution is given in Figure 6.4(a). In addition, the marginal distributions for θ_1 and θ_2 are given in Figure 6.4(b). Clearly, presenting only the marginal distribution of the parameters individually does not capture the correlation structure between the parameters.

This is the identical issue that we can have when considering only the posterior marginal model probabilities – recall that the model is considered to be a parameter. We consider this issue further in the following example.

Example 6.10. Lapwings
Consider the analysis conducted by King et al. (2008b) relating to the UK lapwing population (see Example 1.3), which investigates their declining number in recent years. Here we omit the details of the analysis and simply present the posterior results obtained for two demographic parameters: adult survival probabilities ϕ_a and productivity rate ρ. The parameters are possibly depen-

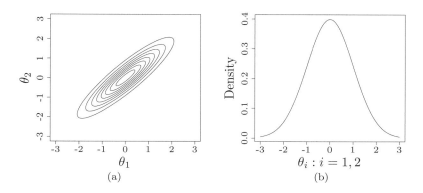

Figure 6.4 (a) The joint posterior distribution of θ_1 and θ_2; and (b) the marginal posterior distribution of θ_1 and θ_2 (they are identical in this case).

dent on two covariates: *fdays* (a measure of the harshness of the winter, in terms of the number of days below feezing at a location in central England; see Figure 2.2) and *year* (both covariates are normalised as usual). Table 6.3 gives the corresponding posterior marginal probability that these parameters are dependent on each of the covariates, using the standard notation that terms in brackets represent the covariate dependence present.

Table 6.3 The Marginal Posterior Probabilities for ϕ_a and ρ

(a) Adult Survival – ϕ_a		(b) Productivity – ρ	
Model	Posterior probability	Model	Posterior probability
$\phi_a(fdays, t)$	0.518	$\rho(t)$	0.457
$\phi_a(fdays)$	0.452	ρ	0.448
$\phi_a(t)$	0.016	$\rho(fdays)$	0.053
ϕ_a	0.015	$\rho(fdays, t)$	0.042

Source: Reproduced with permission from King et al. (2008b, p. 622) published by Wiley-Blackwell.

Thus, there is clearly strong evidence that ϕ_a is dependent on $fdays$. However, there appears to be a reasonable amount of uncertainty as to whether ϕ_a and ρ are time dependent. However, presenting these marginal models masks the intricate detail underneath. Now consider the joint posterior probability for ϕ_a and ρ. The two models with largest support are presented in Table 6.4. When we consider the joint distribution, it is clear that there is strong evidence that either ϕ_a is time dependent *or* ρ is time dependent (but not both). Both of the time dependence relationships (when present) are negative. Thus, using these data alone, the declining population appears to be related to *either* a decline in adult survival probability, *or* a decline in productivity. However, there is not enough information in the data to discriminate between these biological hypotheses. This intricate detail cannot be identified using the marginal distributions given in Table 6.3 alone.

Table 6.4 The Joint Marginal Posterior Probabilities for ϕ_a and ρ

Model	Posterior Probability
$\phi_a(fdays,t)/\rho$	0.443
$\phi_a(fdays)/\rho(t)$	0.418

□

For further illustration of the above ideas and to provide additional examples of applying the methods to different forms of data, we now consider a number of real examples. For each, we discuss the form of the data, the formulation of the set of biologically plausible models, provide brief details of the implementation of the RJMCMC algorithm, and provide sample output results and their corresponding interpretation. We note that the details of these examples may be rather complex, and may be useful for reference if implementing similar approaches in practice, but may not be of primary interest for understanding. As a result, the implementational details have been marked with an asterisk (*).

6.7 Assessing Temporal/Age Dependence

We consider the issue of model discrimination and the RJ algorithm where we have uncertainty in relation to the time dependence of the demographic parameters for mark-recapture-recovery data. We will then extend this to the case where there is additional uncertainty in relation to the age structure of the parameters. To illustrate the ideas we consider the mark-recapture-recovery data relating to the population of shags on the Isle of May, Scotland, discussed in Example 2.4. For this example, there is overall uncertainty in terms of the age and time dependence of the survival, recapture and recovery probabilities. We begin by considering the time dependence of these parameters.

Example 6.11. Shags – Time Dependence
We consider the annual mark-recapture-recovery data from 1982 to 1991. See Example 2.4, Section 2.6.1, Catchpole et al. (1998) and King and Brooks

ASSESSING TEMPORAL/AGE DEPENDENCE

(2002b) for a detailed description of the data. The survival, recapture and recovery probabilities may or may not be dependent on time, and this is unknown a priori. We will initially assume that the age structure of the parameters is known and is as follows:

$$\phi_1, \phi_{2:3}, \phi_{4+}/p_1, p_2, p_3, p_{4+}/\lambda_1, \lambda_{2:3}, \lambda_{4+}.$$

Shags are described as pulli in their first year of life, immature in years 2 and 3, and adults thereafter. Adult birds correspond to the breeding population, whereas immature birds typically spend time away from the Isle of May, before they return to breed. We assume a full age dependence for the recapture probabilities, but a pulli, immature and adult age structure for the survival and recovery probabilities. However, there is uncertainty relating to the time dependence for each of these parameters.

Prior Specification

We begin by considering the set of biologically plausible models. Without any prior information, we allow each individual parameter to be either constant or dependent on each year of the study. This results in a total of $2^{10} = 1024$ possible models. Thus, it is probably just about plausible to consider each individual model, but this would be very time consuming! We shall use the RJ algorithm to explore these possible models, letting the data determine the set of models with largest support. Within the Bayesian framework, we need to specify both priors on the parameters in each model, and the corresponding priors for the models themselves, in terms of the time dependence. Without any strong prior information, we specify an equal prior probability for each model – this once again corresponds to a prior probability of 0.5 that each parameter is time dependent independently of each other. For each model, we specify a $U[0,1]$ for each parameter within the model.

RJMCMC Algorithm*

One possible RJMCMC algorithm for exploring the model and parameter space and estimating the corresponding posterior model probabilities and posterior parameter estimates would be the following. Within each iteration of the Markov chain we:

1. Cycle though each parameter in the model, updating each using a MH random walk (though clearly alternative updates could be used);
2. Consider each parameter $\phi_1, \phi_{2:3}, \ldots, \lambda_{4+}$ in turn and update the corresponding time dependence, i.e. add time dependence if the parameter is constant, otherwise remove the time dependence if it is present, using a RJ step.

Once more we only describe the updating algorithm between different models., i.e. adding/removing time dependence. Without loss of generality we shall consider updating the model for ϕ_1. We initially consider the case where the current model has only a single parameter for ϕ_1. We propose to move to the model with parameters $\phi_1'(t)$ for $t = 1, \ldots, T - 1$, which has a total of $T - 1$ parameters.

One possible updating algorithm is as follows. Suppose that at iteration k of the Markov chain, the current state of the chain is $(\boldsymbol{\theta}, m)_k$. We propose to move to model m' with the set of parameter values $\boldsymbol{\theta}'$, such that the first-year survival probability is time dependent. Thus, we need to propose new values for the first-year survival probability at each time of the study, and we shall use the general RJ algorithm to do this in a similar way to that described in Example 6.7 for the ring-recovery example. For simplicity we assume that all other survival, recapture and recovery parameter values remain the same. For this model move we need to introduce random variables $\boldsymbol{u}$ and $\boldsymbol{u}'$, such that,

$$(\boldsymbol{\theta}', \boldsymbol{u}') = g(\boldsymbol{\theta}, \boldsymbol{u}),$$

where g is a bijective function. In particular we let $\boldsymbol{u} = (u(1), \ldots, u(T-1))$ and simulate values for the set of random variables $\boldsymbol{u}$, from some proposal density q. For example, we may simulate $u(t) \sim U[0, 1]$, independently for $t = 1, \ldots, T - 1$. We then propose the new parameter values,

$$\phi_1'(t) = u(t).$$

Finally, we need to specify the random variables $\boldsymbol{u}'$. For this case, $\boldsymbol{u}' \equiv u'$, where u' is simulated from some distribution with proposal density q' within the reverse move (i.e. from model m' to model m). For example, we may specify $u' \sim U[0, 1]$. The final specification of the model move relates to the relationship between the current first-year survival probability, ϕ_1 and the random variable u':

$$u' = \phi_1.$$

The function g is then completely specified. Recall that all other parameter values remain the same within the proposed model move, so that g essentially corresponds to the identity function for those parameters.

The corresponding acceptance probability for the model move is the standard form of $\min(1, A)$, where,

$$\begin{aligned} A &= \frac{\pi(\boldsymbol{\theta}', m'|\boldsymbol{x})\mathbb{P}(m|m')q'(u')}{\pi(\boldsymbol{\theta}, m|\boldsymbol{x})\mathbb{P}(m'|m)q(\boldsymbol{u})} \left| \frac{\partial(\boldsymbol{\theta}', u')}{\partial(\boldsymbol{\theta}, \boldsymbol{u})} \right| \\ &= \frac{f_{m'}(\boldsymbol{x}|\boldsymbol{\theta}')}{f_m(\boldsymbol{x}|\boldsymbol{\theta})}, \end{aligned}$$

since $\mathbb{P}(m|m') = \mathbb{P}(m'|m) = 1$; $q(\boldsymbol{u}) = 1 = q'(u')$; $p(\boldsymbol{\theta}') = p(\boldsymbol{\theta}) = 1$ and the Jacobian is equal to unity since g is the identify function. If the move is accepted, we set, $(\boldsymbol{\theta}, m)_{k+1} = (\boldsymbol{\theta}', m')$; otherwise we set $(\boldsymbol{\theta}, m)_{k+1} = (\boldsymbol{\theta}, m)_k$.

Alternatively, for the reverse move, suppose that the chain is in model m' (with a time dependent first-year survival probability) with parameters $\boldsymbol{\theta}'$. We propose to remove the time dependence for the first-year survival probability (but keeping all other parameter values the same), and we denote this proposed model by m. We simulate a new parameter value u' from with proposal

ASSESSING TEMPORAL/AGE DEPENDENCE

density q', and set the proposed constant first-year survival probability to be,

$$\phi_1 = u'.$$

Finally, we set,

$$u(t) = \phi_1(t),$$

for $t = 1, \ldots, T-1$. The corresponding acceptance probability for the proposed move is $\min(1, A^{-1})$, where A is given above.

Improving the RJMCMC Algorithm*

The above proposed updating algorithm is somewhat naive, and will typically be very inefficient. The reason for this can be seen if we consider the form of the acceptance rate. The probability of accepting the model move is simply the ratio of the likelihood functions evaluated at the proposed parameter values in model m' and the current parameter values in model m. However, we are proposing $T - 1$ new first-year survival probabilities, each simulated independently from a $U[0, 1]$ distribution. Thus it is likely that the likelihood function (conditional on the other parameter values) is highly peaked in the $(T - 1)$-dimensional parameter space corresponding to the first-year time dependent survival probabilities. Thus, the probability of simultaneously proposing the $T - 1$ values for the time dependent survival probabilities such that they lie in the region of the $(T - 1)$-dimensional space with high likelihood values is very small.

An alternative approach is to consider a different proposal distribution for the new parameter values. In other words, defining a different q and q'. The idea is to specify these proposal densities in such a way that the corresponding simulated proposal values are in areas of high posterior mass. For example, one approach would be to define,

$$u(t) \sim N(\mu(t), \sigma_1^2),$$

independently for each $t = 1, \ldots, T-1$, for some prespecified $\mu(t)$ and σ_1^2. Note that since we set $\phi_1(t) = u(t)$, we could also consider using a truncated normal distribution, in order to ensure that the proposed parameter values lie within the interval $[0, 1]$. Alternatively, we could use a beta distribution with given mean and variance. Otherwise, if we retain the normal distribution, it means that if we simulate any value outside this interval, the corresponding move has zero acceptance probability. Once again there is the trade-off between simplicity and efficiency. However, for this method we need to set the values of $\mu(t)$ and σ_1^2. One approach that often works well in practice is to consider an initial pilot run, where we consider the saturated model (i.e. each parameter fully time dependent). We then use the posterior mean of each parameter as the proposal mean, $\mu(t)$, within the RJ update. Clearly, this assumes that the parameter values are fairly consistent among the different models – an assumption that is often reasonably valid. The proposal variance σ_1^2 can be obtained via pilot tuning, or a similar approach used as for the proposal mean. Typically, defining the variance term σ_1^2 is less important than setting the proposal mean, as it is the location of the parameters that is more important. Finally, we need to specify the proposal density q', for the parameter u'. We consider a similar approach and set,

$$u' \sim N(\mu, \sigma_2^2),$$

for some predefined values μ and σ_2^2. We could use a similar approach as above, and use a pilot run using the smallest model (i.e. no time dependence) to obtain a value for the proposal mean, μ, for each of the parameters. Alternatively, we could simply set,

$$\mu = \frac{1}{T-1} \sum_{t=1}^{T-1} \mu(t),$$

where $\mu(t)$ is the corresponding proposal mean defined above for the demographic parameter.

Once more we consider the move where we propose to add in the time dependence for the first-year survival probabilities. Using the improved proposal distribution, the corresponding acceptance probability of the move is given by $\min(1, A)$, where

$$A = \frac{\pi(\boldsymbol{\theta}', m'|\boldsymbol{x})q'(u')}{\pi(\boldsymbol{\theta}, m|\boldsymbol{x})q(\boldsymbol{u})},$$

such that the proposal distributions, q and q', are the densities of the normal distributions described above. The corresponding acceptance probability for the reverse move follows similarly.

Results

The posterior probabilities that each of the model parameters are dependent on time are provided in Table 6.5. Clearly, we can see that there is strong evidence for time dependence or time independence for the majority of the parameters, with the only exception being adult survival probabilities. We can also estimate the (model-averaged) posterior estimates of the parameters. These are provided in Table 6.6. There are reasonably similar posterior distributions for adult survival probabilities for times 2 through 9 with a significantly lower estimate at time 1. For the marginal model with a constant adult survival probability, the corresponding posterior mean is 0.861 with standard deviation 0.013. This would suggest that the evidence for temporal dependence in the survival probability is primarily led by the difference in survival of adult birds in the first year of the study compared with later years, and hence the reason for the uncertainty in the presence or absence of temporal dependence of this probability. Clearly, if this was of particular interest it could be investigated further. For example, omitting the first year of the data for the adults and re-running the analysis. Alternatively, for the recapture probabilities, with so much marginal posterior mass placed on a single model, it is obvious that model averaging makes little difference from simply using the single marginal model. However, from Table 6.6 we observe that the recapture probabilities

ASSESSING TEMPORAL/AGE DEPENDENCE

for birds in their third year differ quite substantially between years, roughly decreasing and increasing in line with those observed for adults. Finally, there is no evidence of temporal heterogeneity for the recovery probabilities, which is clearly seen within the model-averaged estimates.

Table 6.5 The (Marginal) Posterior Probability that Each of the Model Parameters is Time Dependent

Parameter	Posterior Probability of Time Dependence
ϕ_1	1.000
$\phi_{2,3}$	0.002
ϕ_a	0.537
p_1	0.000
p_2	0.000
p_3	1.000
p_4	1.000
λ_1	0.024
$\lambda_{2,3}$	0.000
λ_a	0.006

There are clearly many alternative proposals that could be used. We have simply presented one possibility that worked well in practice for the shags data set as an illustration of the more general RJ update, and provided some insight into the difficulties that can arise and associated ideas for improving the efficiency of the algorithm. We note further that it is possible to simultaneously update other parameter values within this RJ step. For example, if parameter values are highly correlated, we may want to do this to improve the acceptance rates. If this is done, we simply need to add the corresponding proposal densities to the acceptance probability, analogous to terms within the MH algorithm. We discuss further implementational details relating to the RJMCMC algorithm in Section 6.8.

□

Example 6.12. Shags – Age Dependence
In Example 6.11, we assumed that the underlying age dependence of the demographic parameters was known a priori. We remove this assumption here and extend the above algorithm to the case where we do not know either the age structure or the time dependence of the parameters a priori.

Table 6.6 The Posterior Model-Averaged Mean and Standard Deviation for the Survival, Recapture and Recovery Probabilities for Shags in Years 1–3 and Adults (Denoted by a) and Year

	Survival Probabilities			
	Year of Life			
Year	1	2	3	a
1	0.508 (0.064)	–	–	0.718 (0.141)
2	0.674 (0.055)	0.557 (0.026)	–	0.867 (0.030)
3	0.350 (0.069)	0.557 (0.026)	0.557 (0.026)	0.890 (0.048)
4	0.893 (0.031)	0.557 (0.026)	0.557 (0.026)	0.835 (0.043)
5	0.645 (0.062)	0.557 (0.026)	0.557 (0.026)	0.834 (0.040)
6	0.893 (0.040)	0.557 (0.026)	0.557 (0.026)	0.903 (0.044)
7	0.849 (0.040)	0.557 (0.026)	0.557 (0.026)	0.901 (0.041)
8	0.579 (0.092)	0.557 (0.026)	0.557 (0.026)	0.874 (0.027)
9	0.902 (0.064)	0.557 (0.026)	0.557 (0.026)	0.885 (0.034)

	Recapture Probabilities			
	Year of Life			
Year	1	2	3	a
2	0.0007 (0.0004)	–	–	0.439 (0.086)
3	0.0007 (0.0004)	0.009 (0.002)	–	0.486 (0.040)
4	0.0007 (0.0004)	0.009 (0.002)	0.039 (0.016)	0.064 (0.015)
5	0.0007 (0.0004)	0.009 (0.002)	0.007 (0.005)	0.015 (0.006)
6	0.0007 (0.0004)	0.009 (0.002)	0.020 (0.020)	0.089 (0.012)
7	0.0007 (0.0004)	0.009 (0.002)	0.105 (0.021)	0.283 (0.022)
8	0.0007 (0.0004)	0.009 (0.002)	0.006 (0.006)	0.020 (0.004)
9	0.0007 (0.0004)	0.009 (0.002)	0.073 (0.023)	0.202 (0.020)
10	0.0007 (0.0004)	0.009 (0.002)	0.021 (0.009)	0.137 (0.016)

	Recovery Probabilities			
	Year of Life			
Year	1	2	3	a
1	0.194 (0.038)	–	–	0.093 (0.017)
2	0.200 (0.071)	0.040 (0.005)	–	0.096 (0.046)
3	0.191 (0.024)	0.040 (0.005)	0.040 (0.005)	0.092 (0.019)
4	0.197 (0.063)	0.040 (0.005)	0.040 (0.005)	0.094 (0.027)
5	0.195 (0.046)	0.040 (0.005)	0.040 (0.005)	0.093 (0.025)
6	0.193 (0.051)	0.040 (0.005)	0.040 (0.005)	0.093 (0.027)
7	0.192 (0.044)	0.040 (0.005)	0.040 (0.005)	0.094 (0.030)
8	0.190 (0.034)	0.040 (0.005)	0.040 (0.005)	0.095 (0.042)
9	0.190 (0.041)	0.040 (0.005)	0.040 (0.005)	0.094 (0.029)

ASSESSING TEMPORAL/AGE DEPENDENCE

Prior Specification

We first need to define the set of plausible models, taking into account the fact that there may be both age and time dependence. We do this in a two-step approach. First, we consider the marginal age structure of the parameters. For the recapture and recovery probabilities, we allow all possible combinations of age dependence. This gives a total of 16 possible models for the age structure. However, for the survival probabilities we consider a more restrictive set of plausible models for biological reasons. In particular, we impose the constraint that only what we term *consecutively aged* animals can have a common survival probability. For example, the model with marginal age structure $\phi_1 = \phi_3, \phi_2, \phi_{4+}$ (i.e. shags in their first and third year having a common survival probability, and different from the second-year and adult survival probabilities) is deemed to be implausible. This results in a total of 10 possible age structures for the survival probabilities. Conditional on the age structure for each set of demographic parameters, we allow each of the parameters to be either constant over time or arbitrarily time dependent. We then have a total of 477,144 distinct and biologically plausible models between which we wish to discriminate. Typically, the set of plausible models will be determined via prior biological information, and different structures will be considered plausible for different systems.

In the absence of any prior information, we assume that any biologically plausible model seems no more likely than any other, and place a flat (uniform) prior over all models permissible within the framework described above. (Essentially we specify a prior probability of zero for any other model). For each model, we specify a $U[0, 1]$ prior on each of the parameters.

RJMCMC Algorithm*

As usual, we consider a combination of MH updates for the parameters and RJ updates for the model moves within the constructed Markov chain in order to explore the joint posterior distribution over parameter and model space. Within each iteration of the Markov chain, we perform the same RJMCMC update moves as described above, conditional on the age structure. We consider an additional RJ move for each set of survival, recapture and recovery probabilities, to update the underlying age structure of the parameters. Without loss of generality, here we consider the recapture probabilities.

We propose only to combine age groups that have the same year dependence structure. Thus, to retain the reversibility of the Markov chain, when we split a set of ages; the new sets will have the same year dependence structure as the initial set. This simplifies the between-model move whilst in this example still providing a rapidly mixing chain. There are therefore two types of model moves for age structures: one which replaces a single parameter by two new ones (when we have no year dependence for that particular parameter and age group) and one which replaces $T - 1$ parameters by $2(T - 1)$ new ones (when we have full time dependence for that parameter). Both moves are performed similarly, and for simplicity of notation, we shall consider splitting the age dependence $\{1, 2\}$, i.e. parameter $p_{1,2}$, which we initially assume to be constant over time. This can be regarded as a nested model structure, and so we use the nested RJMCMC-type update.

Within the Markov chain, suppose that at iteration k the state of the chain is $(\boldsymbol{\theta}, m)_k$. We propose to move to model m', with parameter values $\boldsymbol{\theta}'$ using the following algorithm. We propose a *split* move (i.e. add an age group) or a *merge* move (i.e. combine two age groups) with equal probability. Suppose that given the current age group for the recapture probabilities we propose to split the age group into two new age groups, where we choose each possible set of new age groups with equal probability. Alternatively, if we propose a merge move, we choose two age groups to be merged with equal probability. For illustration, we suppose that we choose to make a split move for the parameter $p_{1,2}$. There is only one possible model we can propose, which has a distinct recapture probability for age 1 and age 2, denoted by p'_1 and p'_2. One possible update would be to propose values for these new parameters by setting,

$$p'_1 = p_{1,2} + u;$$
$$p'_2 = p_{1,2} - u,$$

where,

$$u \sim N(0, \sigma^2),$$

for some value of σ^2 chosen via pilot tuning. Essentially we are simply defining the new parameter values to be slightly perturbed from their current value. All the other recapture, recovery and survival probabilities remain the same.

We accept this model move with probability $\min(1, A)$, where,

$$A = \frac{\pi(m', \boldsymbol{\theta}'|\boldsymbol{x})\mathbb{P}(m|m')}{\pi(m, \boldsymbol{\theta}|\boldsymbol{x})\mathbb{P}(m'|m)q(u)} \left|\frac{\partial \boldsymbol{\theta}'}{\partial(\boldsymbol{\theta}, u)}\right|.$$

The probabilities $\mathbb{P}(m|m')$ and $\mathbb{P}(m'|m)$ are calculated from the algorithm used to decide the model move. Additionally, we need to calculate the Jacobian term. We have that,

$$\frac{\partial p'_1}{\partial p_{1,2}} = 1 = \frac{\partial p'_2}{\partial p_{1,2}}.$$

Similarly,

$$\frac{\partial p'_1}{\partial u} = 1; \quad \frac{\partial p'_2}{\partial u} = -1.$$

Thus, when there is no time dependence the Jacobian is,

$$\left|\begin{array}{cc} 1 & 1 \\ -1 & 1 \end{array}\right| = 2.$$

Alternatively, if the initial recapture probability was time dependent (i.e. $p_{1,2}(t)$), we would have a vector of $\boldsymbol{u} = (u(1), \ldots, u(T-1))$ and propose the new parameter values,

$$p_1(t) = p_{1,2}(t) + u(t);$$
$$p_2(t) = p_{1,2}(t) - u(t),$$

where,

$$u(t) \sim N(0, \sigma^2),$$

for $t = 2, \ldots, T$. Using the analogous argument, the corresponding Jacobian is equal to 2^{T-1}.

The reverse move is completely defined by this algorithm. For example, suppose that the current state of the chain is $(\boldsymbol{\theta}', m')$, and that we propose to combine the recapture probability for the first and second year shags. Their current values are p_1' and p_2', respectively. We propose the new parameter value,

$$p_{1,2} = \frac{1}{2}(p_1' + p_2').$$

(The analogous result holds if these parameters were time dependent). The corresponding acceptance rate is again, $\min(1, A^{-1})$ for A given above.

We note that if any model moves are proposed that do not retain the reversibility conditions, then the move is automatically rejected (i.e. accepted with probability 0). For example, proposing to combine the first and second age groups for the recapture probabilities, when the current model has $p_1, p_2(t)$, or when we propose to split an age group that contains only a single age, such as p_1.

Additional types of model moves are possible. For example, we could propose to change the age structure of the recapture probabilities, keeping the total number of age groups the same. If we also retain the time dependence structure for the different age groups, this model move does not change the number of parameters in the model. The corresponding RJ step for such a move simply reduces to the MH algorithm, with standard acceptance probability (see Section 6.3.5 for an example of this).

Results

There are a number of different results that may be of interest, depending on the biological question(s) of interest. For example, identifying the age and time dependence for each of the parameters, the marginal age and time dependence for the survival probabilities (irrespective of the recapture and recovery probabilities) and the (model-averaged) estimates for the survival probabilities. For illustration, we present a variety of the different possible results (given the above questions of interest).

Table 6.7 provides the models with the largest posterior support identified within the Bayesian analysis. It can be clearly seen that the models identified with largest support are all neighbouring. In particular the marginal model for the recapture probabilities is the same for all of the models identified. This is more clearly seen if we consider the marginal models for each of the parameters. These are given in Table 6.8. Typically, models with largest posterior support will be directly neighbouring models, although this is not necessarily the case (for example, see Example 6.10). From Table 6.8, there appears to be some uncertainty relating to the time dependence for adult

Table 6.7 Posterior Model Probabilities for the Most Probable Models, where the Age Dependence is given in Subscripts (for Years of Life 1–3 and then Adults, denoted by a), and Time Dependence Denoted by (t) for the Given Parameters

	Model Parameters		Posterior
Survival	Recapture	Recovery	Probability
$\phi_1(t), \phi_{2,3}, \phi_a(t)$	$p_1, p_2, p_3(t), p_a(t)$	$\lambda_1, \lambda_{2,3}, \lambda_a$	0.309
$\phi_1(t), \phi_{2,3}, \phi_a$	$p_1, p_2, p_3(t), p_a(t)$	$\lambda_1, \lambda_{2,3}, \lambda_a$	0.205
$\phi_1(t), \phi_2, \phi_{3,a}$	$p_1, p_2, p_3(t), p_a(t)$	$\lambda_{1,3}, \lambda_2, \lambda_a$	0.126
$\phi_1(t), \phi_2, \phi_3, \phi_a(t)$	$p_1, p_2, p_3(t), p_a(t)$	$\lambda_1, \lambda_{2,3}, \lambda_a$	0.047

Table 6.8 The Models for the Survival, Recapture and Recovery Parameters with Marginal Posterior Support Greater than 5%

	Posterior
Survival	Model Probability
$\phi_1(t), \phi_{2,3}, \phi_a(t)$	0.366
$\phi_1(t), \phi_{2,3}, \phi_a$	0.219
$\phi_1(t), \phi_2, \phi_{3,a}$	0.142
$\phi_1(t), \phi_2, \phi_3, \phi_a(t)$	0.140
$\phi_1(t), \phi_2, \phi_3, \phi_a$	0.092
Recapture	
$p_1, p_2, p_3(t), p_a(t)$	0.996
Recovery	
$\lambda_1, \lambda_{2,3}, \lambda_a$	0.597
$\lambda_{1,3}, \lambda_2, \lambda_a$	0.182
$\lambda_1, \lambda_2, \lambda_{3,a}$	0.070
$\lambda_1, \lambda_2, \lambda_3, \lambda_a$	0.053

Source: Reproduced with permission from King and Brooks (2002b, p. 846) published by the International Biometric Society.

survival probability, with an overall posterior marginal probability of 55% of time dependence. This is consistent with the previous analysis in Example 6.11, where we assumed an underlying age structure for the model, as we would expect particularly due to the similarity of models identified (in terms of age dependence assumed).

Note that more generally, we may wish to explain the time dependence within the different age groups via a number of covariates. The above ideas can be simply extended, combining the reversible jump-type updates for the different model moves. This is discussed in some detail by King et al. (2006, 2008a) for the Soay sheep data; see Example 1.2.

□

6.8 Improving and Checking Performance

The RJMCMC algorithm is currently the most widely used algorithm to simultaneously explore both parameter and model space, in the presence of model uncertainty. However, typically, the acceptance probabilities for the reversible jump moves are lower than those for MH updates. In the presence of model uncertainty, longer simulations are often needed in order to (i) explore the joint model and parameter space, and (ii) obtain convergence in the posterior estimates of interest, including posterior model probabilities and/or model-averaged estimates of interest. However, a significant bonus is that only a single Markov chain needs to be constructed, irrespective of the number of models, since the chain moves over the model space itself, depending on the observed data. Clearly, with increasing numbers of models, it is likely that convergence will be correspondingly slower and longer simulations will need to be run. As a result of the general increase in length of simulations needed, using efficient reversible jump updating algorithms is important, particularly for large model spaces, or complex and computationally intensive likelihood expressions. We initially describe some convergence diagnostic issues, before discussing the development of efficient RJMCMC algorithms in more detail.

6.8.1 Convergence Diagnostic Tools

The convergence diagnostic tools described in Section 5.4.2 are now more difficult to apply in the presence of model uncertainty. Convergence needs to be achieved over both the parameter and model spaces. In addition, care needs to be taken when looking at parameter convergence, since not all parameters are now common to all models. The BGR statistic can be applied to parameters that are common to all models, and considered for the model-averaged estimate of the parameter. See Brooks and Guidici (2000) for further discussion of the use of convergence diagnostic tests in the presence of model uncertainty. We once again comment that longer simulations are typically needed to ensure convergence, since both the parameter and model spaces need to be explored. Pragmatically, a number of Markov chains should be run starting from over-dispersed parameter values and also from different models, and the posterior results obtained compared. For example, chains can be started in the smallest and largest models from a number of different starting points. In addition, a number of different updating algorithms can be implemented and the results compared.

We again emphasise that there are no tests to prove that convergence has been achieved, any tests applied can only demonstrate that a chain has not converged.

6.8.2 Efficient RJMCMC Algorithms

The development of generic efficient reversible jump algorithms has proven to be somewhat elusive. However, a number of techniques have been developed that may significantly improve the mixing between models. We discuss a few of these ideas here, though there are many others within the literature.

Saturated Proposals

Typically, when models differ by a large number of parameters, constructing the Markov chain such that it is able to move between the different models can be difficult and require extensive pilot tuning. We have already met one idea in Section 6.11, where we use the saturated model to obtain posterior estimates for each of the parameters, and use these within the proposal distribution of the RJ algorithm. However, we note that this algorithm may not always work, in that it assumes that the parameter estimates are similar between models. In addition, poor precision of the posterior estimates in the saturated model may be obtained. When there are only a small number of models, pilot Markov chains can be run in each model to obtain posterior estimates of all of the parameters for each individual model, and these used in the proposal distributions, when proposing to move to a given model. However, clearly, this approach is not feasible for larger model spaces.

Intermediate Models

One alternative to the pilot tuning ideas above is to consider the introduction of *intermediate* models. These are models that are not within the set of plausible models but can be seen to lie between these models. For example, consider the simple case where we are interested in a model where the survival probabilities are regressed on up to two covariates, with a possible further interaction between the covariates. The set of possible models is:

Model 1:   logit $\phi(t) = \alpha$;

Model 2:   logit $\phi(t) = \alpha + \beta y_t$;

Model 3:   logit $\phi(t) = \alpha + \gamma z_t$;

Model 4:   logit $\phi(t) = \alpha + \beta y_t + \gamma z_t$;

Model 5:   logit $\phi(t) = \alpha + \beta y_t + \gamma z_t + \delta(y_t \times z_t)$.

IMPROVING AND CHECKING PERFORMANCE   191

If we consider an RJ updating procedure that only allows the addition/deletion of a single regression coefficient between neighbouring models, then we can only move to model 5 from model 4. Thus, this could in principle lead to poor mixing over the model space within the chain. Note that in practice for such a simple example with reasonable proposal distributions, mixing should be reasonable, but this simple example demonstrates the ideas well. One possible approach is to consider more complex model updates; for example, proposing to add/remove up to two regression coefficients within a single update. However, we consider adding additional intermediate models to improve the mixing. In particular, we could consider the three additional models:

Model 6*:   logit $\phi(t) = \alpha + \beta y_t + \delta(y_t \times z_t)$;
Model 7*:   logit $\phi(t) = \alpha + \gamma z_t + \delta(y_t \times z_t)$;
Model 8*:   logit $\phi(t) = \alpha + \delta(y_t \times z_t)$.

These three models are given stars in order to differentiate them from the plausible models above. Typically, such models are not biologically plausible and so not considered within an analysis (in other words, we specify a prior probability of zero for such models). However, if we introduce them into the set of models that the RJ algorithm can explore, it is now much easier to move between all the models. For example, assuming we only add/remove a single parameter, model 5 can now be proposed when the chain is in models 4, 6* and 7*. In order to obtain estimates of the corresponding posterior distribution of interest, we simply remove the realisations of the chain for which the chain is in models 6*, 7* and 8* and hence only retain the realisations from models 1 through 5. Clearly, there is a trade-off here between the mixing of the Markov chain and the proportion of observations that need to be discarded from the posterior distribution. For example, we may not wish to include model 8* in the set of intermediate models, as mixing may not be significantly improved with the inclusion of this model. Note that this approach is particularly appropriate when considering a variable selection problem within the package WinBUGS (see Chapter 7 and Appendix C), when restrictions cannot be placed on the models. See Example 10.4 and Gimenez et al. (2009a) for a further example of using this approach in WinBUGS where the order of the density dependence is unknown a priori.

Simulated/Importance Tempering

The idea of simulated tempering has been proposed to improve the mixing in poorly-performing Markov chains (Marinari and Parisi 1992; Geyer and Thompson 1995). The basic idea involves introducing an auxiliary variable or *temperature*, denoted by $k \in [0, 1]$, to the posterior distribution and considering the tempered distribution,

$$\pi_k(\boldsymbol{\theta}|\boldsymbol{x}) = \pi(\boldsymbol{\theta}|\boldsymbol{x})^k.$$

Note that for simplicity we write the posterior distribution over only the parameter space, but the idea immediately generalises to additional model space (in the presence of model uncertainty). Clearly, when $k = 1$, this tempered distribution is equal to the posterior distribution of interest. Conversely small values of k, flatten the posterior distribution, so that it is easier to move around the parameter and/or model space. Simulated tempering involves constructing the joint posterior distribution over the parameters (and models) and temperature parameter, so that,

$$\pi(\boldsymbol{\theta}, k|\boldsymbol{x}) \propto \pi_k(\boldsymbol{\theta}|\boldsymbol{x})p(k).$$

Realisations from the posterior distribution $\pi(\boldsymbol{\theta}|\boldsymbol{x})$ are obtained by taking the posterior conditional distribution, when $k = 1$. In practice, simulated tempering involves extensive pilot tuning so that the chain mixes well over the temperature and parameter space. In addition, this method can be fairly inefficient with a large number of realisations discarded (i.e. when $k \neq 1$). An extension of this method, to improve the efficiency of the algorithm, is to use the tempered distribution as the proposal distribution in an importance sampling algorithm, thus not discarding any realisations of the Markov chain (Jennison 1993; Neal 1996; Neal 2001; Gramacy et al. 2009). Note that Gramacy et al. (2009) call this method Importance Tempering and apply this method to the shags example in Section 6.12 and compare the approach with the previous RJMCMC algorithm used.

6.9 Additional Computational Techniques

The development of generic efficient algorithms for RJMCMC has been difficult. Additional approaches not described above also include Brooks et al. (2003b), who considered the form of the acceptance probability itself, in order to choose the proposal parameters. However, this can be difficult to implement, particularly where the likelihood is complex. Alternatively, Al-Awadhi et al. (2004) and Green and Mira (2001) propose stochastic algorithms which do not simply use the initial proposed value of the parameters. In the former case, a secondary Markov chain is implemented to modify the initial proposal before making any acceptance decision, while in the latter, given that a move is rejected, an alternative move is proposed, conditional on the proposal(s) already rejected. However, within all the algorithms there is generally a trade-off between the computational expense of the reversible jump updating procedure and the improvement to the mixing of the chain.

The RJMCMC algorithm is not the only approach for estimating posterior model probabilities. Carlin and Chib (1995) proposed a method where a supermodel is used that contains every possible parameter within the set of proposed models. Each possible parameter is updated at each iteration of the Markov chain. If the parameter is not in the current model, then the posterior distribution of the parameter is simply equal to the corresponding prior distribution. This method can be seen as a special case of the RJMCMC algorithm (in the same way that the MH algorithm is a special case). However, the performance of this method is often poor, and the posterior results obtained can

SUMMARY

be highly influenced by the priors specified on the parameters not even present in the model, due to very poor mixing and lack of convergence. An alternative approach for moving between models has been proposed by Stephens (2000), where a birth-death process is used to move between models of different dimensions. The idea essentially involves each parameter of the model being represented by a point. The parameters in the model are allowed to vary by new parameters being *born*, while existing parameters may *die*, within a continuous time frame. This continuous time process is run for a fixed interval within each MCMC iteration, with the parameters for the model updated as usual in each iteration of the Markov chain. If the birth and death rates of the parameters take a particular form, then the Markov chain has the required joint posterior distribution defined over both the parameters and models.

An alternative approach for discriminating between different competing models has been suggested by Gelfand and Ghosh (1998). Essentially, they propose the use of a posterior predictive loss function in order to discriminate between competing models. The underlying idea for this approach is to choose the model which minimises an expected loss function with respect to the corresponding posterior predictive distribution. Although this approach removes the issue of improper priors on the parameters on which we wish to perform model selection, once again all possible models need to be fitted to the data. In addition, model averaging is again not possible.

6.10 Summary

This chapter addresses the issue of model uncertainty. We show how a simple extension of Bayes' Theorem allows us to incorporate the model uncertainty within a formal and rigorous Bayesian framework. This results in the formation of a joint posterior distribution over both the parameter and model spaces. However, this distribution is typically very complex and little inference can be directly obtained. The most common approach to dealing with this distribution is to once more appeal to the ideas associated with MCMC and use an extension of this algorithm (the reversible jump step) to sample from the distribution and obtain posterior summary statistics of interest. In particular, we estimate the posterior probability associated with each model (or equivalently Bayes factors of competing models), which quantitatively discriminates between the competing models. The appeal of the RJMCMC algorithm is that only a single Markov chain needs to be constructed, irrespective of the number of possible models within the system. It is then the data (and priors) that inform the Markov chain regarding which models are supported and which are not. This approach also allows model-averaged estimates of parameters to be obtained, simultaneously incorporating both the parameter and model uncertainty within the estimate.

The reversible jump algorithm is typically more difficult to implement than the standard MCMC algorithm. This is as a result of (i) having to add/remove parameters when moving between different models so that a different type of move is needed (and often significant pilot tuning), and (ii) keeping track of the models within the algorithm.

There are a number of issues to be aware of when calculating posterior model probabilities and conducting model averaging. These include the sensitivity of the posterior model probabilities on the priors for the parameters, the appropriateness of model averaging on the given parameter of interest, and the efficient implementation of the RJ algorithm. Typically, greater care needs to be taken when implementing an RJ algorithm in the presence of model uncertainty in order to (i) ensure the posterior estimates have converged, and (ii) all results are interpreted correctly. Due to the complexity of the posterior distribution being summarised in the presence of model uncertainty, intricate detail can often be missed, so that additional care should be taken when producing, for example, marginal estimates (see Section 6.6).

6.11 Further Reading

The issue of model uncertainty is a long standing problem. The paper by Chatfield (1995) concludes that "model uncertainty is present in most real problems. Yet statisticians have given the topic little attention ... it is time for statisticians to stop pretending that model uncertainty does not exist, and to give due regard to the computer-based revolution in model-formulation which has taken place." Chatfield also observes that Bayesian model averaging is not without its problems; in particular, an averaged model loses the simplicity of its components. The paper by Hoeting et al. (1999) provides a review of Bayesian model averaging, with a range of interesting examples. Most standard books that include a description of MCMC algorithms, also include the extension to the RJMCMC algorithm and discuss model averaging (see, for example, Gamerman 1997, Givens and Hoeting 2005, Robert 2007, and Robert and Casella 2004). A variety of different examples are provided within such texts, including variable selection problems (for example, in regression problems); mixture models and autoregressive processes. In addition, a review of different methods for comparing competing models is given by O'Hara and Silanpaa (2009) for variable selection problems (including additional proposed methods).

In the classical context, trans-dimensional simulated annealing (TDSA; Brooks et al. 2003a) may also be used to explore large model spaces based on criteria such as the AIC. In particular, Brooks et al. (2003a) apply the TDSA approach to the shags capture-recapture-recovery data presented in Section 6.7 where there is potentially both age and time dependence for the model parameters (survival, recapture and recovery probabilities). Interestingly, the model identified with optimal AIC statistic is the same model with highest posterior support. King and Brooks (2004b) extend these ideas to the case of multi-site data (see Chapter 9) in the presence of possible time and/or location dependence with additional catch-effort information, applied to data from a population of Hector's dolphins.

6.12 Exercises

6.1 Table 6.9 provides the DIC values for four different models fitted to the dippers data, under three different prior specifications. Prior 1 corresponds to $U[0, 1]$ priors specified on each parameter; prior 2 to a $U[0, 1]$ prior on the recapture probabilities and a $Beta(1, 9)$ prior on the survival probabilities; and prior 3 to a $Beta(1, 9)$ prior specified on each survival and recapture probability.

Table 6.9 The DIC Values for the Dippers Data under Four Different Models and Three Different Prior Specifications on Each of the Parameters

Model	Prior 1	Prior 2	Prior 3
T/T	73.0	78.1	105.9
T/C	69.3	74.4	76.0
C/C	66.7	66.9	69.8
$C/C2$	62.0	62.5	65.0

Compare and contrast these results with those presented in Table 6.2 (in Example 6.8) which provides the corresponding posterior model probabilities for each of these models, given the different prior specifications. Discuss the advantages and disadvantages of using these different approaches.

6.2 Consider capture-recapture data with constant recapture probability p and time dependent survival probability ϕ_t, logistically regressed on covariate z_t, at time t, so that,
$$\text{logit } \phi_t = \alpha + \beta z_t.$$
We wish to consider two competing hypotheses, $\mathbb{P}_\pi(\beta < 0)$ (hypothesis 1) against $\mathbb{P}_\pi(\beta > 0)$ (hypothesis 2), where $\mathbb{P}_\pi$ denotes the posterior probability of the given event.

(a) Describe a Monte Carlo estimate for these posterior probabilities, given a set of sampled values from the posterior distribution of β obtained via an MCMC algorithm.

(b) Suppose that we specify a prior distribution on β of the form, $\beta \sim N(\mu, \sigma^2)$. Obtain an expression for the prior probability that $\beta < 0$ and $\beta > 0$, (denoted by $\mathbb{P}_p(\beta < 0)$ and $\mathbb{P}_p(\beta > 0)$), in terms of the normal cumulative distribution function, Φ.

(c) Hence, show that when $\mu = 0$, the Bayes factor reduces to the ratio,
$$BF = \frac{\mathbb{P}_\pi(\beta < 0)}{\mathbb{P}_\pi(\beta > 0)}.$$

6.3 Consider the results presented by Brooks et al. (2000a) for the blue-winged teal ring-recovery data. Four models are considered, and we use the $X/Y/Z$ notation described in Section 3.3.1. We extend the notation to allow for age dependent (adult) survival probabilities, denoted by A within the notation. The four models considered and the corresponding posterior model probabilities are given in Table 6.10. We assume independent $U[0,1]$ priors for each survival and recovery probability and a prior probability of $\frac{1}{4}$ for each model.

Table 6.10 The Posterior Model Probabilities for Four Models Fitted to the Blue-Winged Teal Ring-Recovery Data Ringed from 1961–1967 and Recovered from 1962–1973

Model	$C/A/C$	$T/A/C$	$T/A/T$	$C/C/T$
Posterior probability	0.009	0.054	0.000	0.937

Source: Reproduced with permission from Brooks et al. (2000a, p. 361) published by the Institute of Mathematical Statistics.

(a) Write down the set of parameters for each of the different models.
(b) Calculate the Bayes factor for time dependent recovery probabilities compared to a time independent recovery probability.
(c) Calculate the Bayes factor for age dependent adult survival probability against a constant adult survival probability over all adult individuals.
(d) Suggest an additional model that it might be sensible to fit to the data in light of the results presented for these four models.

6.4 Show that the MH algorithm is a special case of the RJ algorithm, where there is no dimensional change in the parameters.

6.5 Consider the two standard models C/C and C/T for capture-recapture data. An RJ step for moving between the two models is given in terms of the following move from model C/C with parameters p and ϕ to model C/T with parameters $p', \phi'_1, \ldots, \phi'_{T-1}$ as follows. We propose the new parameter values:

$$p' = p$$
$$\phi'_i = \phi \quad \text{where } i \text{ is randomly simulated from the set of values } 1, \ldots, T-1;$$
(in other words $\mathbb{P}(i = k) = \frac{1}{T-1}$ for $k = 1, \ldots, T-1$)
$$\phi'_j = \phi + u_j \quad \text{for } j = 1, \ldots, T-1 \text{ such that } j \neq i,$$
where $u_j \sim N(0, \sigma^2)$.

(a) Calculate the Jacobian for this proposed model move.

EXERCISES

(b) Assuming independent $U[0,1]$ priors are specified on each demographic parameter in each model, calculate the acceptance probability for moving from model C/C to C/T using the above model move.

(c) In order to satisfy the reversibility constraints for the RJ move, describe the reverse move from model C/T to model C/C, and provide the corresponding acceptance probability.

6.6 Consider the following RJ updating algorithm for the dippers data, for moving between the two models C/C and $C/C2$ (compare with the model move proposed in Example 6.3). Suppose that the current state of the Markov chain is model C/C with parameters p and ϕ. We propose to move to model $C/C2$ with parameters, p', ϕ'_n and ϕ'_f (denoting survival probabilities for non-flood and flood years, respectively), using the following bijection:

$$p' = u_1 \qquad u'_1 = p$$
$$\phi'_n = u_2 \qquad u'_2 = \phi$$
$$\phi'_f = u_3,$$

where $\boldsymbol{u} = \{u_1, u_2, u_3\}$ and $\boldsymbol{u}' = \{u'_1, u'_2\}$ are simulated from proposal distributions $q(\boldsymbol{u})$ and $q'(\boldsymbol{u}')$, respectively.

(a) Suggest two possible proposal distributions, q and q', and discuss which may perform better, providing a justification.

(b) Calculate the corresponding acceptance probability for the model move described.

(c) Describe the model move from model $C/C2$ to model C/C and give the corresponding acceptance probability.

CHAPTER 7

MCMC and RJMCMC Computer Programs

In this chapter we present computer programs written in R and WinBUGS for MCMC and RJMCMC analyses, with illustrations provided by the dipper and white stork examples of earlier chapters. In addition we describe the MCMC facilities in the statistical ecology computer package MARK. Detailed introductions to both R and WinBUGS are provided in Appendices B and C, respectively. We recommend that the computer programs are downloaded from the Web site and read in conjunction with the explanations provided within the text for additional clarity and understanding. We begin with an MCMC analysis of model C/C for the dipper data set. Computer timings refer to a 1.8-Ghz computer.

7.1 R Code (MCMC) for Dipper Data

We return to model C/C for the dipper data set of Example 1.1. The MH algorithm is given in Example 5.5 and we now provide the corresponding R-code, using the same single-update MH random walk algorithm with normal proposal distribution for both the recapture and survival probabilities.

The computer code is divided into three distinct sections: reading in the data values along with the prior and proposal parameters, performing the MCMC simulations, and outputting the results. The overall code is simply these parts placed together, in order, to provide a single R function. A detailed explanation, including a line-by-line description, of the program is provided in Section B.4 of Appendix B. We consider these three parts in their natural order.

Figure 7.1 provides the name of the function being defined and reads in the data and prior/proposal values. Thus, *Rdippercode* is a function of two input parameters: nt – the number of iterations; and nburn – the length of the burn-in. In Figure 7.2 we present the main structure of the MCMC simulations, but call two subroutines – one for calculating the log-likelihood value and another for performing the MH updating step. The R commands for doing the postprocessing and outputting of the simulations are given in Figure 7.3 to complete the main part of the code.

The two subroutines are provided in Figures 7.4 and 7.5. The first subroutine, `calclikhood`, calculates the log-likelihood (as given in Equation (2.5)) evaluated at given values for the recapture and recovery probabilities (i.e.

```
###############
# Define the function with input parameters:
# nt = number of iterations; nburn = burn-in

Rdippercode <- function(nt,nburn){

# Define the parameter values:
# ni = number of release years; nj = number of recapture years
# nparam = maximum number of parameters

ni = 6
nj = 6
nparam = 2

# Read in the data:

data <- matrix(c(
11,2,0,0,0,0,9,
0,24,1,0,0,0,35,
0,0,34,2,0,0,42,
0,0,0,45,1,2,32,
0,0,0,0,51,0,37,
0,0,0,0,0,52,46),nrow=ni,byrow=T)

# Read in the priors:

alpha <- c(1,1)
beta <- c(1,1)

# Parameters for MH updates (standard deviation for normal random walk):

delta <- c(0.05,0.05)
```

Figure 7.1 R code for the dipper data: setting up the data and reading in the prior and proposal parameter values.

given parameter values); the second subroutine, **updateparam**, performs the random walk single-update MH update using a normal proposal distribution.

7.1.1 Running the R Dipper Code

To run the dipper code in R for 10,000 iterations, with a burn-in of 1000 iterations, we write:

```
> dip <- dippercodeMHnorm(10000,1000)
```

R CODE (MCMC) FOR DIPPER DATA

```
###############
# Set initial parameter values:

param <- c(0.9,0.5)

# Calculate log-likelihood for initial state using function
# "calclikhood":

likhood <- calclikhood(ni, nj, data, param)

# Define itns - array to store sample from posterior distribution;
# output - vector for parameter values and associated log-likelihood:

itns <- array(0, dim=c(nt, nparam))
output <- dim(nparam+1)

# MCMC updates - MH algorithm - cycle through each iteration:

for (t in 1:nt){

# Update the parameters in the model using function "updateparam":

    output <- updateparam(nparam,param,ni,nj,data,likhood,
                    alpha,beta,delta)

# Set parameter values and log-likelihood to be the output from
# the MH step:

    param <- output[1:nparam]
    likhood <- output[nparam+1]

    for (i in 1:nparam) {
        itns[t,i] <- param[i]
    }
}
```

Figure 7.2 R the main code for the dipper data for the MCMC iterations, calling two separate subroutines for calculating the log-likelihood and performing the MH step.

The 10,000 simulations took approximately 50 seconds to run. The command places the output from the function dippercodeMHnorm into the array dip. The first column of this array contains the set of p values and the second column the set of ϕ values. If the assignment command <- is not used, the output is printed to the screen, which is clearly not desirable in MCMC simulations. However, the posterior mean and standard deviation of the parameters are outputted to the screen, due to the cat commands contained

```
###############
# Remove the burn-in from the simulations and calculate the mean and
# standard deviation of the parameters

subitns <- itns[(nburn+1):nt,]

mn <- array(0,nparam)
std <- array(0,nparam)

for (i in 1:nparam) {
    mn[i] <- mean(subitns[,i])
    std[i] <- sd(subitns[,i]))
}

# Output the posterior mean and standard deviation of the parameters
# following burn-in to the screen:

cat("Posterior summary estimates for each model:   ", "\n")
cat("\n")
cat("mean     (SD)", "\n")
cat("p: ", "\n")
cat(mn[1], "    (", std[1], ")", "\n")
cat("\n")
cat("phi: ", "\n")
cat(mn[2], "    (", std[2], ")", "\n")

# Output the sample from posterior distribution and close the function:

itns
}
```

Figure 7.3 R Code for the Dipper Data Describing the Outputting of the Parameter Values Stored in the Array `itns`.

within the function (see the code in Figure 7.3). When running the above programs we obtain the following output to the screen:

```
Posterior summary estimates for each model:

      mean            (SD)
p: 0.8962453     ( 0.02940489 )

phi: 0.5623169   ( 0.02538913 )
```

It is explained in Section B.4.4 of Appendix B how we can obtain trace plots, as well as histograms and kernel density plots, such as those in Figure 7.6, and also calculate Monte Carlo error.

R CODE (MCMC) FOR DIPPER DATA

```
###############
# This function calculates the log-likelihood of capture-recapture data

calclikhood <- function(ni, nj, data, param){

# Set up the arrays for parameters (phi and p), cell probabilities (q)
# Set parameter values for each year of the study likhood to be zero

    phi <- array(0,nj)
    p <- array(0,nj)
    q <- array(0,dim=c(ni,nj+1))
    for (i in 1:nj) {
        p[i] <- param[1]
        phi[i] <- param[2] }
    likhood <- 0

# Calculate multinomial cell probabilities and log-likelihood value

    for (i in 1:ni){

# For diagonal elements:

        q[i,i] <- phi[i]*p[i]
        likhood <- likhood + data[i,i]*log(q[i,i])

# Calculate the elements above the diagonal:

        if (i <= (nj-1)) {
            for (j in (i+1):nj) {
                q[i,j] <- prod(phi[i:j])*prod(1-p[i:(j-1)])*p[j]
                likhood <- likhood + data[i,j]*log(q[i,j]) } }

# Probability of an animal never being seen again

        q[i,nj+1] <- 1 - sum(q[i,i:nj])
        likhood <- likhood + data[i,nj+1]*log(q[i,nj+1])
    }

# Output the log-likelihood value:

likhood
}
```

Figure 7.4 R code for calculating the log-likelihood value for capture-recapture data. Note that phi[i] $\equiv \phi_i$ and p[i] $\equiv p_{i+1}$ for $i = 1, \ldots, $ nj.

```
###############
# Function for performing single-update MH algorithm:

updateparam <- function(nparam,param,ni,nj,data,likhood,alpha,beta,
                        delta){

    for (i in 1:nparam) {

# Keep a record of the current parameter value being updated
# Propose a new value using a random walk from normal proposal

        oldparam <- param[i]
        param[i] <- rnorm(1, param[i], delta[i])

# If proposed value is in [0,1] calculate acceptable probability

        if (param[i] >= 0 & param[i] <= 1) {

            newlikhood <- calclikhood(ni, nj, data, param)

# In acceptance probability include likelihood and prior contributions
# (proposal contributions cancel since it is a symmetric distribution)

            num <- newlikhood + log(dbeta(param[i],alpha[i],beta[i]))
            den <- likhood + log(dbeta(oldparam,alpha[i],beta[i]))

            A <- min(1,exp(num-den))
        }
        else { A <- 0 }

# Simulate a random number in [0,1] and accept move with probability A;
# else reject move and return parameter value to previous value

        u <- runif(1)
        if (u <= A) { likhood <- newlikhood }
        else { param[i] <- oldparam }
    }

# Set the values to be outputted from the function to be the
# parameter values and log-likelihood value. Then output the values.

    output <- c(param, likhood)
    output
}
```

Figure 7.5 R code for the MH step for updating p and ϕ for capture-recapture data.

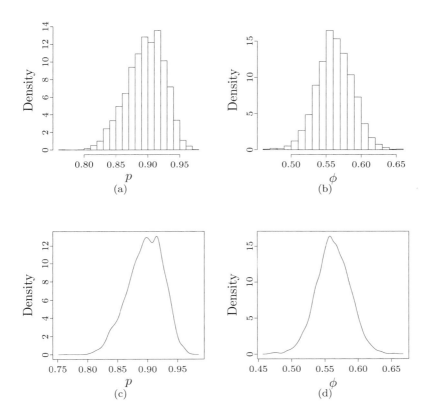

Figure 7.6 Histograms for sampled observations of the posterior distributions of (a) p and (b) ϕ for the dipper data under model C/C. Plots (c) and (d) provide the corresponding kernel density estimates.

7.2 WinBUGS Code (MCMC) for Dipper Data

We now demonstrate how to use the WinBUGS package to fit model C/C to the dipper data set. WinBUGS code is provided in Figures 7.7 and 7.8 for the three components: model specification, data and initial values. Detailed instructions of how to run WinBUGS code are provided in Appendix C.

In Figure 7.7, we define priors for the survival and recapture probabilities, by using the symbol $\sim$ which means *distributed as*. As above we define the size of the data set through the number of years of release (`ni`) and number of years of recapture (`nj`).

Also in Figure 7.7 is the likelihood for the C/C model; this code is similar to that used in R, and so we omit a detailed description.

As shown in Figure 7.8, the data are stored in a `list`, which is an R-type

```
##=############
# WinBUGS model specification (likelihood and priors)

model {

# Set constant recapture and survival probabilities and
# p1 as 1 - recapture probability

for (i in 1 : nj) {
    p[i] <- param[1]
    phi[i] <- param[2]
    p1[i] <- 1-p[i]
}

# Define priors for p = param[1] and phi = param[2] (beta priors)

param[1] ~ dbeta(1,1)
param[2] ~ dbeta(1,1)

# Define the likelihood:
    for (i in 1 : ni){
        m[i, 1 : (nj + 1)] ~ dmulti(q[i, ], r[i])
    }

# Calculate the no. of birds released each year (see input data)
    for (i in 1 : ni){
        r[i] <- sum(m[i, ]) }

# Calculate the multinomial cell probabilities
# Calculate the diagonal, above diagonal and below diagonal elements:

    for (i in 1 : ni){
        q[i, i] <- phi[i]*p[i]
        for(j in (i+1) : nj ){
            q[i,j] <- prod(phi[i:j])*prod(p1[i:(j-1)])*p[j]
        }
        for(j in 1 : (i - 1)){
            q[i, j] <- 0
        }

# Probability of an animal not being seen again
        q[i, nj + 1] <- 1 - sum(q[i, 1:nj])
    }
}
```

Figure 7.7 WinBUGS code providing the model specification for the dipper data set. Note that phi[i] $\equiv \phi_i$ and p[i] $\equiv p_{i+1}$ for $i = 1, \ldots, $ nj.

WINBUGS CODE (MCMC) FOR DIPPER DATA

```
###############
# Data Set: dippers

list(ni=6,nj=6,m=structure(.Data=c(
        11,2,0,0,0,0,9,
        0,24,1,0,0,0,35,
        0,0,34,2,0,0,42,
        0,0,0,45,1,2,32,
        0,0,0,0,51,0,37,
        0,0,0,0,0,52,46
        ), .Dim=c(6,7)))

# Initial values

list(param=c(0.1,0.1))
list(param=c(0.5,0.5))
list(param=c(0.9,0.9))
```

Figure 7.8 WinBUGS code for the data and initial sets of starting values for the dipper data set. Here m denotes the cell entries of the m-array with the final column detailing the number of birds that are released in the given year and not observed again.

object. It has three components, with ni the number of years of release, nj the number of years of recapture, and the capture-recapture m-array which is put into the matrix m. Note that the definition of a matrix requires using the WinBUGS specific function structure and defining the dimension of the matrix with .Dim.

Three independent simulations are initially run for 10,000 iterations each, using the Update Tool (see Section C.1.1 in Appendix C). The corresponding trace plot of the full MCMC simulation is obtained using the history button within the Sample Monitor Tool; see Figure 7.9 for one of the Markov chains. The trace plots suggest that convergence is very rapid (and are very similar for each of the three chains run), and we use a conservative burn-in of 1000 iterations for each chain.

The autocorrelation plots for the survival and recapture probabilities in Figure 7.10 show that the dependence between iterations in the sampled values is negligible. In addition, the BGR convergence diagnostic for both the survival and recapture probabilities is less than 1.2.

To summarise the posterior distribution for the parameter we use the button stats in the Sample Monitor Tool, which provides a table of posterior means, standard deviations and other summary statistics. WinBUGS outputs these summary statistics as in Figure 7.11. Note that within this WinBUGS output, the summary statistics are given for all the chains combined, although it is also possible to obtain the individual output summary statistics for each individual MCMC simulation. The corresponding associated kernel density

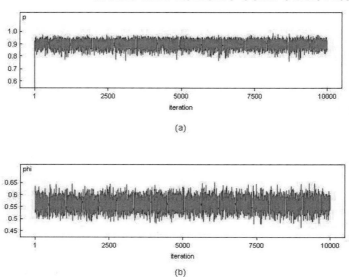

Figure 7.9 The trace plot of the parameters (a) p and (b) ϕ produced in WinBUGS using the history button within the Sample Monitor Tool.

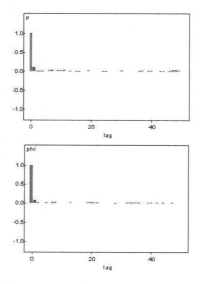

Figure 7.10 The autocorrelation plot of the parameters p (top panel) and ϕ (bottom panel) produced in WinBUGS using the autocor button within the Sample Monitor Tool.

MCMC WITHIN THE COMPUTER PACKAGE MARK

node	mean	sd	MC error	2.5%	median	97.5%	start	sample
p	0.8955	0.02905	1.862E-4	0.8327	0.8979	0.9452	1001	27000
phi	0.5616	0.02493	1.582E-4	0.5131	0.5614	0.6106	1001	27000

Figure 7.11 WinBUGS output summarising posterior distributions for the dipper data set. sd denotes the standard deviation; 2.5% and 97.5% the lower and upper limits of the 95% symmetric credible interval; and MC error the Monte Carlo error.

estimates of the marginal densities of p and ϕ can also be obtained within WinBUGS (using the density button in the Sample Monitor Tool – see Section C.1.1 in Appendix C).

7.3 MCMC within the Computer Package MARK

The computer package MARK (White and Burnham 1999) is a standard tool for analysing capture-recapture-recovery data within the classical framework, as described in Section 3.6.1. The package conducts Bayesian analyses via the MCMC tool. Downloadable help files can also be obtained using corresponding links from the MARK Web site, and also within the Help toolbar command within MARK. We provide a brief overview for using this MCMC facility within MARK, for those generally familiar with the package.

The MCMC function is found within the Run window and can be used for the capture-recapture-recovery models where there is an explicit expression for the likelihood. To perform a Bayesian analysis simply click the MCMC Estimation box when setting up the model, when this is done a ✓ will appear in the box. To continue, press the OK to Run button in the bottom right of the window. This will open up another window containing a number of additional MCMC-related input values, including:

Random Number Seed – the default is zero, which will set the random seed as the function of the time. For a repeatable set of MCMC simulations, with a fixed seed, we simply change this input to a non-zero parameter.

Number of tune-in samples – the number of iterations for which pilot tuning is performed in the MCMC simulations; the default value is 4000.

Number of burn-in samples – the burn-in of the MCMC simulations; default value is 1000 (following the pilot tuning).

Number of samples to store – the number of iterations to run post-burn-in; default value is equal to 10,000.

Name of binary file to store samples – the name of the file that stores the MCMC iteration values which can be used for further post-processing (in particular using coda); the default file name is MCMC.BIN.

Default SD of normal proposal distribution – the MCMC algorithm uses a normal random walk MH updating algorithm with the initial standard deviation inputted (note that this is an initial step size, as it is updated within the initial pilot tuning steps); the default value is 0.5.

The same window also requires a prior specification and possible convergence diagnostics. We consider each in turn, starting with the prior specification. A link function is used for each of the demographic parameters, and it is assumed that in general a logit link function will be used, although alternative link functions, such as log, log log, and complementary log log link functions are available. Priors are specified on the link scale. For example, suppose that we specify,

$$\text{logit } \phi_t = \beta_t.$$

Within MARK, the prior is specified on $\boldsymbol{\beta} = \{\beta_1, \ldots, \beta_T\}$. It is possible to either specify a prior on $\boldsymbol{\beta}$ (Prior Distributions for Parameters not included in Hyperdistributions), or a hyper-prior (i.e. hierarchical prior) on $\boldsymbol{\beta}$ (Hyperdistribution Specification). Taking a standard prior (i.e. non-hyper-prior) there are two possible prior specifications: an (improper) uniform prior over the real line (this is described in MARK as No prior -- Prior Ratio = 1); and a normal prior. The default prior is a $N(0, 1.75^2)$ for each parameter (note that the prior standard deviation is specified within MARK), and is chosen as this induces a near $U[0, 1]$ prior on the corresponding (back-transformed) demographic parameter (i.e. ϕ in the above example). See Sections 8.2.1 and 11.3.1 for further discussion of this issue. However a general prior mean and variance can be specified, either for all parameters, or for each individual parameter using Specify Prior Individually. This opens a new window for which a normal prior mean and variance needs to be specified for each individual parameter. Alternatively, a normal hyper-prior can be specified allowing for random effects. For example, suppose that we specify the logistic link function as given above. The default hyper-prior is of the form, $\beta_t \sim N(\mu, \sigma^2)$, such that $\mu \sim N(m, v)$ and $\sigma^2 \sim \Gamma^{-1}(\alpha, \beta)$, and where m, v, α and β are to be specified. The default is a $N(0, 100)$ for μ and $\Gamma^{-1}(0.001, 0.001)$ for σ^2. However, it is possible to use a more complex multivariate normal hyper-prior, via the specification of a variance-covariance matrix, by checking the box Variance-covariance matrix specified. This permits the specification of a general covariance matrix, with priors specified on the correlation between the (β) variables, typically denoted by rho. A uniform prior is assumed for the correlation rho, where the lower and upper bounds are to be specified. The default is a $U[-1, 1]$ prior. Note that using this prior specification, it is possible to specify a positive (or negative) correlation between variables (by setting either the lower or upper bound to be zero). However, care should be taken if using this approach, since it implies a 100% prior probability (and hence 100% posterior probability) that the correlation between the parameter is positive (or negative), irrespective of the information in the data. Thus, if one is considering using such a prior, a prior sensitivity analysis is highly recommended. Finally, we note that this hyper-prior specification permits an autocorrelation structure to be specified over the β_t values, for example, by specifying a given covariance matrix. Alternatively, a prior can be specified on the correlation between sets of different parameters (such as survival probabilities and productivity rates). See the `help` file on `MCMC VC Matrix` for further discussion and examples of covariance matrices.

Finally, there are the options to conduct convergence diagnostics on the MCMC algorithm. The default option is to run only a single chain and not perform any convergence diagnostics. However, there are two additional options: the use of `gibbsit` (Raftery and Lewis 1996) which can be used to determine the number of iterations necessary to obtain reliable posterior estimates (see Brooks and Roberts 1999 for further discussion), and running multiple replications to obtain a convergence diagnostic statistic (here the number of chains needs to be specified; the default is 10, the minimum number recommended by Gelman 1996). Note that this convergence diagnostic is not the same as the BGR statistic described in Section 5.4.2, but uses the same underlying principles. In particular, the BGR approach calculates the statistic, $\hat{R}$, by comparing the 80% credible intervals between multiple replications, whereas the statistic used in MARK compares posterior means across the multiple replications (see Gelman 1996 for further mathematical details).

MARK also provides output that can be read into the computer packages `SAS` or `R` for additional post-processing. In particular, the `help` file on `MCMC Binary File`, provides detailed instructions and R code (that can be cut-and-pasted in R) for converting the output binary file for use with the `coda` package in R). The `coda` package can be used to obtain additional posterior summary estimates, graphical plots, etc.

Example 7.1. Dippers - Model C/C

We once more consider the dipper data, described in Section 1.1. The package MARK contains the dipper data as an example data set. Note that for the dipper data included in MARK, an additional covariate corresponding to gender is also recorded, so that there are two groups of the recapture and survival parameters, corresponding to these groups. We assume homogeneous model parameters over gender. The corresponding PIMs for the model C/C are provided in Section 3.6.1, where we simply set the PIM corresponding to the survival and recapture parameters to be the same for each group (i.e. gender).

Using the default settings for the MCMC iterations the simulations are run for a total of 15,000 iterations, with the first 4000 iterations used to perform the pilot tuning; the next 1000 iterations are used as burn-in (no posterior samples can be used during the pilot tuning stage or burn-in period). The posterior summary estimates are then obtained from the final 10,000 iterations of the MCMC algorithm.

The summary output file from MARK is automatically opened (as a .TMP file that typically opens in `Notepad`). This output file can then be saved. We present a sample from the output file here, but additional information, such as the input parameter values used, are also included in this output. In particular, posterior estimates of the demographic parameters and parameters on the link function (logit function in this case) are outputted. For example, for the parameters on the link function we obtain the results below:

```
              LOGIT Link Function Parameters of {C/C}
Parameter       Mean      Standard Dev.    Median        Mode      Jumps(%)  Jump Size
---------    ----------  ---------------  ----------  ----------   --------  ---------
1:Phi        0.2516602    0.1022698       0.2500919    0.2458287    57.9      0.15190
2:p          2.1858579    0.3113999       2.1704407    2.1503734    29.8      1.17098
```

The first column provides the parameter and columns 2 through 5 the posterior summary estimates (mean, standard deviation, median and mode). The penultimate column provides the proportion of MH updates accepted and the final column the (updated) normal proposal standard deviation used following the tune-in period of the Markov chain (i.e. these are the pilot tuned values).

Of further interest are the parameter estimates of the survival and recapture parameters. These are also outputted below those for the parameters in the link function. For the above simulation these are:

```
                  Real Function Parameters of {C/C}
Parameter       Mean         Standard Dev.      Median           Mode
-----------  -------------  ---------------  -------------  --------------
   1:Phi      0.5623160       0.0251749       0.5628134      0.5635279
   2:p        0.8951465       0.0296459       0.8976971      0.9100282

Parameter   2.5th Percntile  5th Percntile   10th Percntile  20th Percntile
-----------  --------------- --------------- --------------- ---------------
   1:Phi       0.5111098       0.5202796       0.5300319      0.5409107
   2:p         0.8300242       0.8433405       0.8559984      0.8715489

Parameter   80th Percntile  90th Percntile   95th Percntile  97.5th Percntile
-----------  --------------- --------------- --------------- ---------------
   1:Phi       0.5834665       0.5947611       0.6037405      0.6104707
   2:p         0.9206446       0.9309148       0.9389251      0.9451724
```

These are the posterior mean, standard deviation, median, mode and (2.5, 5, 10, 20, 80, 90, 95, 97.5) percentiles for the survival and recapture probabilities (note that the percentiles are also presented for the link functions, but these are omitted for brevity).

Finally, we can use the R code provided in MARK to convert the binary output file into a format for using the package coda in R. This permits us to obtain trace plots, autocorrelation plots, additional posterior summary estimates, etc. Note that the MCMC values outputted relate only to those post burn-in. For example, the function HPDinterval provides the highest posterior density interval (HPDI) (see Section 4.4). The 95% HPDI for p and ϕ are returned as $(0.845, 0.950)$ and $(0.514, 0.608)$, respectively. Comparing these with the 95% symmetric credible interval obtained from the MARK output, we note that they are very similar. This is as a result of the marginal posterior distributions for the parameters being largely symmetrical in shape.

□

R CODE (RJMCMC) FOR MODEL UNCERTAINTY

The MCMC simulations within MARK are very fast to perform and can be implemented without intricate knowledge of underlying MCMC algorithms themselves. However, as with most computer packages designed to perform particular analyses, there are limitations. For example, the prior specification is restricted to normal distributions (or an unbounded uniform prior) on the parameters of the link function, and can only be performed when the likelihood function can be explicitly calculated. The set of possible models in MARK (with defined likelihood) is extensive (including, for example, for capture-recapture data, ring-recovery data, closed populations and multi-state data), however, the MCMC simulations are limited to the standard models in MARK.

7.4 R code (RJMCMC) for Model Uncertainty

We illustrate R code for RJMCMC through variable selection for the white stork application of Exercise 3.5 and discussed in Example 6.4. The matching MCMC analysis for a fixed model where the survival probability is logistically regressed on a single covariate is given in Section B.5 of Appendix B. The set of possible covariates we consider corresponds to the rainfall at 10 different locations. For the model which includes all the covariates, we set,

$$\text{logit } \phi_t = \beta_0 + \sum_{i=1}^{10} \beta_i cov_{i,t},$$

where $cov_{i,t}$ corresponds to the ith (normalised) covariate value at time t. Allowing for all possible combinations of covariates to be present in the model, there are a total of $2^{10} = 1024$ possible models.

We present the code in sections, relating to the different parts of the RJMCMC algorithm.

7.4.1 Main Code

The basis of the RJMCMC code is very similar to that of the standard MCMC algorithm, with an additional subroutine for performing the RJ step. However, there are also additional minor coding issues that need to be addressed. We consider three distinct parts of the main code: reading in of the data and specification of the prior parameter value, the (RJ)MCMC iterations, and outputting the results.

Part I: Reading in the Data and Prior Parameter Values

The reading in of the capture-recapture m-array and its dimensions is the same as given in Figure B.3 of Appendix B. Thus we omit these from the code given in Figure 7.12, which defines the function, storkcodeRJ, and reads in the data values.

The number of possible covariates (ncov) is defined to be 10, so that the

```
###############
# Define the function with input parameters:
# nt - number of iterations; nburn - burn-in

storkcodeRJ <- function(nt,nburn){

# nparam = maximum number of parameters
# ncov = maximum number of covariates

    nparam = 12
    ncov = 10

# Read in the covariate values:

    cov <- matrix(c(
 0.73,-0.70, 0.61, 0.79, 1.38, 1.60, 0.28, 1.75, 0.17,-0.73,
 0.41, 0.82,-0.23, 2.04, 0.27, 1.50, 0.32, 0.97,-0.81, 0.28,
 0.84, 2.44, 0.03, 1.04, 0.50, 0.41, 0.90,-0.13, 1.89,-0.27,
-0.24, 0.86, 1.77,-0.15,-0.41, 0.79,-0.40, 0.93, 0.79, 0.74,
 0.72,-0.33, 0.41,-0.01, 0.42,-0.50,-0.45,-0.52,-0.92, 0.44,
-0.49,-0.31,-0.13,-0.48, 1.30,-0.77, 1.45,-0.02, 0.13, 0.05,
-0.55,-0.12,-1.10, 1.72, 0.66,-1.56,-0.06, 0.02, 0.07, 3.22,
-1.64, 0.27, 1.38,-0.83,-0.07,-1.21, 1.35,-0.91,-0.01,-0.47,
 1.12, 0.96, 0.02,-0.02,-0.16, 0.80,-0.98, 1.20,-0.13, 0.09,
-0.22, 0.19, 0.58, 0.14,-0.23, 1.08, 0.86, 0.61, 1.11,-0.70,
-0.40,-1.54, 1.22,-0.71, 0.09,-0.44,-1.22, 0.20, 0.01,-0.07,
 0.94,-1.26,-1.57, 0.50, 0.40,-1.43, 1.23, 0.29, 1.38,-1.02,
-2.45, 0.46,-1.82,-0.62,-2.76,-0.65, 0.03,-1.52,-0.72, 0.32,
 1.00,-0.91,-0.53,-0.88,-0.06, 0.09,-0.34,-0.62,-1.48,-0.70,
-0.32, 0.09,-0.16,-1.00, 0.21, 0.40,-1.05,-0.15, 0.33,-0.84,
 0.55,-0.93,-0.48,-1.52,-1.55,-0.11,-1.94,-2.10,-1.81,-0.36),
nrow=16,byrow=T)

    probmodel <- array(0,nparam^2)
    model <- array(0,dim=c(nparam^2,ncov))
```

Figure 7.12 (Partial) R code for the RJMCMC algorithm relating to the white storks.

total number of possible parameters (nparam) is equal to 12, corresponding to p and $\boldsymbol{\beta} = \{\beta_0, \ldots, \beta_{10}\}$. We read the set of covariate values into the array cov. The $(i,j)^{th}$ element of cov corresponds to the amount of rainfall in year i at weather station j. The covariate values have already been normalised per region (i.e. for the columns of cov).

We define two arrays, probmodel, which will be used to calculate the posterior probability for each model, and model, which will be used as a model identifier (i.e. it will identify which covariates are present in the model, for each possible model). These two arrays are defined to simplify the calculation of the posterior model probabilities and in identifying the corresponding model.

We need to specify the priors for each of the parameters. The corresponding R code for this and the proposal parameters for the RJMCMC algorithm are given in Figure 7.13. Note that the prior for p is specified as before (i.e. $p \sim Beta(\alpha_p, \beta_p)$), and omitted from Figure 7.13 for brevity. We consider independent normal priors on each regression coefficient parameter. In particular we specify, $\beta_i \sim N(\mu_i, \sigma_i^2)$, for $i = 0, \ldots, 10$, independently of each other. The values for μ_i are read in to mu, and for σ_i^2 into sig2. For this example, we set $\mu_i = 0$ and $\sigma_i^2 = 0.5$ for $i = 0, \ldots, 10$. Finally, for the prior specification, we define the vector sig to be the prior standard deviations (by taking the square root of the prior variance terms in sig2). In the case of model uncertainty, we also need to specify a prior on each possible model. We do this by specifying the (independent) marginal prior probability that each regression coefficient ($\beta_0, \ldots, \beta_{10}$) is present in the model. These are read into the vector prob. Note that the first element of prob, corresponding to the prior probability that the intercept term, β_0, is non-zero is equal to 1. In other words, we always assume that β_0 is present in the model. (This could be relaxed by changing this prior probability). All other β terms are present with probability $1/2$.

Following the specification of the prior, we set the proposal parameters in the MH and RJ steps. The vector delta corresponds to the proposal parameters for the MH step; and rjmu and rjsig2 (or rjsig) the proposal parameters for the RJ step. We use a single-update MH step with uniform proposal distribution for each parameter. For the parameter p, we propose a candidate value uniformly within ± 0.05 of the current value; for each regression coefficient, we propose a candidate value uniformly within ± 0.1, given the values set for delta. This is the same updating algorithm used in Example 5.7 for the white stork data, but now we allow for more than a single covariate value, a straightforward extension of the given algorithm. Finally, we read in the proposal parameters for the RJ step. In particular, within the RJ step, if we propose to add in the parameter β_i to the model, we propose the candidate value,

$$\beta_i \sim N(\mu_i, \sigma_i^2).$$

The values for $\mu_0, \ldots, \mu_{10}$ are represented by the vector rjmu; and $\sigma_0^2, \ldots, \sigma_{10}^2$ by the vector rjsig2. Note that we have specified the RJ proposal means and variances to include the proposal parameters corresponding to β_0, for generality, even though for our model, we assume that this intercept term is always present.

Part II: RJMCMC Simulations

The second part of the R code given in Figure 7.14 performs the RJMCMC iterations for the given problem. As usual this begins with specifying the initial values for each of the parameters, placing them in the vector param. The first element of param corresponds to p, while the ith element corresponds to the

```
###############
# Read in priors and proposal parameter values for MH and RJ steps
# Normal priors (mean and variance) on regression coefficients:

    mu <- c(0,0,0,0,0,0,0,0,0,0,0)
    sig2 <- c(10,10,10,10,10,10,10,10,10,10,10)
    sig <- sqrt(sig2)

# Specify independent prior probabilities on each
# regression coefficient being present (including beta_0):
# We specify with probability 1 that the intercept beta_0 is present.

    prob <- c(1,0.5,0.5,0.5,0.5,0.5,0.5,0.5,0.5,0.5,0.5)

# Parameters for MH updates (uniform random walk)

    delta <- c(0.05,0.1,0.1,0.1,0.1,0.1,0.1,0.1,0.1,0.1,0.1)

# Parameters for RJ updates for beta_0 to beta_10
# (Normal proposal distribution)

    rjmu <- c(0,0,0,0,0,0,0,0,0,0,0)
    rjsig2 <- c(0.5,0.5,0.5,0.5,0.5,0.5,0.5,0.5,0.5,0.5,0.5)
    rjsig <- sqrt(rjsig2)
```

Figure 7.13 R code providing the prior parameter values for β and the MH and RJ step proposal parameters, relating to the white stork.

parameter β_{i-2} for $i = 2, \ldots, 12$. Thus, for this example, we set the initial parameter values,

$$p = 0.9; \qquad \beta_0 = 0.7; \qquad \beta_4 = 0.3; \qquad \beta_i = 0 \text{ for } i = 1, 2, 3, 5, \ldots, 10.$$

Note that these initial parameter values also implicitly set the initial model, in terms of the covariate dependence on the survival rate. In particular, the non-zero elements of param correspond to the parameters that are present in the model. Conversely, the elements of param set equal to zero correspond to the parameters that are not present in the model. Thus, in this example, we have the model $p/\phi(4)$, i.e. a constant recapture probability and the survival probability as a function of the intercept term (β_0) and covariate 4 (with corresponding parameter β_4). We always assume that the intercept term is present and so do not include this in the model notation. Note that the first element of param (i.e. p) is always non-zero, and more specifically, in the interval $[0, 1]$. In addition, we always assume an intercept term β_0, so that the second element of param is also non-zero. We also define the vectors probparam, sum1 and sum2, which will be used to calculate posterior model

R CODE (RJMCMC) FOR MODEL UNCERTAINTY 217

```
###############
# Set initial parameter values and calculate log-likelihood

    param <- c(0.9,0.7,0,0,0,0.3,0,0,0,0,0,0)
    probparam <- array(0,nparam)
    sum1 <- array(0,nparam)
    sum2 <- array(0,nparam)
    likhood <- calclikhood(ni, nj, data, param, nparam, cov, ncov)
    output <- dim(nparam+1)

# RJMCMC updates: cycle through each iteration:

    for (t in 1:nt){

# Update the parameters in the model using function "updateparam",
# setting parameter and log-likelihood values to current states:

        output <- updateparam(nparam, param, ncov, cov, ni, nj,
                    data, likhood, alphap, betap, mu, sig, delta)
        param <- output[1:nparam]
        likhood <- output[nparam+1]

# Update the covariates in the model using function "updatemodel",
# setting parameter and log-likelihood values to current states:

        output <- updatemodel(nparam, param, ncov, cov, ni, nj,
                    data, likhood, mu, sig, prob, rjmu, rjsig)
        param <- output[1:nparam]
        likhood <- output[nparam+1]

# To calculate what model the chain is in:

        bin <- calcmodel(ncov,param)

# Record the set of parameter values:

        for (i in 1:nparam) {
            itns[t,i] <- param[i]
        }
    }
```

Figure 7.14 R code specifying the initial parameter values and performing the RJMCMC iterations.

probabilities and (model-averaged) posterior means and variances of the parameters (conditional on the parameter being present in the model).

We include the standard loop over the number of iterations to be performed. We call the function **updateparam** which performs the MH updating of the parameters. The additional function, updatemodel, performs the RJ step, updating the model. As usual, after each of these functions, we set param to be the updated parameter values, and likhood to be the corresponding log-likelihood evaluated at these values. We add in another command following the RJ step, where we call the new function calcmodel. This function calculates the model number and is placed into bin, where the models are numbered from $1, \ldots, 1024$ (see Section 7.4.4).

Part III: Outputting the Results

Following the completion of the loop over the number of iterations, we can calculate posterior summary statistics of interest from the set of stored parameter values in itns. The corresponding R code is given in Figure 7.15. Conditional on a parameter being present in the model, we calculate the sum of each parameter value, sum1, and the sum of the square of the parameter value, sum2, in order to calculate the posterior model-averaged mean and variance. In addition, we record the number of times that each parameter is non-zero in probparam (in order to calculate the marginal posterior probability of each covariate being present). Recall that a non-zero parameter value corresponds to the parameter being present in the model, whereas if the parameter value is zero, the parameter is not present in the model. The model indicator has already been stored in bin, and so at the end of each iteration, we increase the number of times the chain has visited the model bin by 1, storing this in the vector probmodel. Finally, for ease of interpretation when outputting the results, we keep a record of the presence or absence of each regression coefficient $\beta_1, \ldots, \beta_{10}$ for each model. This only needs to be calculated the first time the model is visited within the Markov chain, and is stored in model. For example, for the model defined for ϕ from the initial parameter value, i.e. $\phi(4)$ (survival probability only dependent on covariate 4), the corresponding value for model would be,

$$0\ 0\ 0\ 1\ 0\ 0\ 0\ 0\ 0\ 0.$$

As we assume that the intercept, β_0 is always present we do not include this within the model array. However, if we considered model uncertainty on β_0, we simply extend the array model with an additional column at the beginning indicating a 0/1, corresponding to whether β_0 is absent/present in the model.

Figure 7.16 provides the final R commands for the code, outputting the posterior summary statistics and the set of simulated parameter values in itns. The posterior (model-averaged) mean (mn) and standard deviation (std) of the regression coefficients are calculated, conditional on the parameter being present in the model. For example, for the posterior mean, we simply calculate the sample mean, conditional on the parameter being present in the model. This is simply the sum of the parameter values (conditional on the parameter being present in the model, and denoted by sum1[i]), divided by the number of times the chain is in the model (probparam[i]), for $i = 1, \ldots,$ nparam. Similarly, we estimate the posterior standard deviation by the sample standard

R CODE (RJMCMC) FOR MODEL UNCERTAINTY

```
###############
# To calculate posterior marginal probability.

       if (t >= (nburn+1)) {
           for (i in 1:nparam) {
               if (param[i] != 0) {
                   sum1[i] <- sum1[i] + param[i]
                   sum2[i] <- sum2[i] + param[i]^2
                   probparam[i] <- probparam[i] + 1
               }
           }
           probmodel[bin] <- probmodel[bin] + 1

# To update the model probability and record the model:

           if (probmodel[bin] == 1) {
               for (i in 1:ncov) {
                   if (param[i+2] != 0) {
                       model[bin,i] <- 1
                   }
               }
           }
       }
```

Figure 7.15 R code for obtaining summary statistics of interest, including posterior means, variances and model probabilities.

deviation. However, for ease of calculation we use the following (analogous) functional form for calculating the sample standard deviation, but taking into account that we are conditioning on the parameter being present in the model. Generally, for the parameter, θ, say, the sample standard deviation is given by,

$$s = \sqrt{\frac{1}{n-1}\left(\sum_{t=1}^{n}(\theta^t)^2 - n\bar{\theta}^2\right)} \qquad (7.1)$$

where θ^k denotes the value of the parameter θ at iteration k (post-burn-in), and $(\bar{\theta})$ denotes the sample mean of these θ values. Note that for the regression coefficients we consider the posterior mean and standard deviation of the parameter values conditional on them being present in the model (i.e. non-zero).

For each possible parameter, we use the `cat` command to print to the screen the corresponding posterior probability that the parameter is in the model, and the corresponding posterior (model-averaged) mean and standard deviation (conditional on the parameter being present in the model). Finally, we print the representation of the model and the corresponding posterior probability. The models are represented by a series of 0's and 1's, denoting which

```
###############
# Calculate posterior mean and standard deviation of each parameter
# (conditional on being present in the model)

for (i in 1:nparam) {
    mn[i] <- sum1[i]/probparam[i]
    std[i] <- sqrt((sum2[i] - probparam[i]*mn[i]^2)/(probparam[i]-1))
}

# Output the posterior mean and standard deviation of the parameters
# following burn-in to the screen:

cat("Estimates for parameters, conditional on being present:","\n")
cat("\n")
cat("mean(SD)", "\n")
cat("p: ", "\n")
cat(mn[1], "    (", std[1], ")", "\n")

for (i in 2:nparam) {
    cat("\n")
    cat("beta_", (i-2), "     Prob:", probparam[i]/(nt-nburn), "\n")
    if (probparam[i] != 0) {
        cat(mn[i], "    (", std[i], ")", "\n")
    }
}

cat("\n")
cat("Overall model            Posterior probability")
cat("\n")
for (i in 1:(2^ncov)) {
    if ((probmodel[i]/(nt-nburn)) >= 0.01) {
        cat(model[i,], "    :  ", probmodel[i]/(nt-nburn), "\n")
    }
}

# Output the sample from the posterior distribution:

itns
}
```

Figure 7.16 R code for outputting the posterior summary statistics of interest, including the model-averaged posterior mean and standard deviation of each parameter (conditional on being present in the model), marginal posterior model probability for each parameter and finally the posterior probability of each model.

covariates are present in the model. As the number of possible models is fairly large (1024), we only print out those models that have a posterior probability equal to or greater than 0.01.

7.4.2 Subroutine – Calculating Log-Likelihood Function – `calclikhood`

The calculation of the log-likelihood is given in Figure 7.17, and is very similar to the log-likelihood given in Figure B.7 (in Appendix B), where there is only a single covariate for the white stork data. Recall that if the covariate is not in the model, the corresponding regression coefficient is equal to zero, and so does not contribute to the calculation of `expr`. The survival probability is then calculated by taking the inverse of the logit function of `expr`, i.e.

$$\texttt{phi[i]} = \text{logit}^{-1}(\text{expr}) = \frac{1}{1+\exp(-\text{expr})}.$$

The remainder of the log-likelihood, given the specification of the recapture and survival probabilities is given in Figure 7.4, and hence is omitted for brevity. However, we note that we add an additional input parameter `ncov` to the function as an additional argument (corresponding to the number of possible covariates).

```
###############
# This function calculates the log-likelihood of capture-recapture data

calclikhood <- function(ni, nj, data, param, nparam, cov, ncov){

...

# Set the recapture/survival probabilities for each year of the study:

    for (i in 1:nj) {
        expr <- sum(param[2],param[3:nparam]*cov[i,1:ncov])
        phi[i] <- 1/(1+exp(-expr))

        p[i] <- param[1]
    }
...
```

Figure 7.17 R code for calculating the log-likelihood function for a logistic linear regression for the survival probabilities. The remainder of the log-likelihood is already given in Figure 7.4, and is omitted for brevity.

7.4.3 Subroutine – MH Updating Algorithm – `updateparam`

The MH subroutine only updates the parameters that are in the current model. The easiest way to implement this is to cycle through all parameters (as usual), but add in a condition statement (i.e. `if` command) that the parameter is only updated if it is non-zero (and hence present in the model). The corresponding R commands to implement this are illustrated in Figure 7.18 using the `if` statement. The updating of a parameter (conditional on the model) is standard (see Figure B.8 in Appendix B for the single covariate case) and so is omitted

222　　MCMC AND RJMCMC COMPUTER PROGRAMS

```
###############
# MH updating step:

updateparam <- function(nparam, param, ncov, cov, ni, nj, data,
                        likhood, alphap, betap, mu, sig, delta){

# Cycle through each parameter in turn and propose to update using
# random walk MH with uniform proposal density:

    for (i in 1:nparam) {

# Update the parameter if present in model (i.e. not equal to zero)

        if (param[i] != 0) {
...
        }
    }
}
```

Figure 7.18 R code for the MH updating step, in the presence of model uncertainty, so that only the parameters in the model are updated within the step (i.e. all non-zero parameters). The remainder of the MH step is analogous to that presented in Figure B.8.

for brevity. Note that, as for the subroutine `calclikhood`, we do need to include `ncov` in the argument of the subroutine `updateparam`.

7.4.4 Subroutine – Model Identification – `calcmodel`

Within the RJMCMC code, it is convenient to label the possible models from 1,...,1024 in some manner. The easiest way to do this is to express the model using binary form, and then calculate the corresponding numerical interpretation of the binary digits plus 1. For each covariate, $i = 1, \ldots, 10$, we set the ith binary digit to be equal to 0 if the covariate is absent; or 1 if the covariate is present. For example, for the model where the survival probability is dependent on covariates 2 and 4, i.e. $\phi(2,4)$, the binary form is given by,

$$0\ 1\ 0\ 1\ 0\ 0\ 0\ 0\ 0\ 0.$$

This is the same model identification scheme as used in WinBUGS (see Section 7.5). The corresponding numerical interpretation of this model (with units as the left-hand-most element) is given as 6. Thus the model number is $6+1 = 7$. In other words, if we let y_i denote the ith binary element (corresponding to the ith covariate), the corresponding model number is calculated as,

$$1 + \sum_{i=1}^{10} y_i 2^{i-1}.$$

R CODE (RJMCMC) FOR MODEL UNCERTAINTY

We need to include the 1 to correspond to the model with no covariates present. If we omit this 1, then this model has model number 0, so that if we visit this model within the RJMCMC simulations, we try to update the parameter `probparam[0]`, and R will return an error message as elements of vectors and matrices have to be a positive integer. The function in R that performs the calculation of the model identifying number is given in Figure 7.19, and follows the ideas above. However, in the calculation of the model number, we identify whether each covariate is in the model by determining whether the corresponding regression coefficient is non-zero.

```
###############
# Function for calculating the model (uses form of binary conversion)

calcmodel <- function(ncov,param){

    bin <- 1

    for (i in 1:ncov) {
        if (param[i+2] != 0) {
            bin <- bin + 2^(i-1)
        }
    }

# Output model number (as binary conversion)

bin
}
```

Figure 7.19 R code for identifying the corresponding model number given the parameter values.

7.4.5 Subroutine – RJ Updating Algorithm – updatemodel

Finally we consider the RJ subroutine that performs the updating of the model within each iteration of the Markov chain. The RJ updating step is described in Example 6.4. The corresponding R code that implements the RJ step is given in Figure 7.20.

We begin by randomly selecting a parameter to be updated: we randomly sample from the set of integers, 3,...,12 using the function `sample`. This corresponds to sampling from a discrete uniform distribution. (Note that if we wished to perform model selection on the intercept term we would simulate from the set of integers 2,...,12). We let the sampled value be equal to `r`. We propose to add the `r`th parameter to the model if it is absent (i.e. `param[r]` = 0), or remove the parameter from the model if it is present (i.e. `param[r]` $\neq$ 0), but also keeping a record of the current value in `oldparam`. If we propose to add the parameter (i.e. add the corresponding covariate dependence to the

###############
Function for updating the model:

```
updatemodel <- function(nparam, param, ncov, cov, ni, nj, data,
                    likhood, mu, sig, prob, rjmu, rjsig){

# Propose parameter to be added/removed and record current value
# If covariate is not in model, propose candidate value (normal)

    r <- sample(3:12,1)
    oldparam <- param[r]
    if (param[r] == 0) {
        param[r] <- rnorm(1, rjmu[r-1], rjsig[r-1])
    }
    else {
        param[r] <- 0
    }

# Calculate the acceptance probability:

    newlikhood <- calclikhood(ni, nj, data, param, nparam, cov, ncov)

    if (param[r] != 0) {
        num <- newlikhood + log(dnorm(param[r],mu[r-1],sig[r-1]))
        den <- likhood + log(dnorm(param[r],rjmu[r-1],rjsig[r-1]))

        num <- num + log(prob[r-1])     # Prior for covariate present
        den <- den + log(1-prob[r-1])   # Prior for covariate absent
    }
    else {
        num <- newlikhood + log(dnorm(oldparam,rjmu[r-1],rjsig[r-1]))
        den <- likhood + log(dnorm(oldparam,mu[r-1],sig[r-1]))

        num <- num + log(1-prob[r-1])   # Prior for covariate absent
        den <- den + log(prob[r-1])     # Prior for covariate present
    }
# Acceptance probability of RJ step:

    A <- min(1,exp(num-den))
    u <- runif(1)
    if (u <= A) { likhood <- newlikhood }
    else { param[r] <- oldparam }
    output <- c(param, likhood)
    output
}
```

Figure 7.20 R function for performing the RJ step for the white stork data.

R CODE (RJMCMC) FOR MODEL UNCERTAINTY 225

model), we simulate a new proposed parameter value from a normal distribution with mean `mu[r-1]` and standard deviation `rjsig[r-1]`. The indexing of the reversible jump proposal parameters, `mu` and `rjsig`, are both `r-1`, since we do not have any reversible jump proposal parameters for `param[1]` (i.e. p). Otherwise, if the corresponding covariate dependence is present (`param[r]` $\neq$ 0), we set the new proposed parameter value to be equal to zero (and hence effectively remove the covariate dependency).

To perform the accept/reject step, we calculate the corresponding acceptance probability. This involves combining the likelihood, prior and proposal terms. See Example 6.4 for the derivation and an explicit expression of the acceptance probability. In summary, when we propose to add the parameter, the logarithm of the numerator of the acceptance probability (`num`) is the sum of the log-likelihood of the proposed model (`newlikhood`), log prior probability of the covariate being present (`prob[r-1]`) and log prior density for the proposed parameter value (normal distribution). The logarithm of the denominator (`den`) is the sum of the log-likelihood of the current model (`likhood`), log prior probability that the covariate is not present in the model (`1-prob[r-1]`) and log proposal distribution for the proposed parameter (normal distribution). All other terms cancel and so are omitted for brevity. The acceptance probability for the reverse move, where we propose to remove the covariate dependence (i.e. propose the value `param[r]` = 0) is calculated analogously. The acceptance or rejection of the proposed model move, and corresponding parameter values, follows in the same manner as for the standard MH algorithm.

7.4.6 Running the White Stork RJMCMC Code

We run the above code in R for 50,000 iterations, which took approximately 20 minutes on a 1.8-GHz machine; this is substantially longer than earlier MCMC runs. From looking at trace plots of the parameters, we judge that convergence appears to be very fast, and so we use a conservative burn-in of 5000 iterations. The R command used to run the code, and the corresponding output obtained for the given simulations are provided in Figure 7.21.

We note again the excessive number of decimal places in the output. Independent replications of the simulations produced posterior estimates a maximum of ± 0.01 of the posterior estimates given above, so that we assume that the parameter estimates have sufficiently converged.

The first set of output provides the (marginal) posterior probability that each parameter is in the model and the corresponding model-averaged posterior mean and standard deviation of the model parameters, conditional on being in the model. There appears to be strong evidence (Bayes factor of 61) that the survival probability is dependent on covariate 4 (corresponding to the rainfall at weather station Kita). Conversely, there is evidence that survival probability is not related to any other covariate, with a posterior probability of at most 10% of a relationship being present.

```
> storkrj <- storkcodeRJ(50000,5000)
Posterior summary estimates for each parameter, conditional on being
present:
```

		mean	(SD)
p:		0.9119767	(0.01456911)
beta_ 0	Prob: 1	0.6864585	(0.07285264)
beta_ 1	Prob: 0.05511111	0.1090567	(0.09721590)
beta_ 2	Prob: 0.03471111	-0.0015191	(0.09045799)
beta_ 3	Prob: 0.02151111	0.0344788	(0.08002865)
beta_ 4	Prob: 0.9839556	0.3635855	(0.09541476)
beta_ 5	Prob: 0.08615556	-0.1543375	(0.10377240)
beta_ 6	Prob: 0.1033333	0.1329554	(0.08613963)
beta_ 7	Prob: 0.03777778	-0.0801619	(0.08534941)
beta_ 8	Prob: 0.05297778	0.0953845	(0.12908070)
beta_ 9	Prob: 0.04055556	0.0872423	(0.09563914)
beta_ 10	Prob: 0.0648	-0.1074937	(0.09118234)

Overall model	Posterior probability
0 0 0 1 0 0 0 0 0 0	0.6568889
0 0 0 1 0 0 0 0 1 0	0.02446667
1 0 0 1 0 0 0 0 0 0	0.0308
0 0 0 1 0 1 0 0 0 0	0.08393333
0 1 0 1 0 0 0 0 0 0	0.0256
0 0 0 1 0 0 0 1 0 0	0.03228889
0 0 1 1 0 0 0 0 0 0	0.0146
0 0 0 1 1 0 0 0 0 0	0.05902222
0 0 0 1 0 0 1 0 0 0	0.02497778
0 0 0 1 0 0 0 1 0 0	0.03228889

Figure 7.21 R output for the white stork data set where there is uncertainty with respect to the covariate dependence on the survival probability.

The second part of the output provides the posterior model probability for the overall model, in terms of the dependence of the survival probability on the set of covariates. The model with largest posterior support (66%) has the survival probability dependent on the single covariate, labelled 4. The corresponding model-averaged mean (and standard deviation) of the corresponding regression coefficient for this covariate (i.e. β_4) is 0.364 (0.095). Thus there is evidence of a positive relationship between the survival probability and rainfall at the given weather station (Kita).

7.5 WinBUGS Code (RJMCMC) for Model Uncertainty

We now consider how to perform a RJMCMC analysis in WinBUGS, once again using the white stork data set as an illustration. The corresponding WinBUGS code for the fixed model where the survival probability is regressed on a single covariate is provided in Example C.2 of Appendix C. The RJMCMC algorithm has recently been incorporated into WinBUGS, as an additional add-on called *Jump* (Lunn et al. 2006, 2009); see Section C.1.2 (in Appendix C) for further information on how to download this add-on package. However, currently this RJMCMC algorithm is only possible for two situations, namely for variable selection problems (i.e. covariate analyses) in the presence of additional random effects and spline regressions. However, using the package Jump to perform model selection for covariate analyses we need to assume a random effects model for the covariates (see Section 8.5).

For the recapture probability, we assume a $U[0,1]$ prior. For the regression coefficients, we assume independent priors. Conditional on the parameter being present in the model, we specify a $N(0,10)$ prior on the parameter. Within WinBUGS, a hierarchical prior (or random effects models, see Section 8.5) also needs to be specified on the survival probabilities. Thus we specify,

$$\text{logit } \phi_t \sim N(\mu_t, \sigma^2),$$

where μ_t denotes the mean specified as a linear function of the covariates present in the model, and $\sigma^2 \sim \Gamma^{-1}(a,b)$. We initially consider the prior parameters $a = b = 0.001$. Note that we conduct a prior sensitivity analysis of these priors on the posterior distribution.

The specification of the model in terms of the likelihood and prior specification on the recapture probability are provided in Figure C.5, in Appendix C, assuming the survival probability is regressed on a single covariate. However, these parts of the WinBUGS code remain the same for the case where there is model uncertainty in terms of the covariates that the survival probabilities are regressed on, and so are not repeated for brevity. We now consider the specification of the survival probabilities and the additional reversible jump step that are needed in WinBUGS for the case of model uncertainty. Technically in WinBUGS, the survival probabilities are modelled as in Figure 7.22. The function `jump.lin.pred` specifies the linear predictors in the form of a matrix X which contains the covariates, with the number of rows and number of columns equal to the number of years and the number of covariates, respectively. Note that the intercept is always included in the model so we do not need to have a column of ones in this matrix to include an intercept.

The function `jump.model.id` allows the model configuration to be stored efficiently in the variable `id`, which needs to be monitored in order to obtain the posterior model probabilities (using the Jump menu). Parameter `psi[i]` is a vector representing the current values of the linear predictor.

The number K in Figure 7.22 indexes the number of regression parameters (excluding the intercept) in the model for a particular draw from the chain (i.e. K is the dimension of the model). The indicator variable `id` indicates which particular model is in a particular draw; for example, if `id` was 0001000000 that would indicate that the parameter 4, corresponding to an effect of rainfall on survival at the Kita meteorological station, was in the model for that draw.

Finally, we need to specify the priors on the models. Without any prior

```
###############
# Cycle through each release time

for (i in 1:nj) {

# Set the (logit) link function for the survival probabilities

  logit(phi[i]) <- mu[i]

# Specify the prior on the parameter (normal)

  mu[i] ~ dnorm(psi[i],taueps)}

# Define the current values for the linear predictor term
  psi[1:nj] <- jump.lin.pred(X[1:nj, 1:Q], K, beta.prec)

# Set the id function to identify the different models

  id <- jump.model.id(psi[1:nj])
}

# Specify the prior on precision for the random effects (gamma) and
# different models (uniform) and set prior precision for beta terms

taueps ~ dgamma(0.001,0.001)
K ~ dbin(0.5, Q)
beta.prec <- 0.1
```

Figure 7.22 WinBUGS code for modelling the survival probabilities of storks as a linear function of several covariates. Priors on the precision of the random effect taueps, on the models K and on the precision of the regression parameters beta.prec are also specified.

information, we specify equal prior probabilities for all models, in terms of the inclusion/exclusion of each covariate. This equates to specifying a prior probability of 0.5 that each covariate is in the model, independently of all other covariates. In the Jump protocol, this is achieved by specifying a prior distribution on K as in Figure 7.22.

Given the specification of the model we then read in the data and initial values. These are provided in Figures 7.23 and 7.24. The simulations are initially run for 50,000 iterations, with a burn-in of 5000 iterations. The simulations appear to converge very quickly within WinBUGS.

Posterior model probabilities can be calculated from the proportion of time the chain visited each model of interest, as before. There appears to be a single dominating model identified, corresponding to the survival probability being regressed on only covariate 4 (posterior probability of 0.517). The model ranked second, with a posterior probability of 0.080, has no covariate

WINBUGS CODE (RJMCMC) FOR MODEL UNCERTAINTY 229

```
###############
# Data set: white storks and covariate values

list(ni=16,nj=16,Q=10,m=structure(.Data = c(
19,2,0,0,0,0,0,0,0,0,0,0,0,0,0,0,5,
0,33,3,0,0,0,0,0,0,0,0,0,0,0,0,0,14,
0,0,35,4,0,0,0,0,0,0,0,0,0,0,0,0,14,
0,0,0,42,1,0,0,0,0,0,0,0,0,0,0,0,26,
0,0,0,0,42,1,0,0,0,0,0,0,0,0,0,0,30,
0,0,0,0,0,32,2,1,0,0,0,0,0,0,0,0,36,
0,0,0,0,0,0,46,2,0,0,0,0,0,0,0,0,16,
0,0,0,0,0,0,0,33,3,0,0,0,0,0,0,0,28,
0,0,0,0,0,0,0,0,44,2,0,0,0,0,0,0,20,
0,0,0,0,0,0,0,0,0,43,1,0,0,1,0,0,10,
0,0,0,0,0,0,0,0,0,0,34,1,0,0,0,0,25,
0,0,0,0,0,0,0,0,0,0,0,36,1,0,0,0,16,
0,0,0,0,0,0,0,0,0,0,0,0,27,2,0,0,22,
0,0,0,0,0,0,0,0,0,0,0,0,0,22,0,0,16,
0,0,0,0,0,0,0,0,0,0,0,0,0,0,15,1,17,
0,0,0,0,0,0,0,0,0,0,0,0,0,0,0,15,8),.Dim =c(16,17)),

X=structure(.Data = c(
0.73,-0.70,0.61,0.79,1.38,1.60,0.28,1.75,0.17,-0.73,
0.41,0.82,-0.23,2.04,0.27,1.50,0.32,0.97,-0.81,0.28,
0.84,2.44,0.03,1.04,0.50,0.41,0.90,-0.13,1.89,-0.27,
-0.24,0.86,1.77,-0.15,-0.41,0.79,-0.40,0.93,0.79,0.74,
0.72,-0.33,0.41,-0.01,0.42,-0.50,-0.45,-0.52,-0.92,0.44,
-0.49,-0.31,-0.13,-0.48,1.30,-0.77,1.45,-0.02,0.13,0.05,
-0.55,-0.12,-1.10,1.72,0.66,-1.56,-0.06,0.02,0.07,3.22,
-1.64,0.27,1.38,-0.83,-0.07,-1.21,1.35,-0.91,-0.01,-0.47,
1.12,0.96,0.02,-0.02,-0.16,0.80,-0.98,1.20,-0.13,0.09,
-0.22,0.19,0.58,0.14,-0.23,1.08,0.86,0.61,1.11,-0.70,
-0.40,-1.54,1.22,-0.71,0.09,-0.44,-1.22,0.20,0.01,-0.07,
0.94,-1.26,-1.57,0.50,0.40,-1.43,1.23,0.29,1.38,-1.02,
-2.45,0.46,-1.82,-0.62,-2.76,-0.65,0.03,-1.52,-0.72,0.32,
1.00,-0.91,-0.53,-0.88,-0.06,0.09,-0.34,-0.62,-1.48,-0.70,
-0.32,0.09,-0.16,-1.00,0.21,0.40,-1.05,-0.15,0.33,-0.84,
0.55,-0.93,-0.48,-1.52,-1.55,-0.11,-1.94,-2.10,-1.81,-0.36),
.Dim = c(16,10)))
```

Figure 7.23 Example WinBUGS code of the data for the RJMCMC analysis of the storks data set. Here m denotes the cell entries of the m-array with the final column detailing the number of birds that are released in the given year and not observed again, X the matrix of covariates values, and Q the number of covariates.

dependence, so that the posterior distribution of the underlying mean survival probability (μ_t) is a constant. This corresponds to a Bayes factor of 6.46 in favour of the top-ranked model compared to the second-ranked model. The

```
###############
# Initial values
list(taueps=0.01,K=0,p=0.5)
```

Figure 7.24 Example WinBUGS code for initial starting values for the RJMCMC analysis of the storks data set, where `taueps` denotes the precision of the random effect probability, `K` the index for the number of regression parameters and `p` the recapture probability. Note that we start with no covariates in the model.

marginal posterior model probabilities given by WinBUGS are presented in Table 7.1. There appears to be positive evidence that covariate 4 influences the survival probability, but little evidence for the other covariates. It is possible to plot the corresponding posterior distribution of the regression coefficients. Note that WinBUGS plots the posterior distribution as a mixture distribution, which we do not recommend. One component corresponds to a point mass on 0 relating to the probability that the covariate is not in the model; the second component is the posterior distribution of the parameter, conditional on the covariate being present.

Table 7.1 The Marginal Posterior Probability of Inclusion of the Different Covariates for the White Stork Data

Covariate	Posterior Probability of Presence
1	0.064
2	0.034
3	0.034
4	0.811
5	0.066
6	0.087
7	0.041
8	0.076
9	0.049
10	0.053

In addition, the corresponding posterior mean (or median) of the random effect variance is 0.056 (0.084). This estimate, coupled with the posterior probabilities would suggest that there is little temporal heterogeneity that is not explained by the covariates considered. Note that we assess this directly in Section 8.5.

We now consider a prior sensitivity analysis. In particular, we decrease the prior variances for each of the parameters and specify independent $N(0,1)$ priors on the regression coefficients present in the model (prior 2). The corre-

Table 7.2 The Marginal Posterior Inclusion Probability of the Different Covariates for the White Stork Data Using (a) $N(0,1)$ Prior on Regression Coefficients and $\Gamma^{-1}(0.001, 0.001)$ on σ^2 – Prior 2; (b) $N(0,10)$ Prior on Regression Coefficients and $\Gamma^{-1}(3,2)$ on σ^2 – Prior 3

	Posterior Probability	
Covariate	Prior 2	Prior 3
1	0.159	0.089
2	0.108	0.063
3	0.102	0.060
4	0.893	0.260
5	0.202	0.064
6	0.209	0.091
7	0.121	0.064
8	0.169	0.111
9	0.144	0.071
10	0.143	0.062

sponding posterior marginal probabilities obtained within WinBUGS for each covariate is given in Table 7.2. Clearly, we can see that the majority of covariates have an increased posterior probability of being present (see Section 6.4 for a discussion of this). The posterior mean (and standard deviation) of σ^2 is 0.048 (0.071), similar to the results obtained with the first prior. Alternatively, we change the prior on the σ^2, and specify, $\sigma^2 \sim \Gamma^{-1}(3,2)$, (so that σ^2 has a prior mean and variance equal to one), and retain the $N(0,10)$ prior on the regression coefficients (prior 3). The corresponding marginal posterior probability for each covariate is presented in Table 7.2. It is interesting that although the priors on the regression coefficients of the covariates have not changed, the change in the prior variance for σ^2 has significantly decreased the posterior probability that covariate 4 influences the survival probability. Essentially this is a direct result of inflating the posterior distribution of σ^2. The posterior mean (and standard deviation) of σ^2 is 0.400 (0.157), appreciably higher than the posterior mean in the initial simulation. Since this prior results in a larger variance on the survival probabilities, this explains the temporal heterogeneity instead of the covariates themselves.

7.6 Summary

The computer programs illustrated in this chapter and in Appendices B and C may be readily adapted for use in other capture-recapture-recovery data analyses. We see from the illustrations in this chapter and Appendices B and C the relative simplicity of WinBUGS compared with R, for those cases where we can use WinBUGS. However, programming in R, especially for RJMCMC analyses, provides complete control and flexibility.

7.7 Further Reading

For a general introduction into the statistical package R, Crawley (2007) provides a very good overview. In addition, the R Web site given in Section B.1 of Appendix B provides useful documentation for the computer package, including an introductory manual for R and references. However, more specifically, there are a number of recently published books that include the use of R for performing Bayesian analyses, such as Rizzo (2007) and Bolker (2008) (for ecological examples). In particular, Rizzo (2007) provides a good and concise overview of the R language and many of the main principles relating to MCMC. For example, the book includes a description of the statistical distributions, the simulation of random numbers from different distributions (both the theory and the commands in R; see also Morgan 1984), and the MH algorithm and Gibbs sampler, including associated R code for a number of examples. In addition, the book by Albert (2007) discusses how Bayesian analyses can be performed within R using the LearnBayes package for performing the MCMC updates, and which is available via a link from the R Web site. In particular, this package includes two functions for performing MH updates (random walk normal proposal; and an independent sampler). Additional packages in R also include (amongst others), MCMCpack (which performs Bayesian inference for a number of different models) and mcmc (performs MH algorithm using a random walk multivariate normal proposal distribution for continuous parameters). Kerman and Gelman (2006) discuss two further R packages: Umacs (Universal Markov chain sampler) and rv (random variable) that are ongoing packages being developed to perform different aspects of Bayesian inference. Additional packages may also be useful for post-processing the data. For example, the packages boa and coda provide convergence diagnostic tools, functions for summarising posterior output and graphical tools. For further details of these, and additional, see the R Web site.

The WinBUGS Web site provides useful documentation for the computer package, including a tutorial illustrating the various steps of a simple Bayesian analysis using MCMC simulations, as well as additional resources such as courses, tutorial material, book references and Web links. There are also a number of recently published books that include the use of WinBUGS for performing Bayesian analyses, such as Congdon (2003, 2006), Gelman and Hill (2007) and McCarthy (2007) for ecological examples.

In particular, McCarthy (2007) provides a good introduction to WinBUGS for ecologists using standard generalised linear models (linear and logistic regressions, ANOVA, ANCOVA and contingency tables) for analysing data. The book also includes several case studies, with a chapter on capture-recapture analysis using the dipper data, and an illustration of the use of prior information. The book by Gelman and Hill (2007) provides an extensive overview

EXERCISES   233

of hierarchical models, including associated WinBUGS code for a number of examples. This book also provides useful guidelines for reparameterising models to improve the convergence and useful advice on how to debug WinBUGS code.

WinBUGS can be called from other programs such as R (see Section C.2 in Appendix C), SAS, Stata, MATLAB and GenStat. This may help in reading data or displaying the results using specific features of these programs. Note also that we use BUGS for Windows, but BUGS can also be run in other operating systems such as Linux or Macintosh. WinBUGS also has additional features, such as an add-on called GeoBUGS that allows fitting spatial models and building maps using Geographical Information System tools. There are also two further tools for advanced users: the WinBUGS Development Interface (WBDev) allows one to implement specialised functions and distributions through coding in the programming language Component Pascal. This makes the execution faster and more flexible than the built-in WinBUGS functions. The WinBUGS Differential Interface (WBDiff) allows one to solve systems of ordinary differential equations. For further details, see the WinBUGS Web site.

Additional R and WinBUGS examples are provided on the book's Web site; for example, for ring-recovery data (provided for the UK lapwing dataset). In some cases MCMC codes are written in a higher-level language (such as C/C++ or Fortran). The MCMC codes written in these higher-level languages are often significantly faster to run. However, to write programs in these languages requires additional coding skills, particularly without the standard built-in random number generator functions provided in R.

7.8 Exercises

7.1 Use the Help command in R to explore the use of the plot function for producing graphics (for example, determine what the arguments xlab, ylab, xlim, ylim, main, type and col do).

7.2 Use the help.search command to search for the autocorrelation function in R. Determine the command for outputting the values of the autocorrelation function to the screen for lags 1,...,40, and also for producing the corresponding autocorrelation function plot.

7.3 Write an MCMC code in R for ring-recovery data with general m-array, with constant survival and recovery probabilities, using a single-update MH algorithm.

   (a) Run the code for the lapwing data presented in Table 2.2. Plot the corresponding posterior density estimates.

   (b) Adapt the code, such that the parameters are simultaneously updated within each iteration of the MCMC algorithm (i.e. using a block update). Suggest an efficient proposal distribution for the block MH step (note that considering the posterior distribution obtained above using the single-update algorithm may be useful here). Assess the performance of the block updating algorithm with the single-update algorithm.

(c) For the lapwing data set, the birds are ringed as chicks, and an initial first-year survival probability is needed. Extend the (single-update MH) code to fit models with separate first-year and adult survival probabilities.

(d) Finally, extend the code such that the survival probabilities are time dependent. Discuss any issues in relation to parameter redundancy for this model.

7.4 Reconsider the R code for the dipper example in Section 7.1 for model C/C.

(a) Adapt the MCMC code to consider the model $C/C2$, as described in Section 3.3.1. As usual, use pilot tuning to determine values for the parameters in the proposal distributions.

(b) Extend the code to an RJMCMC algorithm where there are only two models C/C and $C/C2$, adding a subroutine to perform the reversible jump step (see Example 6.3), outputting the posterior model probabilities with corresponding parameter estimates for each model and making any other necessary modifications to the MCMC code.

(c) Consider the reversible jump step in more detail. Calculate the mean acceptance probability for moving between the two models. Perform a pilot tuning algorithm to find suitable proposal parameters in the reversible jump step. Suggest alternative reversible jump updating algorithms (in terms of proposal distributions used and/or bijective function for moving between the different models) that may improve the efficiency of the reversible jump step, and how this may be assessed.

7.5 Consider the R code for the white stork data in Section B.5 of Appendix B. Extend the model to allow for additional random effects on the survival probabilities, resulting in a mixed effects model of the form:

$$\text{logit } \phi_t = \beta_0 + \beta_1 w_t + \epsilon_t,$$

where, $\epsilon_t \sim N(0, \sigma^2)$ and,

$$\sigma^2 \sim \Gamma^{-1}(\alpha_\epsilon, \beta_\epsilon).$$

The corresponding priors for β_0, β_1 and constant recapture probability p remain unchanged.

(a) Draw the corresponding DAG for this model with a mixed effects model for the survival probability, and a constant recapture probability.

(b) Write down the posterior conditional distributions for each individual parameter, $p, \beta_0, \beta_1, \epsilon$ and σ^2.

(c) Describe an MCMC updating algorithm for obtaining a sample from the joint posterior distribution of all the parameters.

EXERCISES

(d) Extend the code provided in Section B.5.1 of Appendix B to include the additional random effects terms (i.e. for the mixed effects model).

(e) Perform an initial pilot tuning procedure to obtain efficient MCMC algorithm for the parameters. Describe a number of different methods for assessing the mixing of the Markov chain, and implement these within R for this example.

(f) Run the MCMC simulations and obtain posterior estimates of each of the parameters of interest (remembering to remove the burn-in). Comment on the results and any interesting conclusions that can be drawn.

(g) Write a function to calculate the BGR statistic for each of the parameters in the model given sample output from MCMC simulations (recall that multiple simulations are necessary to calculate this statistic).

7.6 Reconsider the WinBUGS code for the dipper example in Section 7.2 for model C/C.

(a) Adapt the model specification in Figure 7.7 to fit the model T/T to the data.

(b) Obtain summary statistics of the recapture and survival probabilities. Plot the posterior density for each of the parameters. Comment, particularly on the parameters p_7 and ϕ_6 (i.e. parameters p[6] and phi[6] in the WinBUGS code), and compare these with their prior distributions.

(c) Given that the probabilities p_7 and ϕ_6 can only be estimated as their product, since they are confounded in the likelihood, propose a modification of the WinBUGS code to estimate the product. Hint – consider setting one of the parameter values to a constant value. Comment on the posterior estimate of the product. Do the posterior estimates for the other parameters change? What happens if we change the constant from $p_7 = 1$ to $p_7 = 0.5$? What are the implications on the corresponding prior distribution for the survival probability, ϕ_6?

(d) Perform a prior sensitivity analysis on the results, using a variety of different priors. For example, compare the posterior means and variances for a number of different priors. You may like to consider, for example, $Beta(10, 1)$ and $Beta(1, 10)$ priors. Comment on the results.

(e) Amend the code to calculate the geometric mean of the survival probabilities for the first five years. This statistic is often useful to calculate, since it can be useful in summarising time dependent vital rates (such as survival probabilities) in one single quantity. Plot the posterior density of the geometric mean of the survival probabilities, and obtain posterior summary statistics, including the mean and 95% credible interval. Compare these with the geometric mean of 0.561, obtained in a classical analysis. Note that obtaining an estimate of the standard error, or 95% confidence interval in a classical analysis would involve using the delta method, whereas it is straightforward in a Bayesian analysis.

(f) Fit the models T/C (time varying survival, but constant recapture probability) and C/C (constant survival and recapture probabilities) to the data. Comment on the comparison of the posterior estimates with those obtained for the model T/T. From the results obtained, which model might fit best?

7.7 Use MARK to fit the model $C/C2$ for the dippers data, obtaining posterior estimates and using the in-built convergence diagnostic tools. Recall that the data are already included within MARK as an example, and are provided in the MARK input file DIPPERS.INP within the Examples folder contained within the MARK folder.

7.8 Reconsider Example 6.1 relating to the calculation of posterior model probabilities of four different models fitted to the dippers data. Within this example, Monte Carlo estimates of the posterior model probabilities are calculated by simulating parameter values from their prior distributions.

(a) Write an R code for calculating these posterior model probabilities as described in the given example and check that similar posterior model probabilities are obtained from the constructed code.

(b) Discuss the efficiency of this program; for example, consider the number of simulations needed to obtain a reliable estimate of the posterior model probabilities.

(c) Suggest alternative proposal distributions that may be used to improve the efficiency of the algorithm, using an importance sampling approach.

(d) Implement these suggestions by adapting the R code and compare their efficiency.

7.9 Consider the ring-recovery data from lapwings in the UK, presented in Table 2.2.

(a) Write WinBUGS code to fit a model assuming that the recovery probabilities are logistically regressed on time, while the first-year and adult survival probabilities are logistically regressed on the covariate fdays, which corresponds to the number of days that the mean temperature is below freezing in central England. Use standardised values for the covariate fdays (but not for time).

(b) Run the MCMC iterations for 10,000 iterations (this should take about 30 seconds). Remember to monitor the parameters of interest. Plot the history of the parameter values and suggest a suitable burn-in period for the MCMC iterations.

(c) Following a suitable burn-in period, plot the autocorrelation function for each of the parameters. Comment on the results.

(d) To reduce autocorrelation, standardising the covariates often helps. Alter the WinBUGS code to use a standardised time covariate for the recovery probabilities. Repeat the simulations and plot the autocorrelation function for the parameters. Comment on the results.

EXERCISES

(e) Rerun the MCMC algorithm, and calculate the BGR diagnostic (use at least 3 chains) recall that this needs to be specified prior to compiling the model.
(f) Obtain summary estimates of the parameters of interest (including the survival and recovery probabilities), and note the Monte Carlo error. Rerun the chain and compare the summary statistics. Comment.
(g) Plot the running mean and quantiles of the parameters. Do they appear to have converged? If so, by approximately which iteration? If not, run further simulations until they appear to have.
(h) Interpret the results biologically.
(i) Are the posterior estimates sensitive to the priors that are specified for the model? (Try using, for example, a hierarchical prior, i.e. the use of random effects.)

PART III

Ecological Applications

In this part of the book we apply the Bayesian methods previously described to a number of different case studies. We consider a wide variety of ecological problems and discuss and interpret the corresponding results obtained. These examples are more complex than those previously considered. However, within the Bayesian approach the analysis can typically be decomposed into a series of standard and easier problems which can then be recombined to obtain the estimates of the parameters of interest for the full, complex, data set. For example, increased complexity may be due to the incorporation of random effects, observation error, additional data and/or missing values. Note that adding in the complexity one level at a time can also significantly simplify the coding of the (RJ)MCMC algorithms. For each of the following case studies, we focus on the Bayesian methods used, the details of implementation, and the interpretation of the results, rather than the more technical Bayesian issues such as convergence diagnostics, the calculation of Monte Carlo errors and Bayesian p-values, all of which were covered in Part II of the book.

CHAPTER 8

Covariates, Missing Values and Random Effects

8.1 Introduction

As we have seen in Chapters 2 and 3, demographic parameters of interest often exhibit temporal variability. For instance, for the analysis of the dippers data in Example 6.1 there was evidence that the survival probability depended on the year of the study (flood versus non-flood years). Modelling the survival probabilities with an arbitrary time dependence, however, does not provide any information as to an underlying cause for the temporal heterogeneity. In addition, for long-term studies, over a period of 20 or 30 years, say, this creates a potentially very large number of parameters within the model, possibly creating over-fitting. An alternative approach that addresses these two issues is to consider the use of covariates to explain the temporal variation in the demographic parameters. For example, for the dippers example, we considered two different survival probabilities: one for the years that surrounded a flood (ϕ_f), and a different survival probability for the other years (ϕ_n). Here, the covariate that explains the temporal variability is simply an environmental factor indicating whether there was a flood or not.

Typically, more complex environmental factors, rather than a simple indicator function, are considered to explain the temporal variability in demographic parameters. For example, these may include weather conditions, resource availability, predator abundance, human intervention or habitat type. Biological knowledge can be used to determine the factors influencing the system under study. We have already met this idea in Example 4.11 where we consider the capture-recapture data relating to UK lapwings and logistically regress the survival probabilities on a weather covariate and the recovery probability on time. Alternatively, in Example 5.7, we consider capture-recapture data of white storks and logistically regress the survival probabilities on the rainfall at a given weather station. We extend these ideas in this chapter, discussing in greater detail the use of covariates, and consider in depth additional issues such as missing data and prior specification with particular reference to model selection.

Environmental factors are typically assumed to affect all individuals in the same way. In other words, the survival probabilities of individuals may vary over time, for example, as a function of given environmental covariates, but the survival probabilities do not differ between individuals. Alternatively there may be individual heterogeneity, where the demographic parameters differ between individual animals. In the same way that we can explain temporal variability by covariates, it is also sometimes possible to explain individual heterogeneity by covariates. We have already met this idea in Example 2.4, where we admit different survival probabilities for animals of different ages, to incorporate the fact that individuals typically have a higher mortality rate in their first year of life, and/or in later life. This idea can be extended to allow for any form of individual covariate, such as weight/body mass index (which could be regarded as a surrogate for condition), breeding status or location. Individual covariate information is typically difficult to collect, and often restricted by both financial and time constraints. There are two additional issues that often arise when incorporating covariate information into the statistical analysis, and these are missing data (see Section 8.3) and model discrimination, which has been already discussed in Chapter 6 and is revisited in Section 8.4, with particular reference to variable selection.

Given the covariates that are thought to affect the survival probabilities, there can still be additional variability not explained. For example, there may be an unknown additional covariate that is needed to explain temporal heterogeneity, such as predator abundance, for which there may be no information. Alternatively, there may be some overdispersion in the population that is not incorporated into the modelling process. In these circumstances it is possible to incorporate this further uncertainty by the inclusion of random effects. This essentially allows for additional stochasticity, and will also provide an indication of the amount of variability within the survival probabilities explained by the covariates being studied (see Section 8.5).

8.2 Covariates

We shall consider environmental and individual covariates in turn. The underlying concepts relating to these covariates are essentially identical, but there are also some differences that we shall explain. We use the term *environmental covariates* as an all encompassing term for *global covariates*, which are the same for all individuals in the study, but vary over time. Typically these will refer to environmental factors, such as weather conditions or resource availability, as we have already seen in Example 4.11 using weather covariates to explain temporal heterogeneity for the survival probabilities. However, the term is more general, and could include, for example, human intervention in the form of culling strategies or quantity of hunting licenses.

8.2.1 Environmental Covariates

As we have seen in Examples 4.11 and 5.7, we can use environmental covariates to explain temporal heterogeneity of demographic parameters. In general, the number of covariates is assumed to be equal to $n \geq 1$. We observe covariate values over time for covariate i, and denote the set of observed covariate values by $\boldsymbol{y}^i = \{\boldsymbol{y}^i_t : t = 1, \ldots, T\}$. We assume here that the covariate values

COVARIATES

are continuous variables and discuss the modelling of categorical covariates when we consider individual covariates. Note that the covariates $y^1, \ldots, y^n$ should be normalised over the observed time interval for each covariate for interpretability. We assume (for now) that the survival probabilities can be expressed as functions of these covariates and that there is no additional heterogeneity present, although in general, we may have other demographic parameters (such as recapture probabilities or productivity rates) as a function of covariates. All individuals have the same survival probabilities, but these will vary over time due to dependence on the covariate values. Thus, we need to specify the survival probabilities as a function of the covariates of interest. The survival probability is constrained to take values in the interval $[0, 1]$. Thus, a typical formulation for the covariate dependence is to consider a linear relationship between the survival probability and environmental covariates for an appropriate link function. To incorporate the constraint on the survival probability, we consider the logit link function. In other words, we specify the relationship between the survival probability, ϕ_t, and the environmental covariates, $\boldsymbol{y}_t$, at time t to be of the form of a logistic regression,

$$\text{logit } \phi_t = \log\left(\frac{\phi_t}{1 - \phi_t}\right) = \beta_0 + \boldsymbol{\beta}^T \boldsymbol{y}_t,$$

where $\boldsymbol{\beta}^T = (\beta_1, \ldots, \beta_n)$ denotes the $1 \times n$ vector of regression coefficients for the environmental covariates and $\boldsymbol{y}_t = (y_t^1, \ldots, y_t^n)^T$, the $n \times 1$ vector of covariate values at time t. Thus, the parameters to be estimated within this model are β_0 and $\boldsymbol{\beta}$. Recall that we can then express the survival probabilities as,

$$\phi_t = \frac{1}{1 + \exp(-\beta_0 - \boldsymbol{\beta}^T \boldsymbol{y}_t)}.$$

Clearly, we have $\phi_t \in [0, 1]$.

The constraint that the parameter value is within the interval $[0, 1]$ does not arise for all parameters of interest. For example, suppose that we wish to use covariates $\boldsymbol{y}$ to explain temporal variability within productivity rates, denoted by $\boldsymbol{\rho} = \{\rho_1, \ldots, \rho_T\}$. The productivity rate is constrained to be positive, so we could use a log link function, and specify,

$$\log \rho_t = \beta_0 + \boldsymbol{\beta}^T \boldsymbol{y}_t.$$

Clearly, this idea can be extended to a non-linear relationship on the link function, if there is biological evidence of a more complex relationship, such as a quadratic form for the covariates. Alternatively, interactions may be present between the different covariates. These can be simply added to the above model in the standard way. For example, suppose that the biological model we wish to represent has an interaction between covariates $\boldsymbol{y}^1$ and $\boldsymbol{y}^2$. We define the interaction term $\boldsymbol{y}^1 \times \boldsymbol{y}^2 = \{y_1^1 \times y_1^2, \ldots, y_T^1 \times y_T^2\}$, and simply add this term within the regression equation. Typically, only hierarchical models are considered, so that if an interaction is present, then the individual covariates are also present in the model. This implies that if an interaction $\boldsymbol{y}^1 \times \boldsymbol{y}^2$ term is present, the covariates $\boldsymbol{y}^1$ and $\boldsymbol{y}^2$ are also present in the model. Finally, we note that we assume that all the covariate values are observed, which is typically the case. We address the additional problem of missing covariate values in Section 8.3.

The use of covariates to explain temporal heterogeneity can be used for all types of data. A negative/positive regression coefficient, denotes a negative/positive relationship between the demographic parameter and the covariate. In addition, (since all the covariate values are normalised prior to model fitting), the magnitude of the covariate regression parameters can be compared and/or ranked. In this way, the importance of the covariates can be compared: covariates with a regression coefficient of larger magnitude would reflect a larger impact on the demographic parameter. Finally, we note that the survival probabilities are simply a deterministic function of the covariate values and regression coefficients. Since the environmental factors are specified such that the demographic parameters are the same for all individuals (i.e. there is no individual heterogeneity, and individuals are effectively interchangeable), the corresponding data can be represented by a standard m-array and the standard likelihood expressions used (see for example Section 2.3).

Bayesian Framework

The Bayesian analysis follows the standard framework. We specify priors on all of the parameters in the model. This will correspond to the regression coefficients and any additional parameters. For example, suppose that we have capture-recapture data, $\boldsymbol{x}$, where the survival probabilities are specified as a function of covariates. The set of parameters are the regression coefficients, $\boldsymbol{\beta}_0 = \{\beta_0, \ldots, \beta_n\}$ and recapture probabilities $\boldsymbol{p} = \{p_2, \ldots, p_T\}$ (assuming time dependent recapture probabilities). Using Bayes' Theorem, we have,

$$\pi(\boldsymbol{\theta}|\boldsymbol{x}) \propto f(\boldsymbol{x}|\boldsymbol{\theta})p(\boldsymbol{\theta}),$$

where $\boldsymbol{\theta} = \{\boldsymbol{\beta}_0, \boldsymbol{p}\}$. The likelihood is a standard function of the survival and recapture probabilities, as given in Equation (2.5). This leaves the prior on the parameters to be defined. Without any prior information, we could specify independent priors on each of the parameters, so that,

$$p(\boldsymbol{\theta}) = \prod_{i=2}^{T} p(p_i) \prod_{i=0}^{n} p(\beta_i).$$

Typically, since the recapture probabilities are constrained to the interval $[0, 1]$, we would specify the priors on the recapture probabilities to be of the form,

$$p_i \sim Beta(a_i, b_i), \qquad \text{for } i = 2, \ldots, T.$$

Alternatively, if there is prior information relating to a dependence between the parameters, this can be incorporated by specifying the joint prior distribution over these parameters, removing the independence assumption. As-

COVARIATES

suming that we have no prior information, we might specify $a_i = b_i = 1$ for $i = 2, \ldots, T$, i.e. $p_i \sim U[0, 1]$.

What about the priors on the regression coefficients? Let us consider the logistic link function for the survival probabilities. The regression coefficients can take any real value, i.e. they are not constrained. To allow for this, we could consider,

$$\boldsymbol{\beta} \sim \mathcal{N}_{n+1}(\boldsymbol{\mu}, \mathrm{diag}(\sigma_0^2, \ldots, \sigma_n^2)),$$

where $\mathcal{N}_{n+1}$ denotes the multivariate normal distribution of dimension $n+1$; $\boldsymbol{\mu} = (\mu_0, \ldots, \mu_n)^T$; and $\mathrm{diag}(\sigma_0^2, \ldots, \sigma_n^2)$ denotes the diagonal matrix with i^{th} entry equal to σ_i^2 for $i = 0, \ldots, n$, and all other (off-diagonal) entries are equal to zero, corresponding to the assumption of prior independence.

Uninformative Prior Specification

Without any strong prior information, a typical prior mean for β_0 would be to set $\mu_0 = 0$. Similarly, if there is no prior information corresponding to the sign of the regression coefficients (i.e. a positive or negative relationship between the corresponding covariates and survival probability), then it would be natural to take $\mu_1 = \cdots = \mu_n = 0$. Due to the symmetry properties of the normal distribution, this corresponds to a prior probability of 0.5 that each covariate has a positive or negative relationship with the demographic parameter. The corresponding prior on ϕ_t has a prior mean of 0.5, i.e. $\mathbb{E}_p(\phi_t) = 0.5$.

The final specification of the prior relates to $\boldsymbol{\sigma}^2 = \{\sigma_0^2, \ldots, \sigma_n^2\}$, and this is more complex. With no prior information on the magnitude of the regression coefficients, one obvious approach would be to specify a large value for each σ_i^2, for $i = 0, \ldots, n$. However, this prior does not correspond to an uninformative prior on the survival probability – quite the reverse. As the prior variances on the regression coefficients increase (i.e. $\sigma_i^2 \to \infty$), the corresponding prior on ϕ_t tends to point masses on 0 and 1. Consider the logistic relationship between the survival probability and the regression coefficients. In the model where there are no covariates, we have logit $\phi_t = \beta_0$. Figure 8.1 plots ϕ_t as a function of β_0.

Clearly, we can see that there is a sigmoidal relationship, with relatively little change in ϕ_t for values of $|\beta_0| > 3$. Alternatively, $-3 < \beta_0 < 3$ corresponds to $0.05 < \phi_t < 0.95$ (to 2 decimal places). Thus, we could consider specifying the value of σ_0^2 such that the corresponding prior had a 90% credible interval for β_0 of $[-3, 3]$ (i.e. $\sigma_0 = 1.531$, or $\sigma_0^2 = 2.344$). This ignores the covariate dependence, but we could equivalently interpret this prior as the prior on the survival probability corresponding to an "average" year, where all covariate values take their mean value of zero (recall that we normalise the covariate values over time). We also need to specify the prior variances $\sigma_1^2, \ldots, \sigma_n^2$. We would typically consider the prior variances to be similar to that for σ_0^2. Clearly, for any particular set of covariate values, we could plot the corresponding prior on the survival probability.

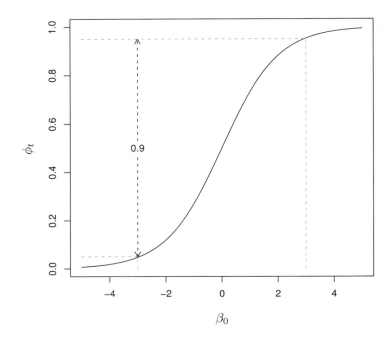

Figure 8.1 The logistic relationship of ϕ_t against β_0 for the function logit $\phi_t = \beta_0$.

We note that within the computer package MARK, the default prior standard deviation $\sigma_0 = 1.75$ (or $\sigma_0^2 = 3.0625$) is used to produce an approximately $U[0,1]$ prior on ϕ_t. Newman (2003) provides a slightly more advanced prior specification, along similar lines. In particular, suppose that there are a total of n covariates on which a parameter is regressed. Assuming, as usual, that the covariate values are normalised, and matching the first two moments of the induced prior on ϕ_t with a $U[0,1]$ distribution, Newman (2003) obtains the proposed prior,

$$\boldsymbol{\beta} \sim \mathcal{N}_{n+1}\left(\mathbf{0}, \frac{\pi^2}{3(n+1)}I\right).$$

So, for example, when $n = 0$, we obtain a prior variance for β_0 of $\sigma_0^2 = 3.29$ (or $\sigma_0 = 1.81$). This is similar to that obtained above using an interval to obtain the prior variance, rather than matching the moments. However, this approach does consider the inclusion of the covariates within the logistic regression expression more explicitly.

Within any Bayesian analysis, a prior sensitivity analysis should be performed, which should identify any sensitivity corresponding to the specification of the σ^2 values. If the posterior results are sensitive to these values, and there is no prior information available, one possible approach would be to consider a hierarchical prior, as discussed in Section 4.2.4. For example, suppose that the posterior distribution of β_j appeared to be sensitive to the prior value of σ_j^2. Then we could specify, $\sigma_j^2 \sim \Gamma^{-1}(a, b)$, the conjugate prior for σ_j^2. The choice of values for a and b should be suitably vague to cover the set of feasible values for the prior variance.

Informative Priors

Conversely, suppose that there is some prior expert information that we wish to incorporate into the prior. For example, this could be as a result of previous analyses of data relating to the population (or similar species) or as a result of experience of similar biological systems. The form of the prior information could, for example, correspond to the sign of the regression coefficients. This can be easily incorporated into the analysis. For instance, suppose that there is prior information for covariate 1, relating to the sign of the coefficient (i.e. positive or negative). Without loss of generality, suppose that the prior information corresponded to a positive relationship. We might then specify,

$$\beta_1 \sim HN^+(0, \sigma_1^2),$$

where HN^+ corresponds to a positive half-normal distribution. However, care should be taken in specifying such a strong relationship, and it should only be used when it is biologically *impossible* for a negative relationship to exist (which is the interpretation of this half-normal prior). An alternative approach would be to specify a non-zero prior mean, μ_1, for β_1. However, perhaps a more elegant and directly interpretable prior would be to consider a mixture model for β_1 of the form,

$$\beta_1 \sim p HN^+(0, \sigma_1^2) + (1-p) HN^-(0, \sigma_1^2),$$

where p denotes the prior probability that β_1 is positive and HN^- denotes the negative half-normal distribution. The expert can provide the value of p, corresponding to the strength of the prior beliefs that the relationship is positive. However, this prior structure still allows the data to provide evidence that the relationship is positive, when $p < 1$. The above cases are then special cases of this structure. For example, setting $p = 0$ (or 1), reduces this prior to the simple negative (or positive) half-normal case; whereas setting $p = 0.5$, reduces to the standard normal prior given earlier. Once more a prior sensitivity analysis should be performed to test the influence of the prior specification on the posterior results, particularly in the absence of any strong prior information.

We now consider an example in more detail, where we assume a linear logistic regression for the survival probabilities, where there are no interactions between the covariates. We return to the lapwing data set.

248   COVARIATES, MISSING VALUES AND RANDOM EFFECTS

Example 8.1. Lapwings
The data are of the form of ring-recovery data from 1963 to 1998. The corresponding likelihood and further description of this type of data are described in Section 2.3.1. To explain the temporal variability of the demographic parameters we consider two different covariates: t corresponding to time; and $fdays$ denoting the number of days that the minimum temperature fell below freezing in central England during the winter leading up to the breeding season in year t. These are available from:

http://www.badc.rl.ac.uk

Here, $fdays$ can be regarded as a (crude) measure of the harshness of the British winter, which is appropriate for birds that are ringed and die over a wide region. We consider the model:

$$\phi_1(fdays); \phi_a(fdays)/\lambda(t).$$

Each of the parameters are logistically regressed on the given covariates. So that, we have,

$$\text{logit } \phi_1 = \beta_0^1 + \beta_1^1 f_t;$$
$$\text{logit } \phi_a = \beta_0^a + \beta_1^a f_t;$$
$$\text{logit } \lambda = \beta_0^\lambda + \beta_1^\lambda u_t,$$

where as usual, f_t and T_t denote the normalised values for the covariates $fdays$ and time, respectively.

We have no prior information, and so we shall use vague priors. In particular, we set,

$$\boldsymbol{\beta}^1 \sim \mathcal{N}_2(\mathbf{0}, 2.334I)$$
$$\boldsymbol{\beta}^a \sim \mathcal{N}_2(\mathbf{0}, 2.334I)$$
$$\boldsymbol{\beta}^\lambda \sim \mathcal{N}_2(\mathbf{0}, 2.334I)$$

where $\boldsymbol{\beta}^1$ denotes the regression coefficients for ϕ_1, $\boldsymbol{\beta}^a$ the coefficients for ϕ_a and $\boldsymbol{\beta}^\lambda$ the coefficients for λ.

The simulations are run for a total of 1 million iterations, with the first 10% discarded as burn-in. The trace plots of the parameters suggested that this burn-in is very conservative. The corresponding posterior estimates for the regression coefficients are given in Table 8.1 and the resulting posterior mean and 95% HPDI of the survival and recovery probabilities are given in Figure 8.2. Clearly we can see that the first-year survival probabilities are significantly lower than those for the adult birds. In addition, the impact of the covariate $fdays$ on the survival probability is clearly demonstrated, with a negative regression coefficient (or slope) for both first years and adults. In other words, when $fdays$ is large (i.e. it is a harsh winter) there is a decrease in the survival probabilities of the birds. Alternatively, it is clear that as time progresses, the

COVARIATES

Table 8.1 The Posterior Mean and Standard Deviation for the Regression Coefficients for the Survival and Recovery Probabilities for the Lapwing Ring-Recovery Data, under the Model $\phi_1(fdays); \phi_a(fdays)/\lambda(t)$

	Parameter	Posterior Mean	Posterior Standard Deviation
ϕ_1	Intercept	0.535	0.069
	Slope ($fdays$)	−0.208	0.062
ϕ_a	Intercept	1.529	0.070
	Slope ($fdays$)	−0.310	0.044
λ	Intercept	−4.566	0.035
	Slope (t)	−0.346	0.039

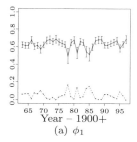
(a) ϕ_1

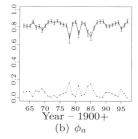
(b) ϕ_a

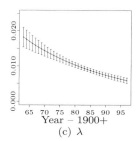
(c) λ

Figure 8.2 The posterior estimates of (a) first-year survival probability, (b) adult survival probability and (c) recovery probability for the lapwing ring-recovery data under model $\phi_1(fdays); \phi_a(fdays)/\lambda(t)$. Also shown are 95% HPDIs. The lower curve in (a) and (b) gives the corresponding $fdays$ covariate values.

recovery probability declines, i.e. there is also a negative relationship between the recovery probability and time.

A prior sensitivity analysis is conducted, changing the prior variances for the regression coefficients of each of the parameters. For example, we consider specifying,

$$\beta^1 \sim \mathcal{N}_2(\mathbf{0}, 10I)$$
$$\beta^a \sim \mathcal{N}_2(\mathbf{0}, 10I)$$
$$\beta^\lambda \sim \mathcal{N}_2(\mathbf{0}, 10I).$$

The increasingly diffuse priors produced essentially identical results for the

8.2.2 Individual Covariates

The previous environmental (or global) covariates may be used to explain temporal heterogeneity. However, there may also be information corresponding to individual-level covariates on the animals in the study, which may affect the demographic parameters at the individual level. For example, there could be a trade-off between the breeding effort of an individual, and their corresponding survival probability – an animal that expends a lot of effort into breeding in a given year may have an increased mortality. We shall consider the particular application of capture-recapture, since individual covariates are often collected at the tagging event and at successive recapture/resightings. Typically, collecting individual-level covariate information can be very time consuming, and often difficult. We distinguish between two different types of covariate information: time invariant and time varying. For example, time invariant individual covariates may include birth weight, parentage and genotype/phenotype information. Alternatively, time varying individual covariates would include weight, wing length, location and breeding effort. Clearly, we could also have additional environmental covariates, and these would simply be added within the logistic regression framework in the standard way. However, for simplicity, we shall simply assume that there are only individual covariates present with the obvious extension for when both environmental and individual covariates feature.

Modelling can continue in the analogous way as for the environmental covariates, but now we must consider the survival probability of each individual animal. We consider the extension to the case where the covariates may be either continuous or categorical (i.e. discrete) variables. For example, suppose that there are r continuous individual covariates and let $\boldsymbol{y}_{it} = \{y_{it}^1, \ldots, y_{it}^r\}$ denote the set of continuous individual covariate values for individual i at time $t = 1, \ldots, \tau$. Similarly, suppose that there are s categorical individual covariates taking values $\boldsymbol{z}_{it} = \{z_{it}^1, \ldots, z_{it}^s\}$ at time t for individual i. We specify the corresponding survival probability of individual i at time t, denoted by ϕ_{it}, as a logistic regression on the individual covariate values:

$$\text{logit } \phi_{it} = \log\left(\frac{\phi_{it}}{1-\phi_{it}}\right) = \gamma_0 + \boldsymbol{\gamma}^T \boldsymbol{y}_{it} + \delta_{it},$$

where γ_0 denotes the intercept term; $\boldsymbol{\gamma}^T = (\gamma_1, \ldots, \gamma_r)$ denotes the corresponding $1 \times r$ vector of regression coefficients; and

$$\delta_{it} = \sum_{k=1}^{s} \delta_{kz_{it}^k},$$

corresponds to the effect of the categorical covariates on the survival proba-

MISSING VALUES

bilities. For identifiability, we need to specify a constraint on the terms corresponding to the categorical covariates, $\delta_{kz_{it}^k}$ for $k = 1, \ldots, s$. Without loss of generality, suppose that $z_{it}^k \in \{0, 1, \ldots, n-1\}$, i.e. the kth covariate can take n discrete values, which we simply label from 0 to $n-1$. One possible constraint is to simply set $\delta_{k0} = 0$ for $k = 1, \ldots, s$, and this is the constraint that we shall adopt throughout. Effectively, the δ_{kj} values correspond to the difference in the survival probabilities (on the logistic scale) for an individual with a value of j for categorical variable k, compared to an individual that has a covariate value of 0 for covariate k.

Likelihood Function

Calculating the likelihood function is typically more complex in the presence of individual covariates, since the survival probabilities of each individual sheep will differ. Thus, we need to calculate the likelihood using the full expression:

$$f(\boldsymbol{x}|\boldsymbol{\theta}) = \prod_{i=1}^{I} \mathbb{P}(\text{capture history of individual } i|\boldsymbol{\theta}),$$

where the data, $\boldsymbol{x}$, is simply the set of capture histories and $\boldsymbol{\theta}$ denotes the set of parameters in the model. This introduces an additional computational expense in the calculation of the likelihood. However, there is another more important issue that arises here, which is the reason we have discriminated between time invariant and time varying covariates. Often within the data collection process it is possible to collect all time invariant individual covariates the first time that an individual is seen within the study; i.e. when it is tagged. (This may not be the case for physical characteristics, for example, that may be difficult to distinguish if tagging occurs for very young individuals.) Alternatively, consider time varying individual covariates, and an individual animal with capture history:

$$1\ 1\ 0\ 1\ 1\ 1\ 1.$$

What is the corresponding contribution to the likelihood for this individual animal? It will be the probability of this capture history, given the set of parameter values (and covariate values). However, we do not observe the individual at time 3, and so we do not know the corresponding value for the time varying covariate at that time. Since we do not know the corresponding covariate value(s), we cannot calculate the survival probability, and hence cannot explicitly calculate the probability for the capture history. In other words, we have a missing data (or value) problem. We shall now discuss this in more detail, and in particular how the Bayesian approach can deal with this problem, using a simple extension to the standard analysis.

8.3 Missing Values

In the presence of missing covariate values, the corresponding likelihood is generally analytically intractable. There are several approaches to dealing with such missing data. For example, in the simplest of cases, where there are only time invariant missing covariates on a relatively small proportion of the data, one approach is to simply ignore those animals, for which there are missing values (Catchpole et al. 2000). This approach can substantially reduce the size of the data set, and ignores any information contained in the data relating to these animals. In addition, this approach cannot generally be implemented when there are missing time varying covariates, since there are typically very few animals (if any) for which all covariate values are known at every capture event. An alternative approach, within a classical analysis, is to use the EM algorithm (Dempster et al. 1977). This algorithm converges to the MLE of the parameter values, given the observed data, but can be computationally very expensive, even for a moderate amount of missing parameter values. Another avenue is to impute the missing covariate values. For example, an underlying model could be specified for the covariates (such as a Gompertz curve), and the missing values simply imputed given this model. However, this approach treats the unknown values as if they were observed at the imputed values, and will underestimate parameter uncertainty. Additionally, specifying a model that does not fit the data very well could severely bias the resulting parameter estimates. In the case of time varying covariates, a similar approach would be to use a *last observation carry forward* method, where we simply impute the missing time varying covariates to be their last observed value. An extension of this approach is a combination of a linear imputation approach (linearly imputing missing covariate values between consecutively observed covariate values) and a last observation carry forward approach (following the final observed covariate value). However, this approach again does not take into account the uncertainty in these missing covariate values, and has been shown to perform poorly in a simulation study when the underlying model for the covariate values are different to the imputed model (Bonner et al. 2009). Catchpole et al. (2008) present an alternative classical approach using a conditional analysis, which does not specify an underlying model for the covariates, but simply conditions on the individuals that are observed at each capture event for which the corresponding covariate values are known.

We present a Bayesian approach where we assume an underlying model for the missing data and which allows us to account for the corresponding uncertainty of the missing values. To describe the approach, we first introduce some further notation. Let the observed capture histories be denoted by $\boldsymbol{h}$ and the observed covariate values by $\boldsymbol{c}_{obs}$. Similarly, let the missing covariate values be denoted by $\boldsymbol{c}_{mis}$. We treat the missing covariate values as parameters, or auxiliary variables, to be estimated. The complete set of parameter values (recapture probabilities, recovery probabilities and survival regression parameters) is denoted by $\boldsymbol{\theta}$. Using Bayes' Theorem, we can express the joint posterior distribution of the parameters $\boldsymbol{\theta}$ and missing covariate values $\boldsymbol{c}_{mis}$, given the capture histories and observed covariate values in the form,

$$\pi(\boldsymbol{\theta}, \boldsymbol{c}_{mis} | \boldsymbol{h}, \boldsymbol{c}_{obs}) \quad \propto \quad f(\boldsymbol{h} | \boldsymbol{c}_{obs}, \boldsymbol{c}_{mis}, \boldsymbol{\theta}) f(\boldsymbol{c}_{obs}, \boldsymbol{c}_{mis} | \boldsymbol{\theta}) p(\boldsymbol{\theta}).$$

MISSING VALUES

The likelihood can be expressed as a product of two terms: the first term, $f(\boldsymbol{h}|\boldsymbol{c}_{obs},\boldsymbol{c}_{mis},\boldsymbol{\theta})$, relates to the likelihood of the capture histories of all the individuals observed in the study given the parameters and all covariate values (known and unknown) and so can be explicitly calculated in the standard way; the second, $f(\boldsymbol{c}_{obs},\boldsymbol{c}_{mis}|\boldsymbol{\theta})$, relates to the underlying model of the covariate values. Thus, we need to specify an underlying model for the covariate values. The function $f(\boldsymbol{c}_{obs},\boldsymbol{c}_{mis}|\boldsymbol{\theta})$ can be thought of as either a *likelihood* term or alternatively as a prior on the underlying model for the covariates. Typically these underlying models for the covariates will be obtained via discussions with ecologists. Assuming that we know the form of this model for each covariate, then we can use an MCMC algorithm to explore the joint posterior distribution and obtain summary estimates of the marginal distribution of interest (see Section 5.4.1),

$$\pi(\boldsymbol{\theta}|\boldsymbol{h},\boldsymbol{c}_{obs}) = \int \pi(\boldsymbol{\theta},\boldsymbol{c}_{mis}|\boldsymbol{h},\boldsymbol{c}_{obs})d\boldsymbol{c}_{mis}.$$

In other words, we take the marginal posterior distribution for $\boldsymbol{\theta}$, integrating out the missing covariate values, $\boldsymbol{c}_{mis}$.

To illustrate this approach we consider a particular example.

Example 8.2. Soay Sheep

We consider the mark-recapture-recovery data from 1986 through 2000 collected via a summer survey for the Soay sheep data set (see Example 1.4), where there is interest in the individual heterogeneity of the survival probabilities. We initially consider the three time invariant covariates, before considering a time varying covariate (namely weight), which will typically have a more complex model structure. In particular, we consider the time invariant covariates:

1. coat type (2 categories);
2. horn type (3 categories); and
3. birth weight.

Here, coat type and horn type are categorical variables, whereas birth weight is continuous (with a lower bound at zero on the raw measurement scale). As usual, we normalise the covariate values for birth weight over all observed values. In some circumstances, we may wish to assume that the covariate values are correlated. In this example, we initially assume that the underlying model specified for each covariate is independent, as there is no information to the contrary, but we also describe how we might incorporate this further complexity for categorical data. We consider possible biological models for each of these covariates in turn.

Coat Type

Let z_i^1 denote the coat type for individual i. We have $z_i^1 \in \{0,1\}$ for all individuals i. In this case it is biologically plausible to assume that each individual has coat type j with probability q_j, for $j = 0, 1$, independently of each other, where $q_0 + q_1 = 1$. In other words,

$$z_i^1 \sim Bernoulli(q_0).$$

However, we do not know q_0, a priori, and so we treat this as an additional parameter to be estimated within the model, and for which we need to specify a prior. In the absence of any strong prior information, a typical prior might be to assume a flat prior, and specify,

$$q_0 \sim U[0,1] \equiv Beta(1,1).$$

For notational purposes, we absorb this additional parameter into the set of model parameters denoted by $\boldsymbol{\theta}$.

If we let $\boldsymbol{z}_{obs}^1$ and $\boldsymbol{z}_{mis}^1$ denote the corresponding set of observed and missing coat types for each individual in the study, we have,

$$\begin{aligned} f(\boldsymbol{z}_{obs}^1, \boldsymbol{z}_{mis}^1 | \boldsymbol{\theta}) &= \prod_{i=1}^{I} q_0^{z_i^1} (1-q_0)^{1-z_i^1} \\ &= q_0^{n_0} (1-q_0)^{I-n_0}, \end{aligned}$$

where n_0 denotes the total number of individuals (either observed with or imputed to have) coat type 0. We are imputing the missing covariate values within the MCMC algorithm, so that this function can be easily evaluated.

Horn Type

Let z_i^2 denote the horn type of individual i, where $z_i^2 \in \{0, 1, 2\}$. Assuming that the distribution of horn type is homogeneous within the population, we can simply extend the above case for coat type where there are more possible categories. In particular, we have that each individual has horn type $k \in \{0, 1, 2\}$ with probability w_k. In other words,

$$z_i^2 \sim Multinomial(1, \boldsymbol{w}),$$

where $\boldsymbol{w} = \{w_1, w_2, w_3\}$ are parameters to be estimated, such that $w_1 + w_2 + w_3 = 1$, and the index of the multinomial distribution is simply 1. Then once more we need to specify a prior for $\boldsymbol{w}$. A natural prior to take is a Dirichlet prior, which is defined over the simplex, and hence constrains the parameters to sum to unity. If we wish to specify a flat prior, we could specify $\boldsymbol{w} \sim Dir(1,1,1)$. We note that the $Beta(1,1)$ distribution specified for the coat type is simply the special case of this Dirichlet distribution where there are only two possible categories for the covariate. The corresponding joint probability of the covariate values is directly analogous as the above case and hence omitted.

Note, we have assumed that the coat type and horn type are uncorrelated. However, it is easy to remove this assumption. In this case there are a total of 2×3 possible combinations of coat type and horn type. We can consider the joint probability of an individual having coat type j and horn type k, with

MISSING VALUES

probability u_{jk}, such that,

$$\sum_{j=0}^{1}\sum_{k=0}^{2} u_{jk} = 1.$$

We then specify,

$$(z_i^1, z_i^2) \sim Multinomial(1, \boldsymbol{u}),$$

where $\boldsymbol{u} = \{u_{jk} : j = 0, 1; k = 0, 1, 2\}$. This extends in the obvious manner for additional covariates or for a larger number of categories.

Birth Weight

Let y_i^1 denote the normalised birth weight for individual i. A number of possible models could be specified for birth weight. For example, perhaps the simplest model would be to assume that,

$$y_i^1 \sim N(\mu, \sigma^2),$$

where μ and σ^2 are again parameters to be estimated. The biological interpretation of this model would be that the birth weight of an individual is normally distributed about some unknown mean μ, which is constant over time, and with some variance σ^2. Letting $\boldsymbol{y}_{obs}^1$ and $\boldsymbol{y}_{mis}^1$ denote the set of observed and missing birth weights, respectively, the corresponding joint probability over all birth weight covariates (those observed and imputed within the MCMC algorithm) is simply a product over normal density functions:

$$f(\boldsymbol{y}_{obs}^1, \boldsymbol{y}_{mis}^1 | \boldsymbol{\theta}) = (2\pi\sigma^2)^{-\frac{I}{2}} \exp\left(-\frac{\sum_{i=1}^{I}(y_i^1 - \mu)^2}{2\sigma^2}\right),$$

where once again we absorb the parameters μ and σ^2 into the full set of parameters $\boldsymbol{\theta}$.

However, this may not be biologically plausible for many systems, for example, since it may be expected that the underlying mean birth weight will be time dependent (or dependent on some given environmental covariates), due to environmental conditions. For example, an alternative model that could represent this would simply be,

$$y_i^1 \sim N(\mu_t, \sigma^2),$$

where individual i is born in year t, so that there is a different underlying mean birth weight dependent on the year of birth. Note that these normal distributions are actually be truncated to ensure that the birth weight cannot be negative (recall y_i^1 is the normalised birth weight). The model should be specified using biological knowledge of the system, where possible. For this illustrative example, we consider the first model for simplicity, but return to the issue of model specification in Section 8.3.1. For this underlying model for the birth weight, we specify the priors, $\mu \sim N(0,1)$ and $\sigma^2 \sim \Gamma^{-1}(0.001, 0.001)$.

It is once again fairly trivial to introduce a correlated structure for continuous covariate values. For example, an obvious approach will be to consider a multivariate normal distribution for the covariate values, with some given covariate matrix, Σ. The simple model corresponding to uncorrelated covariate values is the special case where Σ is a diagonal matrix.

We note that it is also possible to specify a joint model for both categorical and continuous covariates to represent correlated covariate values. For example, by specifying a different underlying model for the continuous covariate values for each possible discrete covariate value. Once again the model representing independence between these covariate values is a special case, where the continuous distribution specified for each covariate value is identical, with common parameter values.

Weight

Typically time varying covariates may have additional structure to be modelled in order to incorporate temporal dependence. For example, it is very plausible that the weight of an individual sheep will be dependent on its corresponding weight in the previous year. Additionally, it is plausible that the age of the individual may also be a relevant factor, as it might be expected that an individual sheep may gain weight in the first few years of life as it matures, before reaching a stable weight. Alternatively, as a consequence of environmental factors, weight may be dependent on the year of the study. A general model to allow for these possibilities would be the following.

Let $y_{i,t}^2$ denote the weight of individual i at time t that is aged a. To incorporate the above assumptions, a typical model would be,

$$y_{i,t}^2 \sim N(y_{i,t-1}^2 + \kappa_a + \nu_t, \tau),$$

where the parameters κ_a, ν_t and τ are to be estimated. Thus, the underlying mean component for the weight of an individual animal is composed of a linear combination of the weight of the individual in the previous year (or birth weight for an individual aged $a = 1$, so that in this case, $y_{i,t-1}^2 = y_i^1$) and terms relating to the age of the animal and the year of the study. We specify the priors, $\kappa_a \sim N(0,1)$, $\nu_t \sim N(0,1)$ and $\tau \sim \Gamma^{-1}(0.001, 0.001)$.

Finally we note that the underlying models specified are simply used as an illustration here and will be highly dependent on the underlying system being modelled.

Results

We use the shorthand notation for covariates given below (note that we use a number of these covariates in later examples relating to the Soay sheep data):

i) Individual covariates:
- Coat type (C)
- Horn type (H)
- Birth weight (BW)

MISSING VALUES

ii) Environmental covariates:
- NAO index (N)
- Population size (P)
- March temperature (T)
- March rainfall (M)
- Autumn rainfall (A)

For further discussion of these covariates, see Catchpole et al. (2000) and King et al. (2006). We consider the following model for the Soay sheep:

$$\phi_1(BW, H, P), \phi_{2:7}(W), \phi_{8+}(P, M, T, A)/p_{1:3}, p_{4:5}, p_{6+}/\lambda_{1:2}(t), \lambda_{3:8}(t), \lambda_{9+}.$$

Thus, we consider an age dependent structure for the demographic parameters. In particular, we assume that the survival probabilities differ for lambs (year of life 1), adults (years of life 2–7) and the most senior sheep (years of life 8+). In addition, the covariate dependence on the survival probabilities differs with age. For first years, birth weight (BW), horn type (H) and population size (P) are included; for adults, weight (W) is included; and for the senior sheep, survival is assumed to depend on population size (P), March rainfall (M) and temperature (T) and autumn rainfall (A). Thus, we have the survival probabilities dependent on both environmental and individual covariates. We have 4.9% of horn type value missing, as well as 21.1% of the birth weight values and 59.2% weight values that need to be imputed. This model is motivated from a previous study of the Soay sheep (see King et al. 2008a). In addition, there are different recapture and recovery probabilities for lambs compared to older individuals. For these probabilities, we take only the adult recovery probabilities to be dependent on the year of the study. These models are once more motivated by the previous study, and are simply used as an illustration of using these methods.

Running the MCMC simulations, taking into account the missing data (and assuming an underlying model for the covariate values), we can obtain estimates of the parameters of interest, such as estimates of the recapture and/or recovery probabilities, and more particularly, estimates for the regression coefficients for different covariates considered within the analysis. To illustrate the results obtained, we present in Table 8.2 the corresponding posterior mean and standard deviation of the regression coefficients for the different covariates, conditional on age.

The interpretation of the posterior estimates of the regression coefficients is quite straightforward, indicating the sign and magnitude of the impact on the survival probability. For example, we can see that the population size has a negative impact on the survival probability for lambs and the oldest sheep, approximately of equal magnitude. The population size can often be regarded as a surrogate for the amount of internal competition for resources, so that in the years with higher population size, and hence competition, the youngest and oldest sheep have a reduced survival probability. Alternatively, for lambs and adults, weight (which could be regarded as a surrogate for condition)

Table 8.2 Soay Sheep: the Posterior Mean and Standard Deviation (in Brackets) of the Regression Coefficients for the Environmental and Individual Covariates for Each Age

		Parameter	Mean (SD)
ϕ_1		Birth Weight	0.483 (0.094)
	Horn Type	(polled)	−1.004 (0.202)
		(classical)	−0.645 (0.210)
		Population Size	−1.212 (0.098)
$\phi_{2:7}$		Weight	0.451 (0.181)
ϕ_{8+}		Population Size	−1.199 (0.185)
		March Rainfall	−0.600 (0.172)
		March Temperature	1.244 (0.191)
		Autumn Rainfall	−0.891 (0.184)

is positively correlated with survival, as we would expect. Finally, we note that for the categorical covariate, horn type, we are comparing the survival probability (on the logistic scale) between polled and classical with the default horn type scurred. It appears that lambs with scurred horns have the largest survival probability, followed by classical, with polled horned sheep having the smallest survival probability. This is in agreement with the classical analysis of Catchpole et al. (2000).

□

8.3.1 Practical Implications

There are a number of practical implications that should be noted when adopting the Bayesian approach to missing data. First, the number of parameters in a model can dramatically increase, since we have additional parameters relating to the underlying models specified for each of the covariates and also since we treat the missing covariate values themselves as parameters (or auxiliary variables). Typically, since recapture probabilities are not very close to unity, the presence of time varying covariates will result in the largest increase in the parameters in the model, but can often be one of the most important factors affecting the survival probabilities of the individuals. Each parameter needs to updated within each iteration of the MCMC algorithm, and so this approach can become computationally intensive. In practice, the updating of the parameter values (using, for example, a pilot tuned MH random walk) is very fast, so that for even relatively large data sets with a reasonably large number of missing values, simulations are feasible. For example, for the previous Soay sheep example and the given model fitted to the data, a total of 1079

ASSESSING COVARIATE DEPENDENCE

individuals are observed within the 15-year study, with a total of 3 individual covariates (horn type, birth weight and weight). In addition to the updating of the parameters in the model, the associated missing covariate values also need to be updated. This corresponds to 53 values for horn type, 228 values for birth weight and 2394 weight values missing. The convergence of the parameter values in this case was also rapid, with only a short burn-in needed and the simulations taking approximately 8.5 hours for 500,000 iterations on a 1.8-Ghz computer. (This is relatively short!).

The above approach allows us to sample from the joint posterior distribution:
$$\pi(\boldsymbol{\theta}, \boldsymbol{c}_{mis}|\boldsymbol{h}, \boldsymbol{c}_{obs}).$$
However, we are primarily interested in the marginal posterior distribution,
$$\pi(\boldsymbol{\theta}|\boldsymbol{h}, \boldsymbol{c}_{obs}) = \int \pi(\boldsymbol{\theta}, \boldsymbol{c}_{mis}|\boldsymbol{h}, \boldsymbol{c}_{obs}) d\boldsymbol{c}_{mis}.$$
Even though we integrate out over the missing covariate values, the marginal posterior distribution is still dependent on the underlying model that we specify for the covariate values (i.e. $f(\boldsymbol{c}_{obs}, \boldsymbol{c}_{mis}|\boldsymbol{\theta})$). Thus, there should be a biological justification for the underlying model specified for the covariates. However, a prior sensitivity analysis can be performed, by considering different models for the covariates or different priors on the underlying model parameters and the corresponding posterior distributions compared for the parameters of interest. For example, a correlated structure could be considered for different covariates; alternatively, we could add in the time dependence assumption for the mean underlying birth weight. Once more for the previous example, the results appear to be data-driven, with the posterior results insensitive to sensible changes in the underlying models.

8.4 Assessing Covariate Dependence

Here we consider capture-recapture-recovery data where we are interested in the dependence of the survival probabilities on different possible covariates. To specify the relationship between the survival probabilities and covariates we use a logistic regression, as described in Section 8.2. In particular, we focus on the Soay sheep capture-recapture-recovery data set.

Example 8.3. Soay Sheep
We consider mark-recapture-recovery data relating to the Soay sheep from 1986 through 1995 for data collected from the spring census (see Example 1.4). This corresponds to a different time period and a different census from that considered in Example 8.2. For simplicity we shall only consider the possible dependence of the survival probabilities on a number of covariates. We also assume that the underlying models for the recapture and recovery probabilities are known, in terms of their age and time dependence structure;

namely, that we have,

$$p_{1:3}, p_{4:5}, p_{6+}/\lambda_{1:2}(t), \lambda_{3:8}(t), \lambda_{9+}.$$

In addition, we assume that the age dependence for the survival probabilities is fixed, and is given by:

$$\phi_1, \phi_{2:7}\phi_{8+}.$$

However, there is uncertainty relating to the dependence of the survival probability on the possible covariates for each of the age groups identified. There are a large number of possible biological models that are deemed to be plausible. Namely, that the survival probability for the different age groups may be dependent on any combination of possible covariates. Thus, for each age group, there are $2^8 = 256$ possible models and overall there are $256^3 \approx 2$ million possible models. It may be just feasible to consider every possible model for a single age group. However, clearly we cannot consider every possible overall model. Thus, the reversible jump algorithm is implemented to explore the posterior distribution and obtain the posterior estimates of interest.

Prior Specification

We need to specify priors on the parameters in each model, and the prior probability of each model. We assume that we have no strong prior information. We begin with the simplest priors, corresponding to the recapture and recovery probabilities. We simply specify an independent $U[0, 1]$ prior for each of these. We specify independent priors on each regression coefficient, conditional on it being present in the model, i.e. conditional on the corresponding covariate being present in the model. Note that given that the parameter is in the model, the prior we specify for the parameter is the same for each of these models. In other words, suppose that we consider the regression parameter θ, then,

$$p(\theta|m) = p(\theta|m'),$$

for all m, m', such that θ is present in models m and m'. In particular, for the regression coefficient θ, conditional on the parameter being present in the model, we specify the prior,

$$\theta|m \sim N(0, 2.334).$$

Finally we need to specify a prior probability for each individual model. We do this by defining the marginal posterior probability for the presence/absence of each covariate. Assuming independent prior probabilities, we construct the corresponding overall prior model probability by multiplying the marginal probabilities. For example, for ϕ_1, we assume that each covariate is present or absent with probability 0.5, in the absence of strong prior beliefs. Since we define the prior probabilities independently for each covariate, the corresponding prior probability for each possible model for ϕ_1 is simply the corresponding product of the marginal prior probabilities of whether each covariate is present or absent. For the above marginal prior specification, this results

ASSESSING COVARIATE DEPENDENCE

in an equal prior probability for each possible model for ϕ_1. We construct the marginal prior model probabilities for $\phi_{2:7}$ and ϕ_{8+}, similarly. Finally, the overall prior model probability for the covariate dependence of ϕ_1, $\phi_{2:7}$ and ϕ_{8+} is simply the product over the marginal prior probabilities for each of the demographic parameters, as a result of the assumed prior independence. This prior construction clearly results in an equal prior probability for each individual model. However, we present the idea here as an example of how prior model probabilities can be constructed in general, since it is feasible that there could be prior information relating to the probability that a covariate does influence the given survival probability. For example, if there is some prior information that ϕ_1 *is* dependent on birth weight, then we could specify a prior probability of greater than 0.5 that birth weight is present for ϕ_1. The actual value of the probability would represent the strength of this belief. Conversely, prior information may be of the form that autumn rainfall does *not* affect ϕ_1. We could specify a prior probability of less than 0.5 that autumn rainfall is present in the model for ϕ_1. Once more, the actual value of the prior probability specified would represent the strength of this belief. Under these circumstances, it may not be clear how this would be incorporated into the prior specification by simply specifying the prior probability for each overall model for the demographic parameters. Clearly, more complex prior structures could be defined, for example, removing the independence assumptions between some of the covariates. For simplicity, we retain the equal prior probability assumption. Note that the Bayes factors for the different model comparisons are independent of the prior specification on the models (see Section 6.2.3). However, the posterior model probabilities are clearly dependent on the prior specification.

RJMCMC Algorithm*

There are missing covariate values for all the individual covariate values considered here. We implement the approach considered in Section 8.2 for the covariates and in particular use the same specifications for their corresponding underlying models as described in Example 8.3. Within each iteration of the Markov chain, we do the following:

1. update each parameter (including any missing covariate values) using the random walk MH algorithm; and

2. cycle through each demographic parameter ϕ_1, $\phi_{2:7}$ and ϕ_{8+} and update each marginal model by proposing to add or remove a single covariate.

We describe in further detail only the second step here. Without loss of generality, suppose that we consider the model update for ϕ_1. Let the state of the chain at iteration k be denoted by $(\boldsymbol{\theta}, m)_k$. We choose each covariate with equal probability. Here there are a total of 8 covariates, so that we choose each covariate with probability $\frac{1}{8}$. If the covariate is present for ϕ_1, we propose to remove it; otherwise, if it is absent, we propose to add the covariate. Thus we only propose model moves between nested models that differ by a single parameter (see Example 6.4 for an analogous model updating algorithm). Initially suppose that we choose covariate j, which is absent from ϕ_1. We propose to add the covariate to the model, where we denote this new model by m'. Let the corresponding proposal values for the regression coefficient for covariate i be denoted by β_i'. We propose the new state of the chain $(\boldsymbol{\theta}', m')$, where $\boldsymbol{\theta}' = (\boldsymbol{\theta}, \beta_i')$. In other words, we do not propose to update the parameter values $\boldsymbol{\theta}$ common to both models, and set $\beta_i' = u$, where we simulate a value of u from some distribution q. A typical choice for the proposal distribution would be,

$$u \sim N(0, \sigma^2),$$

where σ^2 is often chosen via pilot tuning. The proposal distribution is essentially centred at the current value of the regression coefficient. However, many alternative distributions are clearly possible; for example, specifying a non-zero mean, particularly if it is thought that if the covariate is present then it has a positive or negative influence on the survival probabilities. This has totally defined the proposed model move.

We now perform the accept/reject step. We accept the proposed move and set $(\boldsymbol{\theta}, m)_{k+1} = (\boldsymbol{\theta}', m')$ with probability $\min(1, A)$, where,

$$\begin{aligned}A &= \frac{\pi(\boldsymbol{\theta}', m', \boldsymbol{c}_{mis} | \boldsymbol{h}, \boldsymbol{c}_{obs}) \mathbb{P}(m|m')}{\pi(\boldsymbol{\theta}, m, \boldsymbol{c}_{mis} | \boldsymbol{h}, \boldsymbol{c}_{obs}) \mathbb{P}(m'|m) q(u)} \left| \frac{\partial \boldsymbol{\theta}'}{\partial (\boldsymbol{\theta}, u)} \right| \\ &= \frac{\pi(\boldsymbol{\theta}', m', \boldsymbol{c}_{mis} | \boldsymbol{h}, \boldsymbol{c}_{obs})}{\pi(\boldsymbol{\theta}, m, \boldsymbol{c}_{mis} | \boldsymbol{h}, \boldsymbol{c}_{obs}) q(u)}\end{aligned}$$

where $\boldsymbol{h}$ denotes the set of capture histories, $\boldsymbol{c}_{obs}$ the observed covariate values and $\boldsymbol{c}_{mis}$ the missing covariate values. The expression for the acceptance probability has been simplified since $\mathbb{P}(m|m') = \mathbb{P}(m'|m) = \frac{1}{8}$ and the Jacobian term is equal to unity. Otherwise, if we reject the move, we set $(\boldsymbol{\theta}, m)_{k+1} = (\boldsymbol{\theta}, m)_k$.

Now consider the reverse move. This is completely defined by the above. Suppose that at iteration k of the Markov chain the state is $(\boldsymbol{\theta}', m')_k$. We propose to move to model m, by removing the dependence on covariate i. We set $u = \beta_i'$ and keep the current values for the parameters $\boldsymbol{\theta}$. In other words, all parameters except the regression coefficient for covariate i remain the same, and the regression coefficient for covariate i is set to zero (i.e. removed from the model). We then accept the move with probability, $\min(1, A^{-1})$, where A is given above; otherwise the chain remains at the current state for the next iteration.

Results

The simulations are run for a total of 1 million iterations with the first 10% discarded as burn-in. The standard tools described in Sections 5.4.2 and 6.8.1, such as trace plots, suggest that this is adequate. We present a selection of results for illustrative purposes only. Clearly, the posterior estimates presented will reflect the corresponding quantities of interest within the analysis. Here

ASSESSING COVARIATE DEPENDENCE

we assume that we are interested in the posterior support that each survival probability is a function of the different covariates. These are simply marginal posterior probabilities. For example, consider that we are interested in the marginal posterior probability that ϕ_1 is dependent on the NAO index. This is simply calculated by summing the probabilities over all models for ϕ_1 which contain the NAO dependence, and similarly for all other covariates. Table 8.3 gives the marginal posterior probability for each covariate and age group for the survival probabilities.

Table 8.3 The Posterior Probability of Each Covariate for Each Survival Probability for the Soay Sheep Data

Covariate	Posterior Probability		
	ϕ_1	$\phi_{2:7}$	ϕ_{8+}
Coat Type (C)	0.323	0.400	0.224
Horn Type (H)	0.996	0.044	0.084
Birth Weight (BW)	1.000	0.106	0.098
Weight (W)	–	1.000	1.000
NAO Index (N)	0.575	0.165	0.202
Population Size (P)	1.000	1.000	1.000
March Rainfall (M)	0.881	0.999	0.937
March Temperature (T)	0.865	0.995	1.000
Autumn Rainfall (A)	0.774	1.000	1.000

Finally, the sign and magnitude of the regression coefficients may be of interest. Note that for these posterior model-averaged estimates to be meaningful, we must condition on the covariate being present in the model (see Section 6.5). The posterior (model-averaged) mean and standard deviation of the regression coefficients are given in Table 8.4, conditional on the parameter being present in the model. Clearly, we can see a strong correlation between large posterior model probabilities, and parameter estimates away from 0, corresponding to no influence on the survival probabilities. The Bayes factor criterion described in Section 6.2.3 for the models (or marginal models) reduces to the ratio of posterior model probabilities for this example, since the prior model probabilities are all equal. Using the suggested rule of thumb, there is evidence for ϕ_1 to be dependent on horn type, birth weight, population size, March rainfall, March temperature and autumn rainfall. Similarly, for $\phi_{2:7}$ we would identify weight, population size, March rainfall, March temperature and autumn rainfall; and for ϕ_{8+}, weight, population size, March rainfall, March temperature and autumn rainfall. Thus, it appears that weight, environmental conditions (temperature and rainfall covariates) and population size are all related to the survival probabilities throughout the entire life of an individual. □

Table 8.4 The Posterior Mean and Standard Deviation of the Regression Coefficient for Each Survival Probability, Conditional on the Covariate Being Present in the Model, for the Soay Sheep Data

Covariate	Posterior Mean (Standard Deviation)		
	ϕ_1	$\phi_{2:7}$	ϕ_{8+}
Coat Type (C) (light)	0.309 (0.186)	0.389 (0.226)	−0.238 (0.325)
Horn Type (H) (polled)	−0.916 (0.217)	−0.283 (0.271)	0.404 (0.367)
Horn Type (H) (classical)	−0.542 (0.223)	−0.229 (0.224)	−0.043 (0.309)
Birth Weight (BW)	0.578 (0.099)	0.117 (0.122)	0.023 (0.177)
Weight (W)	–	0.940 (0.168)	1.192 (0.260)
NAO Index (N)	−0.424 (0.206)	−0.146 (0.283)	−0.207 (0.316)
Population Size (P)	−1.335 (0.140)	−1.112 (0.152)	−1.488 (0.241)
March Rainfall (M)	−0.574 (0.201)	−1.010 (0.178)	−0.694 (0.213)
March Temperature (T)	0.480 (0.165)	0.831 (0.176)	1.688 (0.253)
Autumn Rainfall (A)	−0.355 (0.124)	−0.961 (0.179)	−1.192 (0.231)

8.5 Random Effects

Random effects can also be introduced within the modelling process in order to allow for additional variability not accounted for by any of the covariates. This could be either temporal variability, individual variability, or both, depending on the underlying assumptions that are made. For simplicity, we shall consider temporal random effects in the context of the logistic regression framework considered within this chapter for the survival probabilities, allowing for environmental covariates. Clearly we could extend this approach to include additional individual covariates within the analysis, by simply adding in additional individual-level covariates within the logistic regression expression. We can express this additional unaccounted for temporal variability (or random effects) by specifying the logistic regression:

$$\text{logit } \phi_t = \beta_0 + \boldsymbol{\beta}^T \boldsymbol{y}_t + \epsilon_t,$$

using the notation introduced for the environmental covariates, and where we have the additional random effects, $\epsilon = \{\epsilon_1, \ldots, \epsilon_T\}$, such that,

$$\epsilon_t \sim N(0, \sigma^2).$$

Here σ^2 is the random effect variance and is a parameter to be estimated. Thus, since σ^2 is a parameter, we need to specify a prior on this parameter within the model. Large posterior values of σ^2 would suggest that there is additional temporal heterogeneity not explained by the covariates, whereas small posterior values of σ^2 would suggest that the temporal variability is explained well by the covariates considered. A common prior that is adopted

RANDOM EFFECTS

is an inverse gamma, which is the conjugate prior for the variance term. However, alternative priors could include a uniform prior over some positive range (typically this is large to reflect prior uncertainty), or a gamma prior. Gelman (2006) suggests the alternative prior specification where the standard deviation (rather than variance) of the random effects is an (improper) uniform distribution on the positive real line. The term ϵ can be regarded as a nuisance parameter that we wish to integrate out. However, the integration is typically analytically intractable. Thus we adopt an alternative approach.

In the presence of random effects, we consider the analogous approach to the problem where we had missing covariate values. Essentially we treat the nuisance parameters, ϵ, as auxiliary variables, or parameters, to be updated within the MCMC algorithm and integrated out. In this case, we have already defined the corresponding underlying model for these parameters, given by the normal density above. In other words, we form the joint posterior distribution,

$$\pi(\boldsymbol{\theta}, \boldsymbol{\epsilon}|\boldsymbol{x}) \propto f(\boldsymbol{x}|\boldsymbol{\theta}, \boldsymbol{\epsilon})f(\boldsymbol{\epsilon}|\boldsymbol{\theta})p(\boldsymbol{\theta}),$$

where $\boldsymbol{x}$ denotes the observed data. Once again we incorporate the parameter σ^2 into the term $\boldsymbol{\theta}$, which denotes all the parameters within the model. The likelihood term, $f(\boldsymbol{x}|\boldsymbol{\theta}, \boldsymbol{\epsilon})$, can once more be easily evaluated when conditioning on the parameters $\boldsymbol{\theta}$ and nuisance parameters $\boldsymbol{\epsilon}$ imputed within the MCMC algorithm. Similarly, the underlying model for the random effects, $f(\boldsymbol{\epsilon}|\boldsymbol{\theta})$, can be evaluated. Thus, we can obtain a sample from this joint posterior distribution and simply marginalise to obtain the posterior distribution of interest, i.e.

$$\pi(\boldsymbol{\theta}|\boldsymbol{x}) = \int \pi(\boldsymbol{\theta}, \boldsymbol{\epsilon}|\boldsymbol{x})d\boldsymbol{\epsilon}.$$

We note that within a Bayesian framework, the difference between a fixed effects model and a random effects model can be expressed in the form of the prior specification. For example, consider the following logistic regression,

$$\text{logit } \phi_t = \beta_0 + \epsilon_t,$$

where,

$$\epsilon_t \sim N(0, \sigma^2).$$

For a fixed effects model, we would simply specify the value of σ^2 (note that this leads to an over-parameterised model, and typically, we would set $\beta_0 = 0$). Alternatively, for a random effects model, we simply place a hierarchical prior on the variance term, $p(\sigma^2)$. Note that if there are additional temporal covariates specified in the regression equation, then the model is over parameterised for the fixed effects assumption on the $\boldsymbol{\epsilon}$. However, we can consider a random effects model, since the parameter of interest is simply the additional parameter σ^2, with the $\boldsymbol{\epsilon}$ terms simply treated as auxiliary variables and integrated out within the MCMC algorithm.

Example 8.4. Soay Sheep
We return to the Soay sheep example, where we assume that there is not

only uncertainty in the covariates present in the model for the survival probabilities, but also whether there are additional random effects (i.e. additional unexplained temporal variability). We once more assume the following underlying age structure (and time dependence) for the demographic parameters:

$$p_{1:3}, p_{4:5}, p_{6+}/\lambda_{1:2}(t), \lambda_{3:8}(t), \lambda_{9+}/\phi_1, \phi_{2:7}, \phi_{8+}.$$

The survival probabilities are then possibly regressed on each of the environmental and individual effects. For simplicity, and for ease of notation, suppose that the survival probabilities are only dependent on continuous covariates $\boldsymbol{y}_{it}$ (which may be environmental or individual). The corresponding survival probability for an individual i in age group $a \in \{1, (2:7), 8+\}$ at time t, in the presence of random effects, is given by,

$$\text{logit } \phi_{it} = \gamma_0 + \boldsymbol{\gamma}^T \boldsymbol{y}_{it} + \epsilon_{a,t},$$

where,

$$\epsilon_{a,t} \sim N(0, \sigma_a^2)$$

denotes the random effect term. For this case study, we assume the prior $\sigma_a^2 \sim \Gamma^{-1}(0.001, 0.001)$, given that random effects are present. All other priors remain the same as specified in Section 8.4 (including a flat prior over all possible models).

In a manner analogous to assessing which covariates influence the survival probabilities, we may wish to assess whether there are additional random effects present. Thus, we can use the standard model discrimination tools, and calculate the posterior probability (or Bayes factor) that random effects are present in addition to the covariates for each of the different age groups. This simply involves an additional step within the previous RJMCMC algorithm described in Section 8.4 corresponding to adding or removing the random effects given that they are absent or present for the given age group.

RJMCMC Algorithm*

Within each iteration of the Markov chain we update the parameters (including auxiliary variables) using the MH algorithm, and the presence of covariates and random effects using the reversible jump algorithm. We discuss here only the reversible jump step on the absence/presence of random effects. The reversible jump step is analogous to the adding/removing temporal dependence on the parameter (as discussed in Section 6.11), but where we have a hierarchical prior specified on the parameters. If random effects are present, we propose to remove them and vice versa. Specifically, at iteration k of the Markov chain, let the current model be denoted by m, with associated parameter values $\boldsymbol{\theta}^k$. Suppose that for a given set of ages i to j, there are no random effects, so that we propose to add them. Dropping the dependence on age for notational convenience, we propose a new model with random effects, $\boldsymbol{\epsilon} = \{\epsilon_1, \ldots, \epsilon_{T-1}\}$,

RANDOM EFFECTS 267

and associated variance σ^2 by simulating,

$$\epsilon_t \sim N(\nu_t, \tau);$$
$$\sigma_\epsilon^2 \sim \Gamma(a, b).$$

The proposal parameters a, b and τ are chosen via pilot tuning (in practice $a = b = 1 = \tau$ appeared to work well), and the $\nu(t)$ are to be defined. This new proposed model, m', has parameters $\boldsymbol{\theta}' = \boldsymbol{\theta}^k \cup \{\boldsymbol{\epsilon}, \sigma_\epsilon^2\}$. The corresponding acceptance probability for this model move reduces to $\min(1, A)$, where,

$$A = \frac{\pi(\boldsymbol{\theta}', m', \boldsymbol{c}_{mis}|\boldsymbol{h}, \boldsymbol{c}_{obs})}{\pi(\boldsymbol{\theta}^k, m, \boldsymbol{c}_{mis}|\boldsymbol{h}, \boldsymbol{c}_{obs})q(\boldsymbol{\epsilon})q(\sigma_\epsilon^2)}.$$

Conversely, in the reverse move, where we remove the random effect terms, the acceptance probability is the reciprocal of the above expression.

In order to improve the acceptance probabilities, by simulating parameter values in an area of high posterior mass, the parameters in the proposal distributions, ν_t, are set via an initial MCMC simulation as follows. We consider the saturated model, where there is full covariate dependence with random effects present for the survival probabilities. We then set ν_t to be the corresponding posterior mean for the random effect terms for the given age group.

Results

The posterior probabilities that the survival probabilities for each age group are dependent on each possible covariate and/or random effects are present are given in Table 8.5. Concentrating initially on the presence/absence of random effects, it is clear that there appears to be additional temporal variability for the adult sheep (ages 2–7), not explained by the covariates considered within the analysis. The corresponding posterior mean (and standard deviation) for the random effect variance is 1.452 (1.600). However, for the oldest sheep there is strong evidence of no additional temporal random effects, whereas for the youngest sheep there is uncertainty as to the presence of random effects.

We can compare these results with those presented in Table 8.3 for the analogous analysis, but assuming that there are no random effects present. There is very little difference between the posterior probabilities for the presence of covariates for the oldest sheep. This is unsurprising, since there is no evidence of random effects being present for these age groups in this latter analysis. However, for the adults (aged 2–7), there are some appreciable differences between the analyses. In particular, the strong dependence on the environmental covariates (population size, March temperature and autumn rainfall), that was previously identified, disappears and is replaced by random effects. There is still some evidence for dependence on March rainfall, but this is reduced, compared to the case with no random effects. This suggests that the temporal variability in the adult survival probability is not well modelled by the environmental covariates considered. However, in the absence of any alternative modelling of the temporal variability (via the random effects), the

Table 8.5 The Posterior Probability that Each Covariate is Present in the Model and Random Effects are Present for Each Survival Probability

	Posterior Probability		
Covariate	ϕ_1	$\phi_{2:7}$	ϕ_{8+}
Coat Type (C)	0.449	0.378	0.215
Horn Type (H)	0.997	0.042	0.085
Birth Weight (BW)	1.000	0.113	0.103
Weight (W)	–	1.000	1.000
NAO Index (N)	0.609	0.296	0.198
Population Size (P)	1.000	0.548	1.000
March Rainfall (M)	0.661	0.734	0.946
March Temperature (T)	0.491	0.246	1.000
Autumn Rainfall (A)	0.378	0.227	1.000
Random Effects (RE)	0.652	1.000	0.025

covariates are identified that provide the "best" fit to the heterogeneity. Finally, for the first-year survival probabilities where there is uncertainty with regard to the presence of random effects, there are both similarities and differences in the covariate identified. For example, horn type, birth weight and population size are still identified as important. However, there is no longer posterior evidence for March rainfall, March temperature and autumn rainfall. The uncertainty in the presence of random effects for the first-year survival probabilities appear to have led to an uncertainty in the dependence on these environmental covariates.

□

Example 8.5. White Storks
We return to Exercise 3.5, relating to white storks, where wish to assess the dependence of the survival probabilities on 10 possible rainfall covariates (see Sections 7.4 and 7.5). Previously, in the analysis in Section 7.5, there appeared some uncertainty as to whether the random effects model was needed, with the posterior estimate for the random effect variance close to zero. Thus, we may wish to consider a model discrimination approach, where there is uncertainty as to whether the model is a fixed effects model or a random effects model, and we calculate the posterior probabilities of the different models.

The presence or absence of random effects can be investigated using the analogous move-type described above. So that when we propose to add random effects to a model, not only do parameter values need to be proposed for the random effects themselves (since they are treated as auxiliary variables within the MCMC approach), but also we need the value for the underlying

RANDOM EFFECTS

variance term. For the white stork data we also allow for the uncertainty on the covariate dependence, using the analogous approach to that presented in Example 8.4 relating to the Soay sheep data. Thus we omit the reversible jump MCMC details and only present the results within this analysis, but also consider an additional prior sensitivity analysis.

We initially specify independent $N(0, 10)$ priors on the regression coefficients, and an $\Gamma^{-1}(0.001, 0.001)$ prior on the random effects variance term, σ^2. For the recapture probability we use a $U[0, 1]$ prior. We refer to this prior specification as prior 1. The posterior marginal probabilities obtained for each covariate and presence/absence of random effects are presented in Table 8.6.

Table 8.6 The Marginal Posterior Probability of the Different Covariates for the White Stork Data and the Presence of Random Effects (REs)

Covariate	Posterior probability		
	Prior 1	Prior 2	Prior 3
1	0.053	0.143	0.054
2	0.031	0.093	0.035
3	0.032	0.086	0.033
4	0.979	0.978	0.958
5	0.086	0.226	0.087
6	0.099	0.203	0.096
7	0.045	0.115	0.045
8	0.059	0.147	0.063
9	0.050	0.128	0.051
10	0.059	0.137	0.060
RE	0.000	0.000	0.020

Note: The analysis uses (a) prior 1: $N(0, 10)$ prior on regression coefficients and $\Gamma^{-1}(0.001, 0.001)$ on σ^2; (b) prior 2: $N(0, 1)$ prior on regression coefficients and $\Gamma^{-1}(0.001, 0.001)$ on σ^2; and (c) prior 3: $N(0, 10)$ prior on regression coefficients and $\Gamma^{-1}(3, 2)$ on σ^2.

We note that the posterior probability that random effects are present is very small < 0.001, so that the rainfall covariates appear to explain all of the temporal heterogeneity within the survival probabilities of the white storks. These results are very similar to those presented in Figure 7.21 assuming no random effects, (allowing for Monte Carlo error). Alternatively, comparing the posterior probability that each covariate influences the survival probabilities with the results presented in Table 7.1, where the random effects model is assumed, the largest difference appears to be in strengthening the dependence of the survival probability on covariate 4.

We consider a prior sensitivity analysis using the same priors in the prior sensitivity analysis of Section 7.5. Prior 2 is specified such that the prior on the regression coefficients are independent $N(0,1)$, and $\sigma^2 \sim \Gamma^{-1}(0.001, 0.001)$. Prior 3 has independent $N(0, 10)$ priors on the regression coefficients, but $\sigma^2 \sim \Gamma^{-1}(3,2)$. The corresponding posterior marginal probabilities are given in Table 8.6. Using prior 2, we see that the posterior probability that each covariate is present in the model is slightly increased, although the corresponding interpretation of the results are unchanged. This is the same response we had assuming the presence of random effects (see Table 7.2). Now, consider prior 3. Previously (assuming random effects were present), changing this prior specification had a large impact on the posterior distribution of the random effect variance term, and consequently on the posterior probability of covariate 4 being present in the model. The same is not true here. The reason is clear – the posterior probability that the random effects are present is very low. Thus, the prior specification on the random effect variance has very little impact on the posterior probabilities, since the fixed effect model has very strong support.

□

8.6 Prediction

We make a small note that the issue of predicting future events is in principle trivial given the work of the previous subsections. In particular, future events can be simply treated as missing data at the end of a given time series, where the actual states of the missing data values are of interest. Since the missing data are treated as parameters and updated within each step of the MCMC algorithm, the posterior distribution of these parameters can easily be extracted from the given MCMC simulation.

8.7 Splines

In the previous sections for modelling covariate data, we make the strong assumption that the effect of the covariate on the survival probability is linear on the logit scale. However, non-linear relationships involving the impact of environmental or individual factors may also occur. It is possible to include additional non-linear factors such as quadratic/cubic covariate values. However, this can appreciably increase the number of possible parameters and still assumes some underlying parametric model. More flexible and alternative models for the survival probability are possible. We consider the alternative approach where we allow the shape of the relationship to be determined by the data without making any prior assumption regarding its form, via the use of splines. Splines are functions flexible in shape made of polynomial segments connected at some points referred to as *knots*. We focus in this section on a specific class of splines called *penalized splines* or *P-splines* (Gimenez et al. 2006a,b).

Suppose that we wish to model a survival probability as a function of the

SPLINES

covariate y. We consider the following regression model for the survival probability:
$$\text{logit } \phi_t \sim N(m(y_t), \sigma^2),$$
where y_t is the value of the covariate at time $t = 1, \ldots, T$ and $m(\cdot)$ is a smooth function, where the term *smooth* refers to the curvature of the function (in mathematical terms, it means that $m(\cdot)$ has derivatives of all orders).

The function specifies a nonparametric flexible relationship between the survival probability and the covariate that allows non-linear environmental trends to be detected. There are many possible functions that can be specified for m. We use a truncated polynomial basis to describe m:

$$m(y|\boldsymbol{\theta}) = \beta_0 + \beta_1 y + \cdots + \beta_P y^P + \sum_{k=1}^{K} b_k (y - \kappa_k)_+^P \qquad (8.1)$$

where $P \geq 1$ is an integer, $\boldsymbol{\theta} = (\beta_1, \ldots, \beta_P, b_1, \ldots, b_K)^{\text{T}}$ is a vector of regression coefficients, and the $(u)_+^P = u^P$ if $(u \geq 0)$ and 0 otherwise and are usually called *spline basis functions* (Figure 8.3), and $\kappa_1 < \kappa_2 < \ldots < \kappa_K$ are fixed knots.

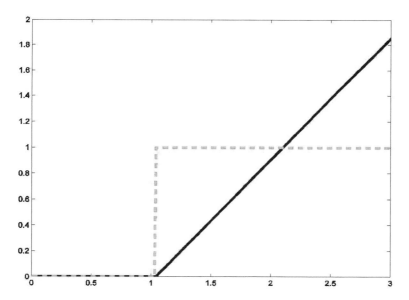

Figure 8.3 Example of a linear spline basis function $(y - 1)_+$ (solid line) with its derivative (dotted line). It consists of two lines that are connected at $y = 1$. The function takes value 0 on the left of 1 and then is a line with positive slope from 1 onward.

This simple function for m, given in Equation (8.1), is exceptionally flexible, allowing an extremely rich set of possible relationships between the mean survival probability and the set of covariates. Essentially, the function can be regarded as a smoothing function of the covariate values. The crucial problem in using the relation above is the choice of the number and the position of the knots. A small number of knots may result in a smoothing function that is not flexible enough to capture variation in the data, whereas a large number of knots may lead to overfitting. Similarly, the position of the knots will influence estimation. We use a P-splines approach inspired by the smoothing splines of Green and Silverman (1994). First, the number of knots is chosen to ensure enough flexibility. We consider $K = \min\left(\frac{1}{4}T, 35\right)$ and let κ_k be *equally spaced sample quantiles*, i.e. the sample quantiles of the y_i's corresponding to probabilities $k/(K+1)$ (Ruppert 2002). Then a quadratic penalty is placed on **b**, which is here the set of jumps in the Pth derivative of $m(\cdot|\beta)$. In brief, this penalty component penalizes the rough functions so that the resulting function is smooth. With the previous equation, we associate the constraint $\mathbf{b}^T\mathbf{b} \leq \lambda$ where λ is called the *smoothing parameter*. This parameter provides a balance between the fit to the data and the penalty on the measure of curviness of m.

The two equations put together lead to the so-called *P-splines approach* (Lang and Brezger 2004). Because roughness is controlled by the penalty term, once a minimum number of knots is reached, the fit given by a P-spline is almost independent of the knot number and location (Ruppert 2002). Note that P-spline models can be fruitfully expressed as Generalized Linear Mixed Models (GLMM, Ruppert 2002; Crainiceanu et al. 2005), which facilitates their implementation in standard software (Ngo and Wand 2004; Crainiceanu et al. 2005), and above all provides a unified framework for generalizations of the nonparametric approach (Gimenez et al. 2006b).

Let $\boldsymbol{\phi} = (\phi_1, \ldots, \phi_T)^T$, $\mathbf{Y}$ be the matrix with i^{th} row $\mathbf{Y}_i = (1, y_i, \ldots, y_i^P)^T$, and $\mathbf{Z}$ be the matrix with i^{th} row $\mathbf{Z}_i = \{(y_i-\kappa_1)_+^P, \ldots, (y_i-\kappa_K)_+^P\}^T$. We define $\sigma_b^2 = \lambda \sigma_\varepsilon^2$, and consider the vector $\boldsymbol{\beta}$ as fixed parameters and the vector **b** as a set of random parameters with $E(\mathbf{b}) = 0$ and $cov(\mathbf{b}) = \sigma_b^2 \mathbf{I}_K$. If $(\mathbf{b}^T, \boldsymbol{\varepsilon}^T)^T$ is a normal random vector and **b** and $\boldsymbol{\varepsilon}$ are independent, then an equivalent model representation of the P-spline model in the form of a GLMM is

$$\text{logit}(\boldsymbol{\phi}) = \mathbf{Y}\boldsymbol{\beta} + \mathbf{Z}\mathbf{b} + \boldsymbol{\varepsilon}, \quad \text{cov}\begin{pmatrix} \mathbf{b} \\ \boldsymbol{\varepsilon} \end{pmatrix} = \begin{pmatrix} \sigma_b^2 \mathbf{I}_K & \mathbf{0} \\ \mathbf{0} & \sigma_\varepsilon^2 \mathbf{I}_I \end{pmatrix}.$$

Note that the connection between the P-splines model and the mixed model allows us to extend the nonparametric approach to incorporate other parametric and nonparametric components as well (Ruppert et al. 2003; Gimenez et al. 2006b).

Example 8.6. White Storks

We revisit the white stork data initially described in Exercise 3.5. We specify the survival probabilities as a function of the rainfall at the Kita meteorological station (covariate 4 in the notation of Example 8.5) which was found to be of particular importance (Kanyamibwa et al. 1990). We use $K = 4$ knots using

SUMMARY

the recommendation above, and linear P-splines ($P = 1$). The relationship between survival and rainfall using P-splines is given in Figure 8.4. With the possibility of fitting flexible relationships, one is obviously interested in testing for the presence of non-linearities in the survival probability regression. We address this question by visual comparison of the model with a linear effect of rainfall as well as a random effect with its nonparametric counterpart. Figure 8.4 shows that the relationship between rainfall in Sahel and white stork survival can be taken as linear.

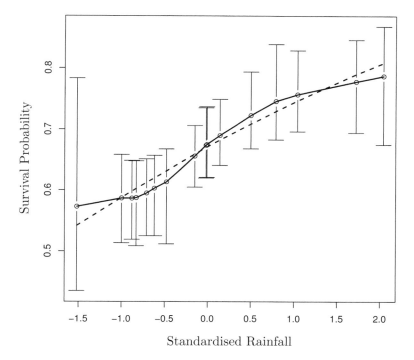

Figure 8.4 Annual variations in white stork survival as a function of the standardized rainfall using a nonparametric approach. Medians (solid line) with 95% credible intervals (vertical solid lines) are shown, along with the estimated linear effect on the logistic scale (dotted line). Reproduced with permission from Gimenez et al. (2009a, p. 896) published by Springer.

8.8 Summary

This chapter focussed on the application of Bayesian methods to analyse ecological data in the presence of covariate information. Typically, two issues commonly arise when relating demographic parameters to different covariates of interest: (i) which covariates to include in the analysis; and (ii) how to deal with missing covariate values. The first issue can be dealt with using the ideas presented in Chapter 6 for discriminating between competing models in terms of the covariates present in the model. This is essentially a straightforward variable selection problem, and hence can be implemented within WinBUGS, assuming random effects are present (see Section 7.5). The second issue is usually a difficult issue within classical analyses, particularly with continuous covariate values. However, within the Bayesian framework, missing values are easily dealt with by simply considering the missing covariate values to be parameters (or auxiliary variables) within the model and the standard Bayesian approach taken. However, we note that using this approach, an underlying model needs to be specified on the underlying model for the covariate values. In addition, particularly in the presence of individual time varying covariates, since we need to update the missing covariate values at each iteration of the Markov chain, the length of time that simulations take will increase.

Random effects can also be easily incorporated within a Bayesian analysis. We note that these can be simply interpreted as a hierarchical prior within this framework. In order to fit a random effects model, we once more treat the random effect terms as parameters (or auxiliary variables) within the analysis. This means that (particularly for individual random effects models), the number of parameters can once more be greatly increased, and hence result in greater computation expense.

Finally, we considered an alternative to a parametric model for the covariates, in the form of splines. These are a non-parametric approach that does not assume a given relationship between the demographic parameter and covariates (for example, logistic relationship). Instead, a smooth relationship via a spline is constructed between the demographic parameter and the covariate(s) of interest, allowing a greater flexibility in the relationship between the covariate and parameter.

8.9 Further Reading

Dealing with missing individual (time varying) covariates is an important subject of current interest, not least, since these factors can often be some of the primary factors driving the system. For discrete-valued covariates, a multi-state approach can be taken (see Chapter 9), and an explicit expression for the likelihood calculated (see for example Brownie et al. 1993 and King and Brooks 2003a). Alternatively, for continuous-valued covariates, the necessary integration is analytically intractable to provide an explicit expression for the likelihood. One possible approach (within the classical framework) would be to use the EM algorithm (Dempster et al. 1977) to obtain the corresponding MLEs of the parameters, however this is generally computationally very intensive, even for a moderate amount of missing parameter values. This approach becomes infeasible when a number of models are to be fitted to the data. An alternative approach has been proposed by Catchpole et al. (2008), which essentially calculates the likelihood conditional on the individuals that are observed at each recapture event with known covariate values. This approach has the advantage of not only being very simple to implement, but

FURTHER READING

also does not make any underlying modelling assumptions for the covariates of interest; however, it can potentially throw away a significant amount of information contained in the data. This is in contrast to typical Bayesian approaches, using an auxiliary variable approach, that assume a model for the underlying covariates. For example, the paper by Bonner and Schwarz (2006) adopts a diffusion model to impute missing information, followed by a Bayesian analysis. See Bonner et al. (2009) for a comparison of these different methods.

When implementing a Bayesian analysis, care does need to be taken when specifying the priors on the parameters, as discussed in Section 8.2.1. We recommend that a prior sensitivity analysis is always performed in a Bayesian analysis, but particular care typically needs to be specified in the presence of model uncertainty, due to the sensitivity that often arises. For logistic models, Newman (2003), Gelman (2006), King and Brooks (2008) and Gelman et al. (2008) discuss the specification of different possible priors when using a logistic link function.

CHAPTER 9

Multi-State Models

9.1 Introduction

We considered multi-state mark-recapture-recovery data and the corresponding Arnason-Schwarz model in Section 2.4. In particular, sufficient statistics can be derived and the corresponding likelihood expression calculated when animals could be observed in different states (Brownie et al. 1993, King and Brooks 2003a). However, the expression for the likelihood is still very complex, and can be computationally very time consuming (particularly when there is a large number of possible states). In addition, this complex likelihood needs to be evaluated repeatedly within each iteration of an (RJ)MCMC algorithm, (i.e. for the updating of each parameter in the model). Here we consider an alternative (and equivalent) approach for analysing data of this form using an auxiliary variable-type approach, analogous to that adopted in Section 8.3 for the missing data problem for categorical variables, considered in the last chapter. However, for multi-state models the covariate corresponds to the state of the individual, which is not only categorical but also time varying. In addition, this approach will also provide posterior estimates of the probability that an individual is in a given state any time, which may prove useful.

9.2 Missing Covariate/Auxiliary Variable Approach

We assume that we have time varying discrete covariates which specify the state of the individual animal. Without loss of generality, we assume that the state of an individual refers to their location at each capture event, where the study area, denoted by K, is composed of k distinct regions. For simplicity, we arbitrarily label the elements of K to be simply the integers $1, \ldots, k$ (i.e. $K = \{1, \ldots, k\}$). The data are composed of series of life-histories for each individual observed within the study, indicating when they were observed within the study, and their corresponding locations at those times. Thus, if an individual is observed, their location is known; however, if the individual is not observed, then their location is unknown. In this case, the migration probability between the different regions is often of interest, and in general the recapture probability may also be dependent on the location of the individual at each capture time. For convenience, we refer to an animal's migration from the area K as a death, with the death state being denoted by $\dagger$. We let $K_\dagger = K \cup \dagger$, which is the set of all possible states in which an animal can exist. Note that the state $\dagger$ may not be observable in the data collection process. However, if an animal can be recovered dead, then we make the usual assumption that this is only possible in the capture event following the death of the individual.

The corresponding likelihood of the data (i.e. the capture histories) can be calculated, but is complex, as it has to account for the fact that an individual's location is unknown when it is not observed at any point within the study. However, an alternative approach is to consider the unknown locations as missing data, in the same manner as discrete missing covariate values.

9.2.1 Notation

Formally, generalising the life history notation established in Section 2.3.3, we let $\boldsymbol{h}$ denote the set of capture histories for the individuals in the study. Let $\boldsymbol{h}^i$ denote the individual capture history of individual i, such that the elements of $\boldsymbol{h}^i$ indicate whether or not animal i was observed at the different capture times and, if it was seen, in which region it was observed. We let $h_t^i \in \{0\} \cup K_\dagger$ denote the tth element of the capture-history vector, where the 0 corresponds to the animal not being observed. Thus, since animals will not typically be observed at all capture times, the location of animal i will not be known at capture time t if $h_t^i = 0$. These missing data can be imputed by introducing auxiliary variables d_t^i, which take some value in $K_\dagger$ whenever $h_t^i = 0$ and take the value zero, if animal i is observed at capture time t, i.e. when $h_t^i \neq 0$. The non-zero elements of $\boldsymbol{d}^i = \{d_1^i, \ldots, d_T^i\}$ therefore correspond to the locations of animal i at times when that animal is unobserved and become auxiliary variables (or model parameters) which need to be estimated. There are, of course, restrictions on the values that the d_t^i can take. For example, if $d_t^i = \dagger$ then $d_l^i = \dagger$ for all $l = t, \ldots, T$, since we assume that the state $\dagger$ is synonymous with "death" (including permanent migration from the study area). Similarly, $d_t^i \neq \dagger$ if there exists $t \leq l \leq T$ such that $h_l^i \in K$. For notational convenience, it is also useful to define the auxiliary variables $z_t^i \in K_\dagger$ to be the location of animal i at time t, regardless of whether it is observed or not, so that, mathematically,

$$z_t^i = \begin{cases} h_t^i & \text{if } h_t^i \neq 0 \\ d_t^i & \text{if } h_t^i = 0. \end{cases}$$

Finally we let $\mathcal{F}_t$ denote the set of animals that have been captured by or at time t, i.e. $\mathcal{F}_t = \{i : h_l^i \neq 0 \text{ for some } l \leq t\}$.

9.2.2 Joint Probability of Capture Histories and Location

We can now form the joint probability of the capture histories and locations (i.e. discrete time varying covariate values). However, this is most easily written as a function of a number of sufficient statistics. We define these next, for $r \in K$ and $t = 1, \ldots, T$:

MISSING COVARIATE/AUXILIARY VARIABLE APPROACH

$u_t(r)$ = the number of animals that are recaptured at time t in location r;

$v_t(r)$ = the number of animals that are unobserved at time t but are imputed to be in location r at this time;

$u_t^\dagger(r)$ = the number of animals in location r at time $t-1$ that die before time t and are recovered in the interval $(t-1, t)$;

$v_t^\dagger(r)$ = the number of animals in location r at time $t-1$ that die before time t and are not recovered in the interval $(t-1, t)$; and

$w_t(r, s)$ = the number of animals that migrate from location r at time t to location s at time $t+1$, for $s \in K_\dagger$.

We note that $w_t(r, \dagger)$ denotes the number of individuals that die in the interval $(t, t+1)$, given that they are in location r at time t. Finally, for notational convenience, we define,

$$w_t(r, \cdot) = \sum_{s \in K} w_t(r, s),$$

to be the number of animals that are in location r at time t and which survive until the following capture time.

Mathematically, we can express the sufficient statistics defined above in the form:

$$u_t(r) = \sum_{i \in \mathcal{F}_{t-1}} I(h_t^i = r); \qquad v_t(r) = \sum_{i \in \mathcal{F}_{t-1}} I(d_t^i = r);$$

$$u_t^\dagger(r) = \sum_{i \in \mathcal{F}_t} I(z_t^i = r, h_{t+1}^i = \dagger); \qquad v_t^\dagger(r) = \sum_{i \in \mathcal{F}_t} I(z_t^i = r, d_{t+1}^i = \dagger);$$

$$w_t(r, s) = \sum_{i \in \mathcal{F}_t} I(z_t^i = r, z_{t+1}^i = s),$$

where $I(\cdot)$ denotes the indicator function.

We can then express the joint probability density of the observed capture history, $\boldsymbol{h}$, and unobserved locations, $\boldsymbol{d}$, for general mark-recapture-recovery data in the form,

$$\begin{aligned}
f(\boldsymbol{h}, \boldsymbol{d} | \boldsymbol{\theta}) &= \prod_{t=1}^{T-1} \prod_{r \in K} p_{t+1}(r)^{u_{t+1}(r)} (1 - p_{t+1}(r))^{v_{t+1}(r)} \\
&\quad \times \lambda_{t+1}(r)^{u_{t+1}^\dagger(r)} (1 - \lambda_{t+1}(r))^{v_{t+1}^\dagger(r)} \phi_t(r)^{w_t(r, \cdot)} (1 - \phi_t(r))^{w_t(r, \dagger)} \\
&\quad \times \prod_{s \in K} \psi_t(r, s)^{w_t(r, s)}. \tag{9.1}
\end{aligned}$$

where $p_t(r)$, $\lambda_t(r)$ and $\phi_t(r)$ are the respective recapture, recovery and survival probabilities of an individual in location r in year t. Similarly, $\psi_t(r, s)$ denotes the migration probability from location r to s in the interval $[t, t+1)$. The first line of the joint probability relates to the actual live resightings; the second line to dead recoveries and survival of the individual; and the third line the migration between different locations. Note that if we only have capture-recapture data (i.e. no dead recoveries), then we simply omit the λ terms from the likelihood.

9.2.3 Posterior Distribution

We denote the set of parameters by $\boldsymbol{\theta} = \{\boldsymbol{p}, \boldsymbol{\phi}, \boldsymbol{\lambda}, \boldsymbol{\psi}\}$. The corresponding joint posterior distribution over the parameters, $\boldsymbol{\theta}$, and auxiliary variables (or missing locations), $\boldsymbol{d}$, is expressed as,

$$\pi(\boldsymbol{\theta}, \boldsymbol{d}|\boldsymbol{h}) \propto f(\boldsymbol{h}, \boldsymbol{d}|\boldsymbol{\theta})p(\boldsymbol{\theta}),$$

where $p(\boldsymbol{\theta})$ denotes the prior on the parameters $\boldsymbol{\theta}$. Typically, (conjugate) beta priors are specified on the survival, recapture and recovery probabilities (so that they are constrained to the interval $[0,1]$), and a (conjugate) Dirichlet prior is specified on the migration probabilities (to ensure that the migration probabilities are restricted to the interval $[0,1]$ and sum to unity). In particular, for $t = 1, \ldots, T-1$, we set,

$$\begin{aligned}
p_{t+1}(r) &\sim Beta(\alpha_p, \beta_p) \\
\lambda_{t+1}(r) &\sim Beta(\alpha_\lambda, \beta_\lambda) \\
\phi_t(r) &\sim Beta(\alpha_\phi, \beta_\phi) \\
\psi_t(r, 1), \ldots, \psi_t(r, k) &\sim Dir(\alpha_1, \ldots, \alpha_k).
\end{aligned}$$

For simplicity, we assume that the same prior is specified over all times and regions. However, this can clearly be relaxed in the presence of expert prior information.

9.2.4 MCMC Algorithm

Within the Bayesian MCMC implementation of this approach, we need to update every parameter and auxiliary variable at each iteration of the Markov chain. We consider each of these in turn:

Step 1: Updating the Model Parameters

Due to the construction of the joint posterior distribution over all parameters and auxiliary variables, it can be shown that the posterior conditional distribution of each is of standard form, so that the Gibbs sampler can be easily implemented. In particular, for $t = 1, \ldots, T-1$, we have the following posterior conditional distributions for the model parameters:

$$\begin{aligned}
p_{t+1}(r)|\boldsymbol{h}, \boldsymbol{d} &\sim Beta\left(\alpha_p + \mathrm{u}_{t+1}(r), \beta_p + \mathrm{v}_{t+1}(r)\right) \\
\phi_t(r)|\boldsymbol{h}, \boldsymbol{d} &\sim Beta(\alpha_\phi + \mathrm{w}_t(r, \cdot), \beta_\phi + \mathrm{w}_t(r, \dagger)) \\
\lambda_{t+1}(r)|\boldsymbol{h}, \boldsymbol{d} &\sim Beta(\alpha_\lambda + \mathrm{u}^\dagger_{t+1}(r), \beta_\lambda + \mathrm{v}^\dagger_{t+1}(r)) \\
\psi_t(r,1), \ldots, \psi_t(r,k)|\boldsymbol{h}, \boldsymbol{d} &\sim Dir(\alpha_1 + \mathrm{w}_t(r,1), \cdots, \alpha_k + \mathrm{w}_t(r,k)).
\end{aligned}$$

Note that we have described the updating procedure for the saturated model, where the model parameters are fully time and location dependent. Submodels may be fitted to the data corresponding to different restrictions placed on the parameters. For example, suppose that the recapture probability is independent of location, so that,

$$p_t(r) = p_t \quad \forall r \in K.$$

The corresponding posterior conditional distribution for p_t $(t = 2, \ldots, T)$ is,

$$p_t|\boldsymbol{h}, \boldsymbol{d} \sim Beta\left(\alpha_p + \sum_{r \in K} u_t(r), \beta_p + \sum_{r \in K} v_t(r)\right).$$

Similarly, suppose that we consider a model with the survival probability independent of time, so that,

$$\phi_t(r) = \phi(r) \quad t = 1, \ldots, T-1.$$

The corresponding posterior conditional distribution is of the form,

$$\phi_t(r)|\boldsymbol{h}, \boldsymbol{d} \sim Beta\left(\alpha_\phi + \sum_{t=1}^{T-1} \mathrm{w}_t(r, \cdot), \beta_\phi + \sum_{t=1}^{T-1} \mathrm{w}_t(r, \dagger)\right).$$

In other words, the terms of the beta distribution simply involve summing over the given sufficient statistics corresponding to the parameters that are specified to be equal within the submodel. For further details, see King and Brooks (2002a), who discuss more general models, and describe the corresponding extension of the updating algorithm to these more general cases.

Step 2: Updating the Auxiliary Variables

Updating the auxiliary variables, $\boldsymbol{d}$, (corresponding to unknown locations when an individual is not observed), involves considering each of the non-zero d_t^i and using a Gibbs sampler update to determine a new location for animal $i = 1, \cdots, I$ at time $t = f_{i+1}, \cdots, T$, since animal i is first observed at time f_i and hence the location of the individual i at time f_i is observed. Since the d_t^i take only a small range of discrete values, the posterior conditional can be easily computed by calculating the conditional probabilities for each possible location at this time, given all of the other parameters.

We note that from the Markovian assumptions of the Arnason-Schwarz model, we need only condition on the previous and following locations of the given animals in updating the auxiliary variables, relating to the location of the animals when they are not observed. We now consider different cases for the update of d_t^i relating to the simulated location of an animal in the study. There are four cases that we consider, relating to those animals either observed alive, or simulated alive, later in the study: those found dead at the following capture time, those simulated to be dead at the following capture time, and those unseen at the final capture time of the study.

Case I: $z^i_{t+1} \in K$, i.e. Animals Known to Be Alive at the Following Capture Time

In this case, since animals are either observed, or simulated to be, alive at the following capture time (i.e. $z^i_{t+1} \in K$), this implies that they must be alive at time t. We simulate their location using a Gibbs sampler. The probability of an animal being in location $k \in K$ at time t can be expressed as a product of recapture, survival and migration probabilities. For $r, s \in K$, and conditioning on the location of animal i at times $t-1$ and $t+1$, the probability that animal i is in location $k \in K$ at time t can be expressed as,

$$\mathbb{P}(d^i_t = k | z^i_{t-1} = r; z^i_{t+1} = s) \propto \phi_{t-1}(r)\psi_{t-1}(r,k)(1-p_t(k))\phi_t(k)\psi_t(k,s),$$

so that the animal survives time $t-1$ when in location r, migrates to location k in the interval $(t-1,t)$ and is not observed at this time, then survives time t and migrates to location s by time $t+1$. The constant of proportionality is denoted by c, where,

$$\begin{aligned} c^{-1} &= \sum_{k \in K} \mathbb{P}(d^i_t = k | z^i_{t-1} = r; z^i_{t+1} = s) \\ &= \sum_{j \in K} \phi_{t-1}(r)\psi_{t-1}(r,j)(1-p_t(j))\phi_t(j)\psi_t(j,s). \end{aligned}$$

Clearly, the constant of proportionality is simply defined to be the reciprocal of sum of the probability of each possible location that the animal can be in, to ensure that the probabilities sum to unity. Thus, we have the explicit conditional probability,

$$\mathbb{P}(d^i_t = k | z^i_{t-1} = r; z^i_{t+1} = s) = \frac{\phi_{t-1}(r)\psi_{t-1}(r,k)(1-p_t(k))\phi_t(k)\psi_t(k,s)}{\sum_{j \in K} \phi_{t-1}(r)\psi_{t-1}(r,j)(1-p_t(j))\phi_t(j)\psi_t(j,s)}.$$

Case II: $h^i_{t+1} = \dagger$, i.e. Animals Found Dead at the Following Capture Time

For this case, the animal is recovered dead by time $t+1$, so that the animal must be alive at time t, but then subsequently die and be recovered in the interval $(t, t+1]$. Conditioning on the location of the animal at time $t-1$ (i.e. z^i_{t-1}) and $h^i_{t+1} = \dagger$, we have for $r \in K$,

$$\mathbb{P}(d^i_t = k | z^i_{t-1} = r; h^i_{t+1} = \dagger) \propto \phi_{t-1}(r)\psi_{t-1}(r,k)(1-p_t(k))(1-\phi_t(k))\lambda_{t+1}(k),$$

so that the individual survives time $t-1$ when they are in location r, migrates to location k by time t where they are not observed, before dying in the interval $(t, t+1]$ and are recovered dead. Once more the constant of proportionality (to ensure all the probabilities sum to unity over all possible locations), is given by

$$\left[\sum_{j \in K} \phi_{t-1}(r)\psi_{t-1}(r,j)(1-p_t(j))(1-\phi_t(j))\lambda_{t+1}(j) \right]^{-1}.$$

Case III: $d^i_{t+1} = †; z^i_{t-1} \in K$, *i.e. Animals That Are Simulated to Be Dead at the Following Capture Time, but Alive at Time* $t - 1$

In this case, since the animal is only simulated to be dead at time $t + 1$, it is possible for the animal to be either dead at time t but not recovered, or simply alive and unobserved in a region $k \in K$. Thus we need to consider both possibilities. For $k \in K$ we have,

$$\mathbb{P}(d^i_t = k | z^i_{t-1} = r; d^i_{t+1} = †) \propto \phi_{t-1}(r)\psi_{t-1}(r,k)(1-p_t(k))(1-\phi_t(k))(1-\lambda_{t+1}(k)),$$

so that the individual survives time $t - 1$, given it is in location r, migrates to location k by time t and is not observed, then dies in the interval $(t, t+1]$ and is not recovered dead.

Alternatively, the probability that the animal is dead by time t, given its location at time $t - 1$, is given by,

$$\mathbb{P}(d^i_t = † | z^i_{t-1} = r; d^i_{t+1} = †) \propto (1 - \phi_{t-1}(r))(1 - \lambda_t(r)),$$

so that the animal dies in the interval $(t-1, t]$, given that it was in location r at time $t-1$, and is not recovered. The constant of proportionality is calculated in the standard manner by taking the reciprocal of the sum over the probabilities of the possible simulated locations for $d^i_t \in K_†$.

Case IV: $t = T$ *and* $z_{T-1} \in K$ *i.e. Animals Not Observed at the Final Capture Time, but Alive at Time* $T - 1$

For this case, there is no subsequent information concerning the animals so that animals may be either alive or dead at this time, and we can only condition on the location of the animal at time $T - 1$. Initially we consider the animal to be simulated alive, so that for $k \in K$, and $r \in K$,

$$\mathbb{P}(d^i_T = k | z^i_{T-1}(r)) \propto \phi_{T-1}(r)\psi_{T-1}(r,k)(1-p_T(k)),$$

so that the individual survives time $T - 1$, given it was in location r, migrates to location k by time T but is not observed.

Similarly, for $r \in K$, the probability that the animal dies is given by,

$$\mathbb{P}(d^i_T = † | z^i_{T-1}(r)) \propto (1 - \phi_{T-1}(r))(1 - \lambda_{T-1}(r)),$$

so that the animal dies in the interval $(T-1, T)$, given that it was alive and in location r at time $T - 1$ and is not recovered. The constant of proportionality is once more calculated in the standard way to ensure that the probabilities sum to unity.

Clearly there are many other types of updates that can be used for the auxiliary variables. For example, we could consider a procedure whereby animals may also be simulated to die at any time following their final capture. Previously this was not possible as an animal was assumed to be alive, if it was simulated to be alive at the following capture time. Once more, a Gibbs sampler can be implemented for this. However, the probability of the full capture history after this time needs to be taken into account, as the death of an animal alters all following locations of the animal (as it remains in the state of death). Alternatively, we could consider a procedure that initially updates the time of death of an animal (if at all) using an MH random walk. Given the time of death of an animal (or its survival until the end of the study), we can employ a Gibbs update to simulate the location of the animal at each time it is known to be alive, yet unobserved. Finally we could of course use "larger" updates that simulate the location of the animal at each time before its final capture using the Gibbs sampler described above, and then all of the locations of the animal after its final sighting using an MH update. The simple procedure described in detail above appears to work adequately in practice for examples considered by the authors.

The introduction of the auxiliary variables $\boldsymbol{d}$ (and $\boldsymbol{z}$) makes the expression for the posterior distribution very simple (up to proportionality). Additional simplicity is achieved through the introduction of the sufficient statistic terms defined above. However, the updates for the auxiliary variables, recording the locations of the animals at unobserved times, does require the raw life history data. Thus, in general, there is a trade-off between updating a large number of parameters within the auxiliary variable approach and the computation of the complex likelihood expression for the sufficient statistic approach (see for example, King and Brooks 2003a). The relative efficiency of the two approaches will typically be data dependent.

Example 9.1. Lizards

We now consider data presented by Massot et al. (1992) concerning the common lizard, *Lacerta vivipara*, and discussed by Dupuis (1995). The data arise from a three year study in which animals were recaptured in one of three regions during the years 1989 to 1991, (animals were not recovered dead within this study). Within each year, there are two capture events, one in June and the other in August, so that there are a total of six distinct capture times. These data were presented in Exercise 2.5. Individuals can be observed in three distinct locations, which we label A, B and C, so that $K = \{A, B, C\}$.

Dupuis (1995) describes additional information relating to the different model parameters, including differing effort in collecting the data and some knowledge of the annual behaviour of the lizards. Based upon information of this sort, Dupuis (1995) specifies informative priors for each parameter within the saturated model (full time and location dependence). However, we assume that we have no prior information, and specify,

$$p_t(r) \sim U[0,1]; \quad \text{and} \quad \phi_t(r) \sim U[0,1],$$

for all times t and regions $r \in \{A, B, C\}$. Alternatively, for the migration probabilities, we need to specify a prior which preserves the sum to unity constraint. Thus, we specify a Dirichlet prior for the migration probabilities from a given location. In particular, we specify a uniform prior on the simplex, such that,

$$\{\psi_t(r, A), \psi_t(r, B), \psi_t(r, C)\} \sim Dir(1, 1, 1).$$

The MCMC algorithm described above is implemented and the corresponding posterior mean and standard deviation of each recapture, survival and migration probability presented in Table 9.1.

Recall that the times 1, 3 and 5 correspond to the June capture events; and times 2, 4 and 6 to the August capture events. Straightaway we can see

Table 9.1 Posterior Means and Standard Deviations (in Brackets) for the Recapture, Survival and Migration Probabilities under the Saturated Model for Lizard Data

(a) Recapture Probabilities

Stratum	Time				
	2	3	4	5	6
A	0.83 (0.08)	0.61 (0.12)	0.78 (0.13)	0.38 (0.22)	0.66 (0.22)
B	0.78 (0.09)	0.75 (0.13)	0.76 (0.14)	0.34 (0.23)	0.62 (0.24)
C	0.85 (0.08)	0.72 (0.14)	0.81 (0.14)	0.18 (0.19)	0.60 (0.24)

(b) Survival Probabilities

Stratum	Time				
	1	2	3	4	5
A	0.82 (0.08)	0.67 (0.12)	0.86 (0.11)	0.54 (0.21)	0.62 (0.23)
B	0.93 (0.06)	0.52 (0.11)	0.81 (0.13)	0.32 (0.20)	0.66 (0.23)
C	0.87 (0.07)	0.51 (0.11)	0.81 (0.13)	0.47 (0.22)	0.54 (0.28)

(c) Migration Probabilities

Migration Regions	Time				
	1	2	3	4	5
$A \to A$	0.88 (0.06)	0.88 (0.08)	0.76 (0.11)	0.58 (0.21)	0.54 (0.21)
$A \to B$	0.08 (0.05)	0.06 (0.06)	0.18 (0.10)	0.18 (0.16)	0.21 (0.17)
$A \to C$	0.04 (0.04)	0.06 (0.05)	0.06 (0.06)	0.23 (0.19)	0.25 (0.19)
$B \to A$	0.11 (0.06)	0.14 (0.09)	0.08 (0.08)	0.23 (0.19)	0.24 (0.18)
$B \to B$	0.82 (0.07)	0.73 (0.12)	0.75 (0.12)	0.49 (0.23)	0.48 (0.22)
$B \to C$	0.07 (0.05)	0.13 (0.09)	0.17 (0.11)	0.28 (0.21)	0.28 (0.20)
$C \to A$	0.03 (0.03)	0.06 (0.06)	0.08 (0.07)	0.27 (0.22)	0.30 (0.22)
$C \to B$	0.14 (0.07)	0.23 (0.11)	0.17 (0.11)	0.31 (0.24)	0.28 (0.21)
$C \to C$	0.83 (0.07)	0.71 (0.12)	0.75 (0.12)	0.42 (0.26)	0.41 (0.24)

from the posterior results that the survival probabilities between consecutive capture events appear to be lower for times 3 and 5. This is unsurprising, since the survival to the following capture event is over a longer time period, and over the winter months, where we may expect more deaths to occur. Alternatively, the recapture probabilities appear to be lower at times 2 and 4, shortly following their mating season in May. Additionally, there appears to be a greater amount of uncertainty in relation to the survival and migration probabilities from time 4 onward, and for the recapture probabilities from time 5 onward. This is most likely as a result of the significantly lower recapture probabilities at time 5, so that fewer individuals are observed.

□

9.3 Model Discrimination and Averaging

From the results obtained in the previous example, the issue of model discrimination clearly arises. For example, the migration probabilities appear to be very similar for times 1 through 3 (and possibly for times 4 and 5 with the reasonable uncertainty in the parameter estimates at these times). As a result, could they be modelled by a single set of migration probabilities between each site, but constant over time? Mathematically, this would be modelled by,

$$\psi_t(r,s) = \psi(r,s),$$

for all $t = 1, \ldots, 5$. Alternatively, the question may arise as to whether there is a seasonal aspect to the survival probabilities, which is constant over years. This biological hypothesis would be represented by the parameter restriction,

$$\phi_1(r) = \phi_3(r) = \phi_5(r), \quad \text{and} \quad \phi_2(r) = \phi_4(r),$$

for all $r \in \{A, B, C\}$. We address the issue of model discrimination for multi-site data next by considering two specific examples for illustration using the ideas presented in Chapter 6.

Example 9.2. Lizards Revisited

We reconsider the above example corresponding to lizards, but focus on the issue of determining the underlying time and/or location dependence of the demographic parameters. See King and Brooks (2002a) for further discussion and details of this example, including the specification of an informative prior in the presence if model uncertainty. Firstly, we need to define the set of possible submodels that we will consider.

Submodels

Often we may be interested in submodels of the saturated Arnason-Schwarz model given above. For the migration probabilities, we consider only a possible time dependence, but for the recapture (or recovery) and survival probabilities there is possible time and location dependence. We restrict attention here to the case where the location dependence is conditional on the time dependence. This is analogous to the type of models discussed in Section 6.7, where there was potentially both age and time dependence with the time dependence conditional on the age dependence.

To illustrate the set of possible submodels and the notation that we will use, consider the case where there are 6 capture events in 3 locations, A, B, C (as we have for the lizard data). The set of parameters in a typical model for the recapture probability may be,

$$p_{2,3,4}(A, B, C); p_{5,6}(A, B); p_{5,6}(C),$$

where the subscript denotes the time dependence and the brackets the location dependence. So, this model has a common recapture probability at times 2, 3 and 4, which is also constant over location, and another recapture probability for times 5 and 6, where the recapture probability is different in location C, compared to A and B. We allow all possible combinations of times, and conditional on the time dependence for the parameter (recapture and/or survival probabilities), all possible combinations of location dependence.

Prior Specification

We need to specify priors on the parameters for each possible model. Recall that in Example 9.1, for the saturated model, we specified uninformative priors. For each possible submodel, we specify the analogous $U[0,1]$ prior for each recapture and survival probabilities, and for the migration probabilities a $Dir(1,1,1)$ prior. Consistent informative priors across different models can be difficult to construct. One possible approach is to define the informative priors for the saturated model, and use these as a basis to form the priors for the parameters in submodels. For example, Dupuis (1995) describes an informative prior for the parameters in the saturated model; and King and Brooks (2002a) extend this to construct consistent (informative) priors across all submodels.

With the additional model uncertainty, we need to place a prior probability on each possible model. We may believe that some groupings of capture events are more likely a priori than others for biological reasons or the design of the experiment. For example, we might believe that the grouping of capture times might correspond to a year effect ($p_2, p_{3,4}, p_{5,6}$) or a seasonal effect ($p_{2,4,6}, p_{3,5}$) and place higher probability mass on such arrangements. However, in the absence of quantifiable information as to how much more likely such arrangements might be than other groupings of times, we simply adopt a flat prior over all possible models.

RJMCMC Algorithm*

Within the constructed Markov chain we perform the following steps at each iteration:

i) update each parameter in the model using the Gibbs sampler;

288   MULTI-STATE MODELS

ii) update the auxiliary variables corresponding to the location of the individuals when they are unobserved;

iii) update the time dependence for the recapture, survival and migration probabilities;

iv) update the location dependence for the recapture and survival probabilities.

Moves (i) and (ii) are discussed in Section 9.2.4 initially for the saturated model, but also for any possible submodel. Thus, we omit the description of these moves here and discuss only the model updates in turn.

(iii) Updating Time Dependence

We cycle through each set of demographic parameters (recapture, survival and migration probabilities) and propose to update the time dependence in turn. Let m denote the current model and m' the proposed model. We begin with considering the recapture probabilities. With probability $\frac{1}{2}$ we propose to increase the number of time groups by one, and with probability $\frac{1}{2}$, decrease the number of time groups by one. Increasing the number of time groups involves splitting an existing time group into two; whereas decreasing the number of time groups involves merging two time groups. Note that if there is only a single time group and we propose a merge move, we automatically reject the proposed move. Similarly, if the marginal model for the times is the saturated model (i.e. a different recapture probability for each time) and we propose a split move, we again automatically reject the proposed move. We begin by considering a split move.

Without loss of generality, suppose that there are V time groups for the recapture probabilities which have more than a single element. We choose each of these groups with equal probability, i.e. with probability $\frac{1}{V}$. Suppose that the time group chosen is denoted by $\boldsymbol{t}$ and contains a total of n times. If there is only a single element in the given time group (i.e. $n = 1$), we cannot split this and the model move is automatically rejected. Otherwise, we propose to split the time group $\boldsymbol{t}$ into two non-empty sets, $\boldsymbol{t}_1$ and $\boldsymbol{t}_2$. We choose each possible set of time groups with equal probability. There are a total of $2^{(n-1)} - 1$ possible combinations for splitting the set $\boldsymbol{t}$ into two non-empty sets. Suppose that the current model is denoted by m and as usual we propose to move to model m' using the above procedure. This gives,

$$\mathbb{P}(m'|m) = \frac{1}{2} \times \frac{1}{V} \times \frac{1}{2^{(n-1)} - 1},$$

where the term $\frac{1}{2}$ corresponds to the probability of proposing a split move. We need to propose recapture probability values for the new time groups $\boldsymbol{t}_1$ and $\boldsymbol{t}_2$. This will depend on the location dependence for the given time group $\boldsymbol{t}$. Let the location dependence be denoted by $\boldsymbol{s}$ for time group $\boldsymbol{t}$. For each element of $\boldsymbol{s}$, $s \in \boldsymbol{s}$, we set,

$$\begin{aligned} p'_{t_1}(s) &= p_{\boldsymbol{t}}(s) + \epsilon(s) \\ p'_{t_2}(s) &= p_{\boldsymbol{t}}(s) - \epsilon(s), \end{aligned}$$

MODEL DISCRIMINATION AND AVERAGING

where $\epsilon(s) \sim U[-\eta(s), \eta(s)]$, such that, $\eta(s) = \min[\delta_1, p_t(s), 1 - p_t(s)]$, ($\delta_1$ chosen via pilot tuning), constraining the proposed recapture probabilities to the interval $[0, 1]$. For example, suppose that $s = \{\{A, B\}, \{C\}\}$, then we consider $s = \{A, B\}$ and $s = \{C\}$ in the above proposed parameter updates.

In order to calculate the Jacobian, we simply need to know the number of location groupings in s. Suppose that there are n_s different recapture probabilities (in terms of location) for time group t. So for the above example, where $s = \{\{A, B\}, \{C\}\}$, $n_s = 2$. The corresponding Jacobian for this model move is simply $|J| = 2^{n_s}$.

The final term we need to calculate is $\mathbb{P}(m|m')$. This is determined by the way that we combine the different age groupings. For our example, to retain the reversibility constraints, we only propose to combine the time groups that have the same location dependence. Suppose that in model m' there are a total of n_g different location groupings, s, common to more than one time group. If $n_g = 0$ we simply skip the model updating step, since there are no time groups that can be merged. We choose each of these possible location groupings with equal probability. For each of the time groups with the given location grouping chosen, we propose to combine any two of the time groups with equal probability. If there are R times with the chosen location grouping in model m', there are a total of $^{R}C_2$ possible time groups that can be combined. Thus, we have that,

$$\mathbb{P}(m|m') = \frac{1}{2} \times \frac{1}{n_g} \times \frac{1}{{}^{R}C_2},$$

where, once again, the $\frac{1}{2}$ corresponds to the probability of proposing a merge move. The model move is then accepted with the standard acceptance probability, $\min(1, A)$, with,

$$A = \frac{\pi(\boldsymbol{p}', \boldsymbol{\phi}, \boldsymbol{\psi}|\boldsymbol{h}, \boldsymbol{d})\mathbb{P}(m|m')}{\pi(\boldsymbol{p}', \boldsymbol{\phi}, \boldsymbol{\psi}|\boldsymbol{h}, \boldsymbol{d})\mathbb{P}(m'|m)q(\boldsymbol{\epsilon})}|J|,$$

where $\boldsymbol{\epsilon} = \{\epsilon(s) : s \in \boldsymbol{s}\}$ and $q(\boldsymbol{\epsilon})$ denotes the corresponding proposal density.

In the reverse move, we propose to reduce the number of time groups by one. As usual, this model move is defined given the above algorithm. In particular, given the two groups that we propose to merge, we simply set the recapture probabilities in the new time group to be the mean of the current recapture probabilities for each location. The corresponding acceptance probability is simply $\min(1, A^{-1})$, where A is given above.

Note that it is possible to simplify the choosing of two time groups to be combined. For example, simply choose any two time groups to be combined. If the location dependence is different, automatically reject the proposed model move; otherwise, if the location dependence is the same in the two groups, continue as usual. This simplifies the updating algorithm, and in particular the form of $\mathbb{P}(m|m')$, but will decrease the mean acceptance probability of moving between models, since it is likely that a large number of model moves will be automatically rejected, and decrease the mixing.

The analogous model move updates are used for the survival probabilities. Conversely, for the migration probabilities, the model moving algorithm is simplified since there is no location dependence within the time grouping structure. We consider the analogous split and merge model moves, increasing or decreasing the number of time groups by one. Within each step of the Markov chain, we choose each of these model moves with equal probability. We begin with the split move. Suppose that there are a total of V time groupings with more than a single element. We choose each of these with equal probability, and use the same procedure for randomly splitting the age group into two non-empty groups, each with equal probability.

Suppose that we choose to split time grouping t into t_1 and t_2. We propose new migration probabilities,

$$\psi'_{t_1}(r,s) = \psi_{t_1}(r,s) + \omega(r,s) \quad \text{for all } r \in \{A,B,C\};\ s \in \{A,B\}$$
$$\psi'_{t_2}(r,s) = \psi_{t_2}(r,s) \quad \text{for all } r \in \{A,B,C\};\ s \in \{A,B\},$$

with probability $\frac{1}{2}$; otherwise we set

$$\psi'_{t_1}(r,s) = \psi_{t_1}(r,s) \quad \text{for all } r \in \{A,B,C\};\ s \in \{A,B\}$$
$$\psi'_{t_2}(r,s) = \psi_{t_2}(r,s) + \omega(r,s) \quad \text{for all } r \in \{A,B,C\};\ s \in \{A,B\}.$$

To retain the sum-to-unity constraint on the migration probabilities, for $j = 1,2$ we set,

$$\psi'_{t_j}(r,C) = 1 - \psi'_{t_j}(r,A) - \psi'_{t_j}(r,B).$$

The parameters $\omega(r,s)$ are simulated from some proposal distribution. In our case, we assume $\omega(r,s) \sim N(0,\sigma^2)$ for $r \in \{A,B,C\}$ and $s \in \{A,B\}$, where σ^2 is chosen via pilot tuning. The migration probabilities for the times in each of the two new sets of times remain unchanged with probability $\frac{1}{2}$. Clearly, there are many different possible updating procedures that we could implement for the migration parameters. However, this updating procedure appeared to perform best in practice, in terms of both the acceptance probabilities of the proposed moves and the mixing of the chain.

The corresponding Jacobian for these moves is easily shown to be equal to unity, since the map from $(\boldsymbol{\psi},\boldsymbol{\omega})$ to $(\boldsymbol{\psi}')$ is the identity. Finally, in order to calculate the acceptance probability for this move type, we need to calculate an expression for the probability of proposing the different moves. Let n denote the number of elements in the chosen age group, t. Following the same argument as above for the recapture probabilities, we have,

$$\mathbb{P}(m'|m) = \frac{1}{2} \times \frac{1}{V} \times \frac{1}{2^{(n-1)} - 1}.$$

Similarly, in the reverse move, we propose to combine any two time groups. Suppose that there are a total of $R = (V+1)$ time groupings in model m'. The corresponding probability of moving to model m, conditional on the chain being in model m' is given by,

$$\mathbb{P}(m|m') = \frac{1}{2} \times \frac{1}{{}^{R}C_2},$$

MODEL DISCRIMINATION AND AVERAGING

using the analogous argument as above for the recapture probabilities (ignoring the location dependence that is absent for the migration probabilities). The move is accepted with the standard acceptance probability. The reverse move follows similarly.

(iv) Updating Location Dependence

Within each iteration of the Markov chain we cycle through the recapture and survival probabilities and propose to update the location dependence for each individual time grouping for these demographic parameters. Without loss of generality, suppose that we consider the recapture probabilities and time grouping t. There are a total of 5 possible location dependence structures,

$$\{\{A, B, C\}\}, \quad \{\{A\}, \{B, C\}\},$$
$$\{\{A, B\}, \{C\}\}, \quad \{\{A, C\}, \{B\}\},$$
$$\{\{A\}, \{B\}, \{C\}\}.$$

With probability $\frac{1}{2}$ we propose a split move, increasing the number of age groups by one; otherwise we propose a merge move, decreasing the number of age groups by one. If we propose a split move when the current model is $\{\{A\}, \{B\}, \{C\}\}$, the move is automatically rejected. Similarly, if we propose a merge move when the location dependence is $\{\{A, B, C\}\}$, the move is once again rejected. We now consider the case when the move is not automatically rejected.

Suppose that we propose a split move from model m to model m'. In this instance (as there are only three locations), there can only be one set of locations that can be split, so we automatically choose this element. We denote this element by s, so that $s \in \{\{A, B, C\}, \{A, B\}, \{A, C\}, \{B, C\}\}$. If $s = \{A, B, C\}$, then there are 3 possible models that can be proposed:

$$\{\{A\}, \{B, C\}\}, \quad \{\{A, C\}, \{B\}\} \quad \text{and} \quad \{\{A, B\}, \{C\}\}.$$

We choose each of these with equal probability. Alternatively, for any other set of locations that can be split, there are only two elements, and so only one possible model that can be proposed, namely, $\{\{A\}, \{B\}, \{C\}\}$. Letting, n_s denote the number of elements in s, we have that,

$$\mathbb{P}(m'|m) = \frac{1}{2} \times \frac{1}{2^{(n_s-1)} - 1}.$$

Conversely, consider moving from model m' to model m. Following a similar argument to that above, we have that,

$$\mathbb{P}(m|m') = \frac{1}{2} \times \frac{1}{{}^{(n_s+1)}C_2}.$$

Suppose that we split s into new sets s_1 and s_2. In order to generate the new parameter values associated with this move, we adopt a similar approach to that used in step (iii) above and set

$$p'_t(s_1) = p_t(s) + \zeta, \quad p'_t(s_2) = p_t(s) - \zeta,$$

where

$$\zeta \sim U[-Z, Z], \quad Z = \min[\delta_2, p_t(s), 1 - p_t(s)],$$

for some predefined δ_2 chosen via pilot tuning. The corresponding Jacobian for this defined model move is simply equal to 2. All of these terms can be substituted into the standard acceptance probability in order to obtain the corresponding acceptance probability of the model move.

In the reverse move, the recapture probabilities for the new merged location groups is simply taken to be the mean of recapture probabilities of the individual groups.

This model move can easily be extended to allow for a greater number of locations, with the moves very similar to those described above for the time dependence. King and Brooks (2002a) also describe the more complex case for any given number of possible locations.

Results

We concentrate on the marginal models for the time and location dependence. Table 9.2 provides the posterior probabilities for the time dependence structure of the recapture, survival and migration probabilities; Table 9.3 provides the posterior probabilities for the location dependence for the recapture and survival probabilities. There appears to be a reasonable amount of uncertainty with respect to the time dependence structure for the recapture and survival probabilities, but very strong posterior support for a single model for the migration probabilities, corresponding to constant migration over each time period.

There may be some evidence for a weak seasonal effect for the survival probabilities, with the June (odd times) and August (even times) grouped together in many of the top few models. Thus, there appears to be no annual effect, but some weak support for a seasonal effect, though it is plausible that the survival probabilities simply differ at each capture time. Further data would be required to investigate this aspect further. Unsurprisingly, the posterior mean survival probabilities for the June through August period (odd times) are all significantly higher than the August through June (even times). For example, the posterior mean survival probability for times 1 and 3 are all approximately 90%, whereas for times 2 and 4, the corresponding posterior mean survival probability vary between 37 and 56%. Note that the survival probability at time 5 has very little posterior precision as a result of the reduced amount of data at the end of the study. In addition, there is a significantly smaller recapture probabilities at time 5 (with posterior mean between 20 and 31%, dependent on location), compared to other time periods (posterior means between 67 and 82%, dependent on time and location), resulting in less individuals being observed and hence less information regarding their future state (dead or alive). This is probably also the reason for the general uncertainty as to the model relating to ϕ_5 seen in Table 9.2. Conversely, there is very strong evidence that the migration probabilities are constant over all re-

Table 9.2 The Posterior Marginal Probabilities of the Most Probable Groupings of Times for the Recapture, Survival and Migration Probabilities for Lizards

(a) Recapture Probabilities

Model	Posterior Probability
$p_{2,3,4}, p_{5,6}$	0.150
$p_{2,3,4}, p_5, p_6$	0.146
$p_{3,4}, p_2, p_5, p_6$	0.086
p_2, p_3, p_4, p_5, p_6	0.085
$p_{2,3}, p_4, p_5, p_6$	0.079
$p_{2,3,4,6}, p_5$	0.061
$p_{2,4}, p_{3,6}, p_5$	0.061
$p_{2,4,6}, p_3, p_5$	0.059

(b) Survival Probabilities

Model	Posterior Probability
$\phi_{1,3}, \phi_2, \phi_4, \phi_5$	0.250
$\phi_1, \phi_2, \phi_3, \phi_4, \phi_5$	0.128
$\phi_{1,3,5}, \phi_2, \phi_4$	0.113
$\phi_{1,3}, \phi_{2,5}, \phi_4$	0.082
$\phi_{1,3}, \phi_{2,4}, \phi_5$	0.075
$\phi_{1,5}, \phi_2, \phi_3, \phi_4$	0.058
$\phi_1, \phi_2, \phi_{3,5}, \phi_4$	0.056

(c) Migration Probabilities

Model	Posterior Probability
$\psi_{1,2,3,4,5}$	0.962

capture times. Under this constant migration probability model, the posterior mean (and standard deviation) of the probability of remaining in the same location between recapture events is 0.919 (0.034), 0.853 (0.048), 0.846 (0.047) for locations A, B and C, respectively.

Table 9.3 Posterior Marginal Probabilities for the Arrangement of Locations at Different Times for the Recapture and Survival Probabilities for the Lizard Data

(a) Recapture Probabilities

	Location Groupings				
Time	A,B,C	AB, C	AC, B	A, BC	ABC
2	0.045	0.160	0.141	0.129	0.525
3	0.062	0.162	0.145	0.165	0.467
4	0.058	0.172	0.154	0.138	0.478
5	0.144	0.211	0.158	0.204	0.283
6	0.114	0.182	0.178	0.173	0.354

(b) Survival Probabilities

	Location Groupings				
Time	A,B,C	AB,C	AC,B	A,BC	ABC
1	0.045	0.133	0.144	0.147	0.532
2	0.102	0.194	0.155	0.185	0.363
3	0.048	0.148	0.142	0.137	0.525
4	0.144	0.179	0.183	0.207	0.286
5	0.122	0.184	0.173	0.175	0.346

□

Example 9.3. Hector's Dolphins – Catch-Effort Models

We now consider another multi-site capture-recapture data set, relating to a population of Hector's dolphins off the coast of Akaroa, New Zealand, collected annually between 1985 and 1993. The study site is divided into three locations, which we label 1, 2 and 3. The data-collection process involves a boat going out to observe the dolphins within the inshore waters of the different areas over a number of days each year. Within each trip, individual dolphins sighted are uniquely identified via markings on their dorsal fin and/or body (see Hammond et al. 1990, for example). The data comprise the capture histories of each of 102 individuals, detailing the years and locations that each dolphin is observed. The data set is presented in Table 2.4; for further details see Cameron et al. (1999) and King and Brooks (2004a,b). Particular interest lies in whether a managerial conservation policy introduced in 1988 affected the dolphin population.

MODEL DISCRIMINATION AND AVERAGING

Model Specification

Within this example we focus on identifying whether there is any change in the survival, recapture, or migration probabilities over time, following the introduction of a new conservation policy. To investigate whether this policy had an impact we consider a number of different models that represent competing biological hypotheses. We represent the different competing models, in terms of the dependence of the parameters upon the time and/or location, by placing different restrictions upon the parameters. For example, the biological hypothesis that the survival probability remains constant throughout the study, and is common to all areas would be represented by the restriction,

$$\phi_t(r) = \phi, \qquad \text{for all } r \in \mathcal{R} \text{ and } t = 1985, \ldots, 1992.$$

Clearly, this model implies that the policy had no effect upon the survival probability of the dolphins. Conversely, if we believed that the survival probability did change at the time that the policy was introduced, then we may wish to consider the model with a change-point, so that

$$\begin{aligned}\phi_t(r) &= \phi_a(r), &&\text{for } t = 1985, \ldots, 1987; \\ \phi_t(r) &= \phi_b(r), &&\text{for } t = 1988, \ldots, 1992,\end{aligned}$$

for example.

We are particularly interested in whether or not the policy that was introduced in 1988 had a significant impact upon the survival probabilities of the dolphins. Thus, we impose the constraint that there may be at most one change-point, which may or may not occur at the time that the sanctuary was introduced. Our class of models therefore comprises models for which there is either no change-point over time; or where there are the restrictions,

$$\begin{aligned}\phi_t(r) &= \phi_a(r), &&\text{for } t = 1985, \ldots, v-1; \\ \phi_t(r) &= \phi_b(r), &&\text{for } t = v, \ldots, 1992,\end{aligned}$$

for $v = 1986, \ldots, 1992$.

We also need to define the possible location dependencies. We allow all possible restrictions upon the locations, for example, a common survival probability for all locations, distinct survival in each area, or a common survival probability for two different areas which is distinct from the third. We identify these models by defining the sets of areas with common survival probabilities. For example, the model with a common survival probability over areas is denoted by {1,2,3}, while the model {1}, {2,3} denotes that there is a common survival probability in areas 2 and 3, which is distinct to that in area 1, and so on. Since we consider change-point models here, we allow the location restrictions to be placed independently on the survival probabilities both before and after the change-point, if there is one. Thus, the dependence of the parameters upon location is conditional on the year.

Catch-Effort Information

We have additional covariate information relating to the catch-effort of each year. This is given in Table 9.4.

Table 9.4 The Number of Days Spent Observing Dolphins in Each Location for Years 1986–1992

				Year				
Region	1986	1987	1988	1989	1990	1991	1992	1993
1	3	1	9	5	0	1	10	12
2	0	30	34	41	15	13	12	10
3	4	5	5	2	0	5	9	14

Source: Reprinted with permission from *The Journal of Agricultural, Biological, and Environmental Statistics*. Copyright 1999 by the American Statistical Association. All rights reserved. Published in Cameron et al. (1999, p. 128).

We specify the recapture probabilities as a function of the corresponding effort taken in that year and location. In particular, we assume that sightings in year $t = 1986, \ldots, 1992$ and location $r \in 1, 2, 3$, occur as a Poisson process with general underlying recapture intensity rate $\beta_t(r)$, so that the recapture probability is specified in the form,

$$p_t(r) = 1 - \exp[-\beta_t(r)x_t(r)],$$

where $x_t(r)$ denotes the catch-effort in year t and location r. Note that this can be reparameterised in the form,

$$p_t(r) = 1 - [1 - \alpha_t(r)]^{x_t(r)},$$

where $\alpha_t(r)$ is directly interpretable as the underlying recapture probability per unit time. We retain the recapture intensity notation.

We consider special cases for the $\beta_t(r)$ parameter, representing different possible models directly analogous to those considered above for the survival probabilities, (i.e. a maximum of a single change-point and all possible location dependencies). For example, the model,

$$\beta_t(r) = \beta(r), \qquad \text{for } t = 1986, \ldots, 1993,$$

represents the system where the recapture intensity rates depend only upon the location of the dolphin, suggesting that the dolphins may be inherently more observable in some areas than others, or that the observers themselves have more (or less) information relating to the areas where dolphins are most likely to be seen in each of the designated areas. Note that for all models we implicitly assume that the underlying recapture probability for any given year is homogeneous over days within that year.

MODEL DISCRIMINATION AND AVERAGING

Finally, we impose the same year constraints upon the migration probabilities, as for the survival probabilities for similar reasons as those discussed above. See King and Brooks (2002a) for further discussion of model structures. Discriminating between these competing models tells us about the underlying dynamics of the system in terms of the possible effect of the introduction of the sanctuary upon the survival and behaviour of the dolphins.

Prior Specification

For each of the parameters, we place an equal prior probability on each possible age dependence structure. For the recapture intensities and the survival probabilities we place a flat prior across each possible combination of locations independently within each set of ages. Note that for the recapture intensities and survival probabilities, the prior is not flat over each individual model. Placing a flat prior over the whole of the model space, so that each individual model was equally likely a priori, would result in a greater amount of prior mass on models with a single change-point compared to those with a constant probability over time. For example, when there are no change-points, there are a total of 5 possible location dependence structures as illustrated by the column headings of Table 9.6. However, when there is a single change-point, there are 5 location dependent models both for before and after that change-point and so 5^2 models in total. Thus, putting a flat prior across all models would make a change-point model five times more likely than having no change-point, since there are five times more change-point models than there are models without a change-point.

We do not have any prior information and so we specify independent vague priors on each of the parameters. In particular, for each model, we place independent $U[0,1]$ priors on each of the survival probabilities. For the migration probabilities we use a $Dir\left(\frac{1}{2}, \frac{1}{2}, \frac{1}{2}\right)$ prior, corresponding to an uninformative Jeffreys prior (see Section 4.2.3). Finally, we need to specify a prior on the recapture intensity rates. Placing a prior on the recapture intensity rates implicitly imposes a prior on the recapture probabilities, which may be more directly interpretable. Thus, we begin by specifying a prior on the recapture intensity rates and calculate the corresponding prior on the recapture probabilities themselves, to ensure that our priors are consistent with our beliefs concerning the recapture probabilities.

The recapture intensity rates are strictly positive, and so a natural prior for these parameters is $\Gamma(a,b)$. For ease of notation, we assume the model with a common underlying recapture intensity rate for each year and location, denoted by β. The same prior will then be used independently for each recapture intensity rate across all models (i.e. time and location dependence structures). Given $\beta \sim \Gamma(a,b)$, the corresponding prior on the recapture probability in any given year t and location r, $p_t(r)$ is given by,

$$f(p_t(r)) \propto (1-p_t(r))^{\frac{b}{x_t(r)}-1}(-\log(1-p_t(r)))^{a-1}.$$

Placing what is generally regarded as a vague prior on β (such as $\Gamma(0.001, 0.001)$)

does not induce a vague prior on $p_t(r)$. In this case using such a prior on β produces a prior on $p_t(r)$ (assuming $x_t(r) \neq 0$), which essentially places all the prior mass at zero and one. See Section 8.2.1 for discussion of this same prior specification issue relating to parameters expressed in the form of a logistic regression.

However, consider the form of the probability density function for $f(p_t(r))$ in more detail. Setting $a = 1$, we obtain,

$$p_t(r) \sim Beta\left(1, \frac{b}{x_t(r)}\right).$$

This suggests a possible specification for b such as, $b = \bar{x}$, where $\bar{x}$ is the mean catch-effort spent over all times and locations. This induces a flat prior for the recapture probability (i.e. $p_t(r) \sim U[0,1]$) when the mean amount of time is spent observing dolphins in the given year and location (i.e. $b = \bar{x}$). If a larger amount of time is spent ($x_t(r) > \bar{x}$) the prior is skewed to the right, and there is a greater prior mass on larger recapture probabilities. Alternatively, if a less than average amount of time is spent (i.e. $x_t(r) < \bar{x}$), the prior is skewed to the left and has more prior mass on lower recapture probabilities. There are clearly other possible values of b that may be sensible in any given application. For example, taking b to be the median catch-effort value. Prior information might be used to discern the most appropriate form of b. In addition, a prior sensitivity analysis can also be performed to observe the sensitivity of the posterior distribution on the specification of b. For the purposes of this example, we take $b = \bar{x}$.

RJMCMC Algorithm*

The reversible jump MCMC algorithm consists of two different move types: one for updating the parameters; and the other for updating the model itself. The MH random walk is used to update each of the parameters, conditional on the model. We consider the model updates in more detail.

Within each step of the Markov chain, we propose to update:

i) the number of change-points on the survival probabilities, recapture intensities and migration probabilities;

ii) the location of a change-point (if any) for each of the parameters, and

iii) the area dependence on the survival probabilities and recapture intensities for each age group.

We consider the different types of reversible jump updates in turn.

(i) Adding/Removing Change-Point

We initially consider the survival probabilities. If there is a constant survival probability, we propose to add a change-point. Otherwise, if there is already a change-point, we propose to remove it, since we only consider models with a maximum of a single change-point. We need to account for the added complication that the recapture probabilities are also location dependent. Initially,

MODEL DISCRIMINATION AND AVERAGING

suppose that we propose to add a change-point, so that the current model, denoted by m, has a common survival probability over time, and for simplicity we assume that the survival probability is also common over all areas, i.e.

$$\phi_t(r) = \phi \qquad \text{for all } r \in \{1, 2, 3\} \text{ and } t = 1985, \ldots, 1992.$$

Then we propose to move to the new model, m', with survival parameters,

$$\begin{aligned} \phi'_t(r) &= \phi + \epsilon & \text{for all } r \in \mathcal{R} \text{ and } t = 1985, \ldots, v - 1; \\ \phi'_t(r) &= \phi - \epsilon & \text{for all } r \in \mathcal{R} \text{ and } t = v, \ldots, 1992, \end{aligned}$$

where v is randomly chosen in the interval $\{1986, \ldots, 1992\}$, and

$$\epsilon \sim N(0, \sigma^2),$$

for σ^2 chosen via pilot tuning. If any $\phi'_t(r) \notin [0, 1]$ the move is automatically rejected, otherwise it is accepted with probability,

$$\min\left(1, \frac{\pi(\boldsymbol{\phi}', \boldsymbol{\psi}, \boldsymbol{\beta}, m'|\boldsymbol{h}, \boldsymbol{d})|J|}{\pi(\boldsymbol{\phi}, \boldsymbol{\psi}, \boldsymbol{\beta}, m|\boldsymbol{h}, \boldsymbol{d})\mathbb{P}(m'|m)q(\epsilon)}\right), \qquad (9.2)$$

where $\boldsymbol{x}$ denotes the data; $|J|$ is the Jacobian term, which is equal to 2 in this case; $\mathbb{P}(m'|m) = \frac{1}{7}$ is the probability that given we are in model m, we propose to move to model m' with a change-point at time v (for which there are seven possible choices, each chosen with equal probability); and $q(\epsilon)$ is the corresponding normal proposal density. Note that $\mathbb{P}(m|m') = 1$ and hence is omitted in the acceptance probability. Clearly this approach can be generalised for any given location dependence, with the restriction that the new survival probabilities before and after the change-point have this same location dependence, with the corresponding changes to the Jacobian term. Note that in general $|J| = 2^k$, where k is simply the number of distinct survival probabilities over the locations.

Alternatively, in the reverse move, to retain the reversibility conditions, we only propose to remove the change-point if the location dependence is the same over all times. Then the proposed survival probability is simply taken to be the mean of the survival probabilities either side of the change-point for each area. The corresponding acceptance probability is simply the reciprocal of Expression (9.2). The analogous update is used for the recapture intensity, with the restriction that the proposed recapture intensity rates are positive.

However, we need to consider a different updating procedure for the migration probabilities, since we need to retain the restriction that the migration probabilities sum to unity. Again, suppose that we propose to add in a change-point to the current model m with migration probabilities $\psi(r, s)$, $r, s \in \{1, 2, 3\}$. For $r \in \{1, 2, 3\}$ and $s = 1, 2$, with probability $\frac{1}{2}$ we propose the parameters in the new model to be,

$$\begin{aligned} \psi'_t(r, s) &= \psi_t(r, s) + \eta(r, s) & \text{for all } r \in \mathcal{R} \text{ and } t = 1985, \ldots, v - 1; \\ \psi'_t(r, s) &= \psi_t(r, s) & \text{for all } r \in \mathcal{R} \text{ and } t = v, \ldots, 1992; \end{aligned}$$

otherwise we set,

$$\psi'_t(r,s) = \psi_t(r,s) \quad \text{for all } r \in \mathcal{R} \text{ and } t = 1985,\ldots,v-1;$$
$$\psi'_t(r,s) = \psi_t(r,s) + \eta(r,s) \quad \text{for all } r \in \mathcal{R} \text{ and } t = v,\ldots,1992.$$

Here, v is chosen uniformly in $[1986,\ldots,1992]$, and,

$$\eta(r,s) \sim N(0,\sigma^2),$$

where σ^2 is chosen via pilot tuning. Essentially, we are simulating a new set of migration probabilities for either before or after the change-point, which are similar to their current values, while the others remain the same. Note that this is similar to the proposed analogous model move in Example 9.2. We also set

$$\psi'_t(r,3) = 1 - \sum_{r=1}^{2} \psi'_t(r,s),$$

to ensure that the migration probabilities sum to unity. If any $\phi'_t(r,s) \notin [0,1]$, we automatically reject the move, otherwise we accept the move with the acceptance probability,

$$\min\left(1, \frac{\pi(\boldsymbol{\phi},\boldsymbol{\psi}',\boldsymbol{\beta},m'|\boldsymbol{h},\boldsymbol{d})\mathbb{P}(m|m')|J|}{\pi(\boldsymbol{\phi},\boldsymbol{\psi},\boldsymbol{\beta},m|\boldsymbol{h},\boldsymbol{d})\mathbb{P}(m'|m)q(\boldsymbol{\eta})}\right).$$

Here $|J| = 1$, $\mathbb{P}(m'|m) = \frac{1}{7}$, $\mathbb{P}(m|m') = 1$ and $q(\boldsymbol{\eta})$ denotes the normal proposal density for the set of parameters $\boldsymbol{\eta} = \{\eta(r,s) : r \in \{1,2,3\}, s = 1,2\}$.

(ii) Location of Change-Point

Initially, consider the survival probabilities. Within this updating procedure, we propose to update the location of the change-point if there is one present. Suppose that we are in the model m with the change-point at time v, so that,

$$\phi_t(r) = \phi_a(r) \quad \text{for all } r \in \{1,2,3\} \text{ and } t = 1985,\ldots,v-1;$$
$$\phi_t(r) = \phi_b(r) \quad \text{for all } r \in \{1,2,3\} \text{ and } t = v,\ldots,1992.$$

Then we propose to move to model m' by updating the change-point to time $v' = v \pm 1$. If $v' \notin [1985,\ldots,1992]$, we reject the proposal; otherwise we set

$$\phi'_t(r) = \phi_a(r) \quad \text{for all } r \in \{1,2,3\} \text{ and } t = 1985,\ldots,v'-1;$$
$$\phi'_t(r) = \phi_b(r) \quad \text{for all } r \in \{1,2,3\} \text{ and } t = v',\ldots,1992.$$

We accept the proposed move with the standard MH acceptance probability with symmetric proposal distribution, i.e.

$$\min\left(1, \frac{\pi(\boldsymbol{\phi}',\boldsymbol{\psi},\boldsymbol{\beta},m'|\boldsymbol{h},\boldsymbol{d})}{\pi(\boldsymbol{\phi},\boldsymbol{\psi},\boldsymbol{\beta},m|\boldsymbol{h},\boldsymbol{d})}\right).$$

We use the analogous updating procedure for the recapture intensities and migration probabilities.

MODEL DISCRIMINATION AND AVERAGING

(iii) Updating Area Dependence

Initially consider the survival probabilities and assume that there is no change-point; otherwise, if there is a change-point, then we update the location dependence before and after the change-point independently of each other as separate model moves. We assume that there is a common survival probability over time and location, i.e. the area dependence is denoted by $\{1,2,3\}$. We propose to update this area dependence by splitting the group into two: there are three possibilities $\{1\}, \{2,3\}$; $\{1,3\}, \{2\}$; and $\{1,2\}, \{3\}$. We choose each one with equal probability; without loss of generality suppose that we propose to move to model $\{1\}, \{2,3\}$. For the new model m', we propose the parameters,

$$\phi'_t(r) = \phi_t(r) + \omega \quad \text{for } r = 1 \text{ and } t = 1985, \ldots, 1992;$$
$$\phi'_t(r) = \phi_t(r) - \omega \quad \text{for } r = 2, 3 \text{ and } t = 1985, \ldots, 1992,$$

where,

$$\omega \sim N(0, \tau),$$

for τ chosen via pilot tuning. We reject the proposed move if any $\psi'_t(r) \notin [0,1]$, otherwise, we accept the move with probability,

$$\min\left(1, \frac{\pi(\phi', \psi, \beta, m'|x)\mathbb{P}(m|m')|J|}{\pi(\phi, \psi, \beta, m|x)\mathbb{P}(m'|m)q(\omega)}\right),$$

where the Jacobian $|J| = 2$; $\mathbb{P}(m'|m)$ denotes the probability of moving to model m' from model m and $q(\omega)$ is the normal proposal density for the simulated parameter ω. Here, $\mathbb{P}(m'|m) = \frac{1}{3}$, since we can only increase the dimension of the model, and there are three possible location dependence possibilities chosen with equal probability. Similarly, $\mathbb{P}(m|m') = \frac{1}{2}$, as we assume that we propose to split or merge the location dependence with equal probability, and for each of these moves (since the model m' corresponds to $\{1\}, \{2,3\}$) there is only a single possible location dependence ($\{1,2,3\}$ for a merge move and $\{1\}, \{2\}, \{3\}$ for a split move). Once more, to retain the reversibility condition, the reverse move is deterministically defined by this move. In particular, we set the new survival probability to be the mean of the current values, and accept the move with probability equal to the reciprocal of the above expression for the acceptance probability. The analogous move holds when proposing to move to the model with distinct survival probabilities over all areas. We apply the analogous move to the recapture intensities, restricting the parameter values to be simply greater than zero.

Results

We once more present a selection of posterior estimates to illustrate the type of results that are obtained for multi-site data of this form. Table 9.5 gives the marginal posterior probabilities for the time dependence of each of the demographic rates. We concentrate on the survival probabilities. There appears to be a reasonable amount of uncertainty in relation to the time dependence, particularly for the survival probabilities. Table 9.6 provides the corresponding marginal posterior probability for the location dependence for the survival probabilities for each year. There appears to be posterior support that the survival probabilities are independent of location. Finally, the corresponding posterior (model-averaged) estimates for the survival probabilities are provided in Figure 9.1. Clearly, these model-averaged estimates are influenced by the largest posterior support for a location independent model, demonstrated by the large overlap in the survival estimates between the different regions. In addition, there is significant overlap over time in the posterior distributions for the survival probabilities, as a result of the largest posterior support for a constant survival probability over time. There also appears to be a significant decrease in the precision of the survival probability over time. Conditional on a change in the survival probability over time, then this would appear to be a decrease. This would suggest that the conservation policy did not decrease the mortality rates of the dolphins; however, the data are only a relatively short time series, and so further investigation would be advisable.

Table 9.5 The Posterior Marginal Probabilities of No Change-Point and a Change-Point in Each Year for the Survival Probabilities, Recapture Intensities and Migration Probabilities for Dolphins

Change-Point	(a) Survival Probabilities Posterior Probability	(b) Recapture Intensities Posterior Probability	(c) Migration Probabilities Posterior Probability
None	0.314	0.175	0.180
1986	0.088	-	0.036
1987	0.036	0.241	0.155
1988	0.160	0.035	0.549
1989	0.136	0.421	0.041
1990	0.112	0.087	0.024
1991	0.069	0.069	0.010
1992	0.084	0.012	0.005
1993	–	0.012	–

Source: Reproduced with permission from King and Brooks (2004a, p. 347) published by Museu de Ciéncies Naturals de Barcelona.

Note: Recall that the survival and migration probabilities are defined for years 1985–1992, and the recapture intensities for years 1986–1993.

□

Table 9.6 Posterior Marginal Probabilities for the Arrangement of Locations Over Time for the Survival Probabilities for the Dolphins

Year	Location Groupings				
	{1},{2},{3}	{1,2}, {3}	{1,3}, {2}	{1}, {2,3}	{1,2,3}
1985	0.046	0.138	0.144	0.135	0.537
1986	0.035	0.138	0.132	0.133	0.562
1987	0.035	0.140	0.133	0.132	0.561
1988	0.041	0.152	0.134	0.138	0.534
1989	0.051	0.166	0.135	0.145	0.503
1990	0.059	0.170	0.136	0.154	0.482
1991	0.063	0.172	0.137	0.159	0.468
1992	0.070	0.178	0.142	0.164	0.446

Source: Reproduced with permission from King and Brooks (2004a, p. 349) published by Museu de Ciéncies Naturals de Barcelona.

9.4 Summary

This chapter focusses on more complex capture-recapture data, where individuals are not only observed, but their state is also recorded. We consider the case where the state is a discrete (or categorical) value (in such instances where the state is continuous a discretisation might be applied). Models applied to data of this form are typically referred to as multi-state models. We note that they can be regarded as a special case of covariate data, where the states (or covariates) are individual, discrete and time varying. The explicit likelihood expression is very complex, so that typically, the MCMC iterations are relatively slow to perform. Thus, we consider an alternative, auxiliary variable, approach.

Essentially, we consider the missing states (i.e. when an individual is unobserved) as auxiliary variables, analogous to the idea presented in Chapter 8. This greatly simplifies the expression for the likelihood (defined over observed and missing states), so that the Gibbs sampler can be used to update each parameter in turn (and missing states). However, there is a trade-off between the increase in the number of parameters having to be updated within this approach and the simplification of updating the parameters. Thus, the efficiency of using the auxiliary variable approach to the full complex likelihood is likely to be problem dependent.

Finally, in this chapter we also considered the issue of model uncertainty, in terms of the dependence of the demographic parameters on the states and time. This once again uses the methodology presented in Chapter 6, and the RJMCMC algorithm to calculate posterior model probabilities for discriminating between competing models. These reversible jump moves are typically more complex than the ones presented for variable selection problems.

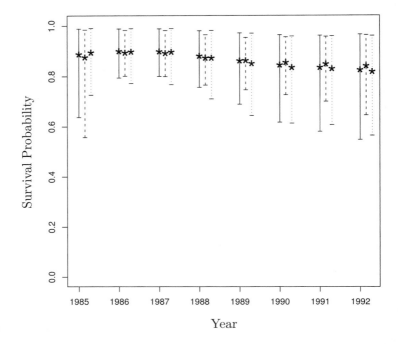

Figure 9.1 The posterior model-averaged mean (*) and 95% HPDI (vertical lines) for the dolphin survival probabilities over time, where (—) denotes area 1, (- -) denotes area 2, and (···) denotes area 3. Reproduced with permission from King and Brooks (2004a, p. 348) published by Museu de Ci'encies Naturals de Barcelona.

9.5 Further Reading

Typically within the analysis of capture-recapture-recovery data the model parameters (recapture, recovery and survival probabilities) are assumed to be time varying and/or dependent on other external factors, such as the age of gender of the individuals. A further factor influencing these parameters might be the animal's location at any given time. Multi-site models are able to allow for possible area influence, and incorporate the possibility of migration between different predefined regions. See Schwarz et al. (1993) and Brownie et al. (1993), for example. Typically, within these models, a first-order Markovian assumption is made, in that the migration probabilities are only dependent on the current time and location of the individual. More general, memory models can be considered (for example, second- or third-order Markovian models), where the movement of individuals between the different states may also be dependent on its previous locations. See Hestbeck et al. (1991), Brownie et al.

FURTHER READING

(1993) and McCrea and Morgan (2009) for further discussion of memory models.

Categorical covariates may be modelled in the same manner as multi-state models. The levels of the covariate are often referred to as states, and individuals belong in a single state at each recapture time but their state can change between recapture events. For example, the breeding state of individuals may be considered (such as breeding, lactating and suckling previous young, and non-breeding), with transition probabilities defined for movement between these different states. For example, Moyes (2007) applies such ideas to red deer; and Chilvers et al. (2009) to a population of New Zealand sea lions, where the transition probabilities between states are also logistically regressed on given covariates (namely age). Sample WinBUGS code is also provided in the latter case.

Continuous covariates can also be discretised to fit into the multi-state framework. For example, Nichols et al. (1992) consider the mass of the meadow vole, *Microtus pennsylvanicus*, in a multi-state approach, by defining "bins" or classes for mass. The states are simply different mass classes, which the animals move between with transition probabilities (equivalent to migration probabilities in the multi-site case). Bonner and Schwarz (2006) reconsider this meadow vole data using the continuous mass data for the individuals, using the ideas presented in Chapter 8 (and in particular Section 8.3), relating to missing covariate data. We note that often there may be some structure (particularly for multi-state models), where there are restrictions as to the movement between the different possible states. For example, there may only be movement in one direction between the states of non-breeding to breeding (once an individual starts to breed, it may continue to do so). Such restrictions can easily be incorporated by setting the given transition probabilities equal to zero.

The analysis of multi-state/site data can also be extended to allow for different cohorts. For example, Dupuis et al. (2002) and King and Brooks (2003b) consider the additional covariate of gender, relating to a population of mouflon (wild sheep). Dupuis et al. (2002) also discuss the specification of informative priors that can be derived, given external data (radio-tagging data in this case). These ideas are extended by King and Brooks (2003b) (and King and Brooks 2002a) for the case of model uncertainty, who derive consistent informative priors on the parameters in the different models, given the informative prior specified on the parameters within the saturated model. The paper by Pradel (2005) extends multi-state capture-recapture models to include cases when states are uncertain.

Finally, we note that the mixture models of Pledger et al. (2003) can be regarded as multi-state models, where individuals cannot move between states (i.e. the off-diagonal elements of the transition probabilities are all equal to zero), and where typically, the state of each individual is unknown. We consider these models again in Chapter 11, in the context of estimating the size of a closed population.

CHAPTER 10

State-Space Modelling

10.1 Introduction

In this chapter we consider data relating to open populations, where we have estimates of the size of the population over time. In other words, we consider data relating to population sizes, where there are births, deaths and movements occurring throughout the data-collection process. For simplicity we shall refer to this type of data as *count data*. Generally, within open populations a number of different questions may be of interest. For example, this may include estimating the size of populations that may not be directly observable (Thomas et al. 2005); estimating demographic parameters, such as the survival probabilities and/or fecundity rates of the population of interest; or determining the underlying factors that may be driving an increasing or decreasing trend in population size. The alternative case of closed populations is considered in Chapter 11.

There are numerous methods for collecting count data, ranging from large-scale aerial surveys (Jamieson and Brooks 2004a) to volunteers visiting different breeding sites (such as the Constant Effort Site scheme; see for example Peach et al. 1998 and Cave et al. 2009a). See Section 2.2.1 for further examples. However, for all methods of data collection, it is typically impossible to enumerate totally all individuals within a population. Thus, the count data that are provided are typically in the form of an estimate of the total population size, where there is some (unknown) observation error. We take this observation error into account by considering a state-space approach which directly incorporates this additional element of stochasticity within the statistical analysis. The use of state-space models has become increasingly popular within statistical ecology, with applications to many different systems (see for example, Newman 1998; Millar and Meyer 2000; Buckland et al. 2004b; Jamieson and Brooks 2004a; Brooks et al. 2004; Baillie et al. 2009; Newman et al. 2006 and Besbeas et al. 2002). A state-space model essentially considers two separate processes: the observation process (taking into account the observational error associated within the data collection process), and the system process (which models the underlying biological system over time). This approach results in a smoothing of the estimated population sizes with respect to the assumed underlying biological processes, rather than using an arbitrary smoothing such as the use of general additive models (Fewster et al. 2000). We now discuss the separate observation and system processes in further detail.

10.1.1 Observation Process

The observation process of the population under study typically involves some form of error. Without loss of generality, we assume that the observation process relates to estimates of the total number of individuals. This is often as a result of the total enumeration of the individuals being infeasible due to their number and/or the data collection process. For example, only the individuals within a part of the colony/site may be enumerated, and a correction factor applied to allow for this (see, for example, Reynolds et al. 2009 and references therein where this is applied due to the inaccessibility of part of the colony). We note that often the count estimates may relate to one or more of the states of the whole population. For example, for terrestrial mammals, the observation process may provide estimates of the total number of breeding females and total number of offspring produced each year in a given area.

Notationally, we let $\boldsymbol{y} = \{\boldsymbol{y}_1, \ldots, \boldsymbol{y}_T\}$ denote the set of observed count data over the period $t = 1, \ldots, T$. However, the true underlying population sizes are denoted by $\boldsymbol{N} = \{\boldsymbol{N}_1, \ldots, \boldsymbol{N}_T\}$. We define some observational process that describes how the observed data $\boldsymbol{y}$ are related to the true underlying population sizes, $\boldsymbol{N}$. For example, suppose that for the population of interest there are two age classes: first years and adults (often the case for avian populations), but where we only have observational count data on the number of adults within the population. Typically, we may assume that the estimates of adult population are unbiased. One possible observational process may be of the form,

$$y_t = (0\ 1) \begin{pmatrix} N_1 \\ N_a \end{pmatrix}_t + e_t,$$

where,

$$e_t \sim N(0, \sigma^2).$$

Here σ^2 denotes the observational error, and is typically a parameter to be estimated. In other words, the observed adult population size is assumed to be normally distributed about the true value with some constant, but unknown, error. This distribution can be easily extended, for example, to allow for an increased uncertainty for larger population sizes by expressing the observational error variance to be proportional to (or some other function of) the population size. Clearly there may be many other distributional forms that may be appropriate (for example, a log-normal error, which would allow for a skewed distribution for the observation error). Finally, we note that some data sources may also provide estimates of the observation error. In these circumstances, this information can be easily incorporated into the observational process by simply specifying a known value for the observation error, which may vary over time (see, for example, King et al. 2008b). Ignoring the observation error, and treating the observed population sizes as the true sizes can produce severely biased results (see, for example, Example 10.3 and in particular Table 10.3).

INTRODUCTION

10.1.2 System Process

The size of a population changes over time as a result of births, deaths and movement within the population. However, there is typically biological information concerning these different underlying processes. Using this information we are able to construct underlying transition equations that govern the changing population size over time, describing the underlying dynamics of the system. For example, the transition equations may represent survival, fecundity and/or movement processes within the population. The processes may vary over different ages, sex and/or other characteristics of the states being modelled. For example, only mature animals may reproduce, only juvenile males may migrate from the study area and/or only individuals within a particular area are culled.

We describe the demographic processes via transition equation(s). For example, consider the (simple) case for the surplus production fisheries model (Schaefer 1954), relating to the modelling of biomass. Millar and Meyer (2000) consider the quadratic model proposed by Polacheck et al. (1993), such that the biomass in year $t+1$, B_{t+1}, is described by,

$$\log B_{t+1} = \log\left(B_t + rB_t\left(1 - \frac{B_t}{K}\right) - C_t\right) + \epsilon_{t+1},$$

where r denotes the intrinsic growth rate of the population, K the carrying capacity, C_t the catch during year t, and

$$\epsilon_t \sim N(0, \sigma^2),$$

and corresponds to the normal stochastic error assumed for the model. In other words, the model assumes a multiplicative log-normal error structure for the biomass, B_t.

A more complex example is where there is an age structure to the population, for individuals in years 1 through 3, with all older individuals regarded as 'mature' (or adults, and hence we denote this age group by a). All adult birds breed, with common reproductive rate/fecundity ρ. We could represent the transition equation for this population as,

$$\begin{pmatrix} N_1 \\ N_2 \\ N_3 \\ N_a \end{pmatrix}_{t+1} = \mathbf{X} \begin{pmatrix} N_1 \\ N_2 \\ N_3 \\ N_a \end{pmatrix}_t + \boldsymbol{\epsilon}_{t+1}$$

$$= \begin{pmatrix} 0 & 0 & 0 & \phi_1\rho \\ \phi_2 & 0 & 0 & 0 \\ 0 & \phi_3 & 0 & 0 \\ 0 & 0 & \phi_a & \phi_a \end{pmatrix} \begin{pmatrix} N_1 \\ N_2 \\ N_3 \\ N_a \end{pmatrix}_t + \boldsymbol{\epsilon}_{t+1},$$

where ϕ_j corresponds to the survival probability of individuals of age $j = \{1, 2, 3, a\}$ (assumed constant over time), $\boldsymbol{\epsilon}_{t+1}$ represents an additive error function (since the transition equations are stochastic), and $\mathbf{X}$ denotes the transition (or process) matrix. We note that this system process implies that the numbers of individuals in each age group (N_j for $j \in \{1,2,3,a\}$) correspond to the time at the end of year t, just before births in year $t+1$. Clearly, further complexity can be added, allowing for a time dependent system process. In addition, the demographic parameters of interest (such as survival probabilities and/or fecundity rates) may often be expressed as a function of different covariates, as described in Section 8.2, or to allow for density dependence within the population.

Finally, we note that an alternative representation often used for the system process would be to specify the expectation

$$\mathbb{E} \begin{pmatrix} N_1 \\ N_2 \\ N_3 \\ N_a \end{pmatrix}_{t+1} = \begin{pmatrix} N_a \phi_1 \rho \\ N_2 \phi_2 \\ N_3 \phi_3 \\ (N_3 + N_a)\phi_a \end{pmatrix}_t,$$

with the same assumed error function.

10.1.3 Process Matrix Decomposition

Often it can be straightforward to explicitly state the projection matrix, $\boldsymbol{X}_t$. However, for more complex situations, where there may be a number of different processes acting on the same population, this matrix can be difficult to write down. Buckland et al., (2004b, 2007) discuss an approach for constructing the projection matrix in a number of steps. The idea essentially involves building up the projection matrix by considering the individual processes, such as births, deaths, movement, age increment, etc., and the ordering in which these processes occur. Process matrices are constructed for each of the individual processes and simply multiplied together to define the overall projection matrix. For example, in the above case there are three processes acting on the system. In order of occurrence these are: breeding, aging and survival. We consider each individual process matrix in turn. We begin with the breeding process into each separate state, and introduce a new state corresponding to newly born animals. We can write the breeding process matrix in the form:

$$\boldsymbol{B} = \begin{pmatrix} 0 & 0 & 0 & \rho \\ 1 & 0 & 0 & 0 \\ 0 & 1 & 0 & 0 \\ 0 & 0 & 1 & 0 \\ 0 & 0 & 0 & 1 \end{pmatrix}.$$

The new state of newborn animals is represented by the first row of $\boldsymbol{B}$. The diagonal elements of rows 2 through 5 represent the current individuals in each age category.

The breeding process is followed by the age incrementation process. This is a deterministic process, and so contains no stochastic parameters. In particular,

INTRODUCTION

we have:

$$A = \begin{pmatrix} 1 & 0 & 0 & 0 & 0 \\ 0 & 1 & 0 & 0 & 0 \\ 0 & 0 & 1 & 0 & 0 \\ 0 & 0 & 0 & 1 & 1 \end{pmatrix}.$$

The aging process increases the age of each individual, except for the adult state, which is an absorbing state (once an individual reaches adulthood, it stays an adult thereafter). This process effectively collapses the 5 states (with the additional newborn chicks) back to the 4 states of years 1 through 3 and adults (since the adult state is an absorbing state). Finally, there is the survival process:

$$S = \begin{pmatrix} \phi_1 & 0 & 0 & 0 \\ 0 & \phi_2 & 0 & 0 \\ 0 & 0 & \phi_3 & 0 \\ 0 & 0 & 0 & \phi_a \end{pmatrix}.$$

The projection matrix X is then easily constructed as a product of these individual process matrices:

$$X = SAB.$$

Note that the order of the matrices is from right to left in order of the operating processes. This clearly follows, given the specification of the system process of the form,

$$N_{t+1} = X_t N_t + \epsilon_{t+1},$$

so that the rightmost matrix operates on N_t first, and so on, and where ϵ_{t+1} denotes the stochastic error associated with the system process, for example, $\epsilon_{t+1} \sim N(\mathbf{0}, \sigma^2 I)$. See Section 10.2 for further discussion.

We note that in the above processes, an additional "intermediate" state (i.e. newborn) is needed. An alternative is to consider a single recruitment process matrix, R_t, say, corresponding to recruitment into all of the ages at time t. This essentially combines the breeding and age incrementation processes into a single process. The corresponding recruitment matrix is given by,

$$R_t = \begin{pmatrix} 0 & 0 & 0 & \rho \\ 1 & 0 & 0 & 0 \\ 0 & 1 & 0 & 0 \\ 0 & 0 & 1 & 1 \end{pmatrix}.$$

The first row corresponds to the number of individuals born and then become age 1; the second and third rows indicate recruitment into ages 2 and 3 (from those aged 1 and 2); and the fourth row gives the recruitment into adulthood from those aged 3 and those who are already adults. Thus, this approach removes the necessity of introducing the intermediate state of newborns, but combines the processes, making a more complicated transition matrix in general. It is trivial to see that $R_t = AB_t$.

10.1.4 Bayesian State-Space Model

The system and observation processes describe the biological system being studied in terms of the changing population size(s) over time, and the associated process in observing the population. Clearly, the set of parameters within the model is determined by the transition processes modelled within the system process. Typically we are interested in obtaining estimates of the demographic parameters within the system process (such as survival probabilities and/or fecundity rates), and the observation error variance (if appropriate). For notational convenience, we let $\boldsymbol{\theta}$ denote the set of these parameter values to be estimated. In addition, we shall typically be interested in obtaining estimates of the true underlying population sizes, $\boldsymbol{N}$.

In the state-space approach, by combining the system and observation process, it is possible to obtain estimates of the number of individuals in each state, which may not be directly observable within the observation process. For example, in Section 10.2 we only have observational data relating to the number of breeding adult female lapwings (defined to be ≥ 2 years of age), yet, using the state-space approach we are able also to obtain estimates of the number of birds aged 1 throughout the study period.

Within the classical framework for analysing state-space models the Kalman filter may be used (Kalman 1960). In order to apply the Kalman filter, we need to make linear and normality assumptions for both the system and observation processes. However, in practice these assumptions are often violated. In some cases, it is possible to use normal approximations for the distributions within the system process. For example, when the system process is modelled via binomial and/or Poisson distributions, a normal approximation can be used, assuming that the standard approximation results are valid; see Besbeas et al. (2002) for a particular example of this technique. However, the normal approximation will not always be valid as a result of small population sizes, for example, or for more complex underlying distributions, such as an exponential distribution. It is harder to use the Kalman filter for non-linear or non-normal system and/or observation processes, including, for example, density dependent population size transitions or a log-normal observation process.

We present a Bayesian approach which can be applied to any system and observation processes. In particular, we implement a *missing data approach* (see Section 8.3). Essentially, we treat the true underlying population sizes $\boldsymbol{N}$ to be missing values that are imputed within the MCMC algorithm. These are then integrated out within the MCMC algorithm to provide estimates of the posterior (marginal) distributions of the parameters $\boldsymbol{\theta}$. This approach also allows the posterior distribution of the true underlying population sizes to be estimated within the same MCMC algorithm.

Bayesian Formulation

Mathematically, we treat the true unknown population sizes as parameters or auxiliary variables to be estimated. We then form the joint posterior distrib-

INTRODUCTION

ution over both the parameters $\boldsymbol{\theta}$ and the true unknown population sizes, $\boldsymbol{N}$. In particular, using Bayes' Theorem, we have,

$$\begin{aligned}\pi(\boldsymbol{N},\boldsymbol{\theta}|\boldsymbol{y}) &\propto f_{obs}(\boldsymbol{y}|\boldsymbol{N},\boldsymbol{\theta})p(\boldsymbol{N},\boldsymbol{\theta}) \\ &= f_{obs}(\boldsymbol{y}|\boldsymbol{N},\boldsymbol{\theta})f_{sys}(\boldsymbol{N}|\boldsymbol{\theta})p(\boldsymbol{\theta}),\end{aligned}$$

where $p(\boldsymbol{\theta})$ denotes the priors on the parameters $\boldsymbol{\theta}$, $f_{obs}(\boldsymbol{y}|\boldsymbol{N},\boldsymbol{\theta})$ is the corresponding likelihood for the observation process, and $f_{sys}(\boldsymbol{N}|\boldsymbol{\theta})$ is the likelihood relating to the system process. We note that f_{sys} is not strictly a true "likelihood," since the expression does not contain any information relating to the observed data, $\boldsymbol{y}$, but describes how the unobserved underlying population size changes over time. Alternatively, we can regard this term as the prior specification of the underlying model for the true population sizes.

With the joint posterior distribution calculated above (up to proportionality), we can use an MCMC algorithm to explore the distribution and obtain posterior estimates of the parameters of interest. Within each step of the MCMC algorithm, we update the model parameters, $\boldsymbol{\theta}$, and the auxiliary variables, $\boldsymbol{N}$ (see King 2010 for further discussion and comparison of different updating algorithms). This allows us to obtain a sample from the joint posterior distribution, $\pi(\boldsymbol{N},\boldsymbol{\theta}|\boldsymbol{y})$, and use this sample to obtain posterior (marginal) estimates of the model parameters and true population sizes (if required). Note that this approach also allows us to obtain estimates of posterior correlations between parameters and/or true population sizes, and/or functions of the true population size, such as annual growth rates. In particular, we can easily obtain posterior credible intervals for such quantities (obtaining error bands for similar quantities in the classical framework can be difficult), using the standard Monte Carlo estimate. For further discussion, see for example Royle (2008). Finally we note that a Bayesian analysis of a normal linear state-space model is often referred to as a *Dynamic Linear Model* (West and Harrison 1997), particularly within time-series analysis and forecasting.

Practical Implications

There are a number of practical implications that can arise when fitting a Bayesian state-space model. We begin by discussing the system process (or prior specification on $\boldsymbol{N}$). In particular, the population size at time t can often be expressed as a function of the population at the previous time $t-1$. In other words, $\boldsymbol{N}_t = f(\boldsymbol{N}_{t-1}, \boldsymbol{\theta})$, for some predefined function of f, describing the biological processes. In particular, we note that the true population size at the first time period of the study, $\boldsymbol{N}_1$, is a function of $\boldsymbol{N}_0$, which is also unknown and for which there is no corresponding observational data. Two possible approaches can be used to solve this problem. First, we can once more treat the $\boldsymbol{N}_0$ as an additional parameter to be estimated, and specify some prior $p(\boldsymbol{N}_0)$. Alternatively, we can truncate the system and observation likelihoods (starting at time $t=2$), and specify a prior on $\boldsymbol{N}_1$. An obvious choice of prior for this latter approach would be to specify a prior, analogous to the observation process at time $t = 1$. For example, suppose that the observation process is specified such that $y_t \sim N(N_t, \sigma^2)$. For $t = 1$, we specify the prior, $N_1 \sim N(y_1, \sigma^2)$. Clearly, if there are a number of possible states, additional priors need to be specified on each state (where all states may or may not be observed). These ideas can be extended where $\boldsymbol{N}_t$ is a function of $\boldsymbol{N}_{t-1}, \ldots, \boldsymbol{N}_{t-\tau}$, i.e. we add the additional parameters $\boldsymbol{N}_{-\tau+1}, \ldots, \boldsymbol{N}_0$ with some specified prior, or truncate the likelihoods to start at time $t = 1 + \tau$ and specify a prior on $\boldsymbol{N}_1, \ldots, \boldsymbol{N}_\tau$.

An additional issue often arises in terms of specifying the starting values for the MCMC algorithm. Initial starting values need to be specified for the parameters in the model and the true underlying population sizes (auxiliary variables). Naive starting values for these parameters may result in essentially incompatible values for the population size over time, given the demographic parameters, or even an impossible set of population sizes. This may produce slow mixing within the Markov chain, with long burn-in periods, or in the case of impossible starting values, a non-functioning computer code. Possible solutions to specifying the initial $\boldsymbol{N}_1, \ldots, \boldsymbol{N}_T$ values may include setting these values to be equal to the observed values $\boldsymbol{y}$. However, there may still be unobservable states that need initial values specified. In addition, setting the states to be equal to the observed values does not necessarily ensure consistent (or even a set of possible) values for the states, particularly, for example, if there is large observation error. Alternatively, another approach would be to condition on the initial values specified for the demographic rates and assuming an initial state for $\boldsymbol{N}_0$, simulate a set of possible values for $\boldsymbol{N}_1, \ldots, \boldsymbol{N}_T$, and use these for the initial values. Clearly these two approaches can also be combined, using the first approach for observed states, and then conditioning on these and the initial demographic rates, simulate possible values for the unobserved states to use as their initial values.

We note that in principle it is possible to fit state-space models within WinBUGS. See, for example, Brooks et al. (2004) and Gimenez et al. (2009a), where sample code is also provided. However, the simulations can be slow to run (particularly if convergence is slow), and lengthy simulations are needed to ensure convergence of the summary statistics post-burn-in.

10.2 Leslie Matrix-Based Models

Often, the transition equations can be expressed in the form of a Leslie matrix (Leslie, 1945, 1948, Caswell 2001). In other words we can express the transition equations in the form,

$$\boldsymbol{N}_{t+1} = \boldsymbol{X}_t \boldsymbol{N}_t + \boldsymbol{\epsilon}_{t+1},$$

where $\boldsymbol{N}_t$ is the $p \times 1$ state vector at time t, $\boldsymbol{X}_t$ is a known $p \times p$ projection matrix and $\boldsymbol{\epsilon}_{t+1}$ denotes the error term. Typically, $\boldsymbol{X}_t$ will be time independent but this need not be the case in general. An example is the culling of animals at certain times over the time period, and so using the general time dependent $\boldsymbol{X}_t$ projection matrix permits such complicated system processes.

LESLIE MATRIX-BASED MODELS

Recall that in Section 10.1.2 we described how these projection matrices can be constructed from matrices describing the individual processes acting on the system. We note that if the projection matrix is used to describe developmental growth of the population (rather than aging), the matrix is typically referred to as a Lefkovitch matrix (Lefkovitch 1965).

Example 10.1. Two-State System Process

Suppose that we have a two-state system, relating to age 1 animals and adults, so that,

$$\boldsymbol{N}_t = \begin{pmatrix} N_1 \\ N_a \end{pmatrix}_t.$$

In order to calculate the corresponding transition matrix for the system, we assume that there are two stochastic demographic processes operating: survival, which is age dependent, and breeding (adults only), and one deterministic process: age incrementation. The ordering of the processes corresponds to the breeding of individuals (in spring, say), their age incremented, followed by their survival to the following year. We denote the survival matrix by $\boldsymbol{S}_t$, the age incrementation matrix by $\boldsymbol{A}$ (assuming time independent age transitions), and the breeding matrix by $\boldsymbol{B}_t$. We consider each process in turn, in order of occurrence, and so begin with the breeding process. We need to introduce a further state (as in the example of Section 10.1.2) relating to newborns, and represented in the first row of the matrix for $\boldsymbol{B}_t$. The process matrix is then given by,

$$\boldsymbol{B}_t = \begin{pmatrix} 0 & \rho \\ 1 & 0 \\ 0 & 1 \end{pmatrix}_t.$$

The age incrementation matrix is then applied, so that age 1 animals become adults, adults remain adults and the newborns become aged 1:

$$\boldsymbol{A} = \begin{pmatrix} 1 & 0 & 0 \\ 0 & 1 & 1 \end{pmatrix}.$$

This matrix also essentially removes the newly introduced state of newborns, as they become animals aged 1, thus reducing the states to the original set: age 1 animals and adults. Finally, the survival process matrix is simply,

$$\boldsymbol{S}_t = \begin{pmatrix} \phi_1 & 0 \\ 0 & \phi_2 \end{pmatrix}_t.$$

The overall projection matrix is simply,

$$\begin{aligned} \boldsymbol{X}_t &= \boldsymbol{S}_t \boldsymbol{A} \boldsymbol{B}_t \\ &= \begin{pmatrix} \phi_1 & 0 \\ 0 & \phi_a \end{pmatrix}_t \begin{pmatrix} 1 & 0 & 0 \\ 0 & 1 & 1 \end{pmatrix} \begin{pmatrix} 0 & \rho \\ 1 & 0 \\ 0 & 1 \end{pmatrix}_t \\ &= \begin{pmatrix} 0 & \rho\phi_1 \\ \phi_a & \phi_a \end{pmatrix}_t. \end{aligned}$$

Alternatively, we could combine the breeding and age matrices into a single recruitment process matrix, $\boldsymbol{R}_t$, say, corresponding to recruitment into both adulthood and age 1 animals (as in the example in Section 10.1.3). This essentially combines the breeding and age incrementation processes into a single process. In particular, for the above example, we have the recruitment matrix,

$$\boldsymbol{R}_t = \begin{pmatrix} 0 & \rho \\ 1 & 1 \end{pmatrix}_t.$$

The matrix $\boldsymbol{R}_t$ generates new births (with productivity rate ρ from only adult animals), existing first year birds are recruited into the adult population, and adults remain adults. It can clearly be seen that $\boldsymbol{R}_t = \boldsymbol{A}\boldsymbol{B}_t$. □

Example 10.2. Lapwings: Count Data

We consider survey data relating to the UK lapwing, *Vanellus vanellus*, described in Example 1.3. The data are derived from the common birds census (CBC) collected via annual counts made at a number of sites around the United Kingdom and collated by the British Trust for Ornithology (BTO). We assume that all lapwings begin breeding at the age of 2 (Cramp and Simmons 1983, Besbeas et al. 2002). Birds of one year of age are classed as (non-breeding) juveniles, and birds of age 2 or greater are classed as adults. Birds are assumed to have a different survival probability in their first year of life, and then a common "adult" survival probability thereafter. All birds aged at least 2 years old are assumed to breed. From the annual count data, an index value is calculated, providing a measure of the population level (for the number of adult breeding females), taking account of the fact that in each year only a small proportion of sites are actually surveyed. We treat this index as survey data, and consider data from 1965 to 1998 inclusive. The data are reproduced in Besbeas et al. (2002) and Brooks et al. (2004). We denote the index value for year t by y_t and, for consistency with the ring-recovery data and further analyses described later (in Section 10.2.1), we associate the year 1963 with the value $t = 1$, so that we actually observe the values $y_3, \ldots, y_{36}$. We consider the observation and system process for this illustrative example in turn.

Observation Process

The index data can be considered to be an estimate of the true female breeding population size at the CBC sites over time, and typically an index proportional to the national population size. We first assume the underlying model for the observation process to be of the form,

$$y_t \sim N(N_{a,t}, \sigma_y^2) \qquad (10.1)$$

where $N_{a,t}$ represents the true underlying numbers of breeding adult females at time t, and σ_y^2 is the observation error variance to be estimated. This describes the observation process by which the estimates y_t are derived from the true underlying process $N_{a,t}$. As observed earlier, in general, alternative observation process models may be plausible, for example, a log-normal distribution, or a scaled variance term in a normal distribution (e.g. scaled by the coefficient of variation squared). Additionally, due to the data collection process and/or processing of the data, it is possible that the variance term σ_y^2 may be known (and possibly data dependent). For example, see King et al. (2008b), who derive annual variance estimates, $\sigma_{y,t}^2$, in addition to the estimates for the population size, y_t, given the underlying raw data corresponding to estimates at the site level. However, for simplicity, in order to illustrate the methodology we adopt the simple normal observation error with unknown variance. Clearly, the form of the error will be problem dependent and should be justified biologically.

System Process

The system process describes the underlying dynamics of the population level over time representing the biological processes. For the lapwing data we need to define the transition equations relating to both the juvenile and adult birds. A natural model would be to assume that

$$N_{1,t} \sim Po(N_{a,t-1}\rho_{t-1}\phi_{1,t-1})$$
$$N_{a,t} \sim Bin(N_{1,t-1} + N_{a,t-1}, \phi_{a,t-1}),$$

where ρ_t denotes the productivity rate in year t, i.e. the average number of female offspring per adult female. Thus, the number of birds aged 1 in year t stems directly from the number of chicks produced the year before, which then survive from $t-1$ to t. Similarly, the number of adults in year t is derived directly from the number of adults in the previous year which survive from $t-1$ to t and the number of juveniles (i.e. age 1 birds) that become adults and survive from year $t-1$ to t. We could represent this model using a Leslie matrix:

$$\begin{pmatrix} N_1 \\ N_a \end{pmatrix}_{t+1} = \begin{pmatrix} 0 & \rho\phi_1 \\ \phi_a & \phi_a \end{pmatrix} \begin{pmatrix} N_1 \\ N_a \end{pmatrix}_t + \begin{pmatrix} \epsilon_1 \\ \epsilon_a \end{pmatrix}_{t+1},$$

where ϵ_1 and ϵ_a denote Poisson and binomial errors. The transition matrix above is identical to the process matrix derived in Example 10.1, and can be constructed in the same manner.

We note that it is clear from the form of the transition equations, and hence corresponding system likelihood, that the productivity rate and first-year survival probability are, in general, confounded without further information, so that it is only possible to estimate their product, i.e. the model is parameter redundant (see Section 2.8). However, annual variation in the population parameters is to be expected, and weather is almost certainly a contributing factor to changes in survival probabilities (see Section 8.2). In particular, the severity of the winter is likely to have a substantial influence on the survival probabilities, with most mortalities occurring during the winter months. As a surrogate for the harshness of the winter, we consider the covariate corresponding to the number of days that the temperature fell below zero in central

England during the winter. For year t, we denote this by $fdays_t$. See Besbeas et al. (2002) for further details. The survival probabilities are constrained to the interval $[0, 1]$, and so we consider a logistic regression of both first-year survival probability and adult survival probability on the covariate $fdays_t$ (see, for example, Section 8.2.1). In other words, we specify:

$$\begin{aligned} \text{logit } \phi_{1,t} &= \alpha_1 + \beta_1 f_t \\ \text{logit } \phi_{a,t} &= \alpha_a + \beta_a f_t, \end{aligned}$$

where f_t denotes the normalised value for the covariate $fdays_t$ (see Example 4.11).

Additionally, the productivity rate is assumed to be dependent on time. The productivity rate is constrained to be positive, so that we consider a log link function. In particular, we set,

$$\log \rho_t = \alpha_\rho + \gamma_\rho T_t,$$

where T_t denotes the normalised value for the time t.

We note that the given specification of the first-year survival probability and productivity rate as a function of the covariates removes the confounding nature of the demographic parameters, with all the regression coefficient parameters specified above estimable. Thus the parameters to be estimated are the regression coefficients for the survival and recovery probabilities, observation error variance, and the true underlying population sizes (treated as auxiliary variables, but also of interest in their own right).

Prior Specification

We need to specify priors on the regression coefficients, observation error variance parameter and the initial population sizes, as discussed in the Practical Implications of Section 10.1.4. In particular, without any prior information, we specify independent $N(0, 10)$ priors for each regression coefficient, and for the observation error variance, an $\Gamma^{-1}(0.001, 0.001)$ prior. Finally, for the state of the initial population size, we place an uninformative prior on $\mathbf{N}_2$. Recall that the observed survey data start at time $t = 3$, so that we place the prior on the population sizes at time $t = 2$. In particular, we set,

$$\begin{aligned} N_{1,2} &\sim Neg - Bin(0.04, 0.002) \\ N_{a,2} &\sim Neg - Bin(1.00, 0.001). \end{aligned}$$

This prior is specified so that $\mathbb{P}_p(N_{1,2}) = 200$, $\mathbb{P}_p(N_{a,2}) = 1000$ and $\text{Var}_p(N_{1,2}) = 10^6 = \text{Var}_p(N_{a,2})$. Thus, we specify independent priors on the initial population sizes for juveniles (age 1) and adults (age 2 and greater). Clearly, the specification of the magnitude of the prior means (and variances) to be uninformative will be data dependent. A sensitivity analysis can clearly be conducted on the prior specified for the model parameters (regression coefficients and observation error variance) and initial population sizes.

Simulations

In order to initialise the Markov chain, we need to specify starting values for each of the model parameters and the underlying population sizes, for first years and adults. Specifying starting values for the model parameters is generally straightforward. However, care needs to be taken for the true underlying population sizes. We simply set $N_{a,t} = y_t$ (i.e. set the true adult breeding population sizes to be equal to the estimated population sizes). However, we also need to specify starting values for the juvenile population sizes. Specifying values of $N_{1,t}$ that are too small can result in impossible values for $N_{a,t}$. In particular, we have the restriction that $N_{a,t+1} \leq N_{1,t} + N_{a,t}$, i.e. $N_{1,t} \geq N_{a,t+1} - N_{a,t}$. If we specify the initial starting values for the adult population size as given above, this means that we need to specify initial starting values for juvenile birds, such that $N_{1,t} \geq y_{t+1} - y_t$. In the MCMC simulations, we simply set $N_{1,t} = 400$ for all t, which satisfies the above conditions. The MCMC simulations were run for 1 million iterations, with the first 10% discarded as burn-in, which appeared to be a very conservative burn-in with respect to the trace-plots of the parameters.

Results

Table 10.1 provides the corresponding estimates for the model parameters. Clearly, we can interpret from these results that there appears to be a negative relationship between first-year survival probabilities and the covariate $fdays$; and a linear decline (on the logarithmic scale) for the productivity rate. The relationship between adult survival probability and $fdays$ appears to be less clear, with the 95% HPDI containing both negative and positive values. In addition, we note the generally low posterior precision of these parameters (we return to this in Section 10.2.1).

Table 10.1 The Posterior Mean, Standard Deviation and 95% HPDI for the Model Parameters for the Lapwing Count Data from 1965–1998

Parameter	Posterior Mean (SD)	95% HPDI
α_1	1.260 (1.512)	(−1.877, 3.865)
β_1	−1.784 (0.966)	(−3.613, −0.169)
α_a	2.261 (0.382)	(1.402, 2.916)
β_a	0.064 (0.171)	(−0.263, 0.412)
α_ρ	−2.053 (0.501)	(−1.070, 0.035)
γ_ρ	−0.613 (0.254)	(−1.134, −0.197)
σ_y^2	21395 (6297)	(11132, 33867)

In addition, Figure 10.1(a) provides the corresponding posterior estimates for the true juvenile and adult population sizes, along with the corresponding point estimates of the population, y_t. Note that although we only have data relating to estimates of the adult population size, we are also able to provide posterior estimates for the number of juvenile birds, due to the inclusion of the system process describing the underlying biological system. Clearly, the Bayesian state-space approach has smoothed over the estimates of the population size, but the explicit system equations have constrained this smoothing to be biological plausible. The Bayesian approach is also very flexible in terms of the functions of parameters that can be estimated. For example, it is straightforward from the constructed MCMC algorithm to obtain posterior estimates of the adult growth rate each year, denoted by $r_{a,t} = N_{a,t+1}/N_{a,t}$ or total growth rate, $r_t = (N_{1,t+1} + N_{a,t+1})/(N_{1,t} + N_{a,t})$. The log growth rate is also a common statistic that is of interest. Posterior summary statistics of these quantities are simply calculated at each iteration of the Markov chain (using the imputed values for the true underlying population sizes), and the corresponding empirical quantities of interest calculated as an estimate of the posterior quantity (i.e. using the Monte Carlo approach). For example, Figure 10.1(b) provides the posterior mean and 95% HPDI for the total growth rate, r_t, for the lapwing count data. This clearly shows an increasing growth rate for earlier years and a declining growth rate in latter years. Finally, we note that we can also calculate the posterior probability that the total growth rate is positive and/or negative for any given year as the proportion of the posterior distribution of the growth rate in year t (r_t), which is greater/less than 1 (or equivalently the log growth rate ($\log r_t$) is greater/less than 0). See King et al. (2008b) for further extensions to this idea.

□

10.2.1 Integrated Data

Often, more than one type of data may be collected on the same population, independently of each other. We consider the case here, where both count data and ring-recovery data are collected on the UK lapwing population. We consider the same count data described above between 1965 and 1998 (see Table 2.2 for a subset of the data). To augment the count data, we also have recovery data from lapwings ringed as chicks between 1963 and 1997 and later found dead and reported between 1964 and 1998. The full data set are reproduced in Besbeas et al. (2002) and King et al. (2008b). By combining data from different sources, we obtain more robust (and self-consistent) parameter estimates that fully reflect the information available.

Example 10.3. Lapwings: Integrated Analysis
We consider the identical state-space model as described above for the count data in Example 10.2. For the ring-recovery data, we once more assume different time varying survival probabilities for first years and those aged at least

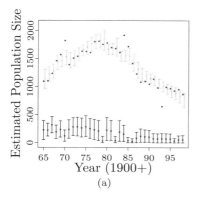
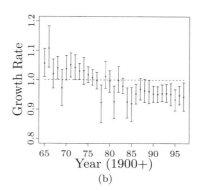

Figure 10.1 (a) The posterior mean (*) and 95% HPDI (vertical lines) of the number of juvenile birds aged 1 (in black) and adults aged at least 2 (in grey). The additional *'s give the data y over time. (b) The posterior mean (*) and 95% HPDI (vertical lines) of the total growth rate r_t in year $t = 1965,\ldots,1997$, where $r_t = (N_{1,t+1} + N_{a,t+1})/(N_{1,t} + N_{a,t})$. The horizontal dashed line corresponds to no change in population size.

1 (logistically regressed on the covariate $fdays$), but common time varying recovery probability λ_t denoting the probability that a bird which dies in year t is recovered. Here we assume that the recovery probability is logistically regressed on time (where the time covariate is once more normalised), so that,

$$\text{logit } \lambda_t = \alpha_\lambda + \gamma_\lambda u_t,$$

where u_t is the normalised time covariate.

The state-space model described above depends upon parameters ρ_t, $\phi_{1,t}$, $\phi_{a,t}$, σ_y^2 and the underlying population levels $\boldsymbol{N}_1$ and $\boldsymbol{N}_a$, which we treat as missing values (or auxiliary variables) to be estimated. This model is described as a joint probability distribution for the observed data $\boldsymbol{y} = (y_3,\ldots,y_{36})$ in terms of these parameters as follows,

$$f(\boldsymbol{y},\boldsymbol{N}|\boldsymbol{\rho},\boldsymbol{\phi},\sigma_y^2) = f_{obs}(\boldsymbol{y}|\boldsymbol{N},\sigma_y^2)f_{sys}(\boldsymbol{N}|\boldsymbol{\rho},\boldsymbol{\phi}),$$

explicitly writing out the set of parameters. Additionally, the ring-recovery model depends upon parameters λ_t, $\phi_{1,t}$ and $\phi_{a,t}$ and has corresponding likelihood $f(\boldsymbol{m}|\boldsymbol{\lambda},\boldsymbol{\phi}_1,\boldsymbol{\phi}_a)$ given in Equation 2.3 of Section 2.3.1.

It is clear that both of these models have parameters in common ($\boldsymbol{\phi}_1$ and $\boldsymbol{\phi}_a$). Thus, in the presence of both count data and ring-recovery data, combining the two data sets and analysing them together pools the information regarding these parameters and this filters into the estimation of the remaining parameters. The combination of these two models is most clearly demonstrated in the Directed Acyclic Graph (DAG, Section 4.5) given in Figure 10.2. Note that the top line of the DAG, corresponding to the prior parameters specified on each parameter, has been omitted to avoid the DAG being too cluttered.

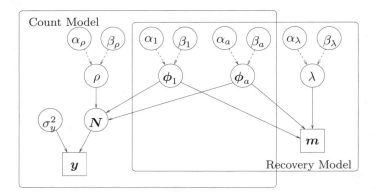

Figure 10.2 Directed Acyclic Graph (DAG) corresponding to the combined model for the count and recovery data. Note that the top level of the DAG relating to the prior parameters has been omitted. Reproduced with permission from Brooks et al. (2004, p. 519) published by Museu de Ciéncies Naturals de Barcelona.

There is generally a large amount of information contained within the ring-recovery data relating to both the survival probabilities and (essentially nuisance parameters) recovery probabilities. Alternatively, for the count data, there is relatively a much smaller amount of information relating to the survival probabilities. Thus, we can essentially "borrow" information from the ring-recovery data relating to the survival probabilities, and use this in the analysis of the count data. Formally, we assume that the data sets are independent of each other, and obtain a corresponding joint probability distribution for the combined data (and auxiliary variables, $\boldsymbol{N}$), as follows

$$f(\boldsymbol{y}, \boldsymbol{N}, \boldsymbol{m}|\boldsymbol{\rho}, \boldsymbol{\phi}, \boldsymbol{\lambda}, \sigma_y^2) = f(\boldsymbol{y}, \boldsymbol{N}|\boldsymbol{\rho}, \boldsymbol{\phi}, \sigma_y^2) f(\boldsymbol{m}|\boldsymbol{\phi}, \boldsymbol{\lambda}),$$

where $f(\boldsymbol{y}, \boldsymbol{N}|\boldsymbol{\rho}, \boldsymbol{\phi}, \sigma_y^2) = f_{obs}(\boldsymbol{y}|\boldsymbol{N}, \sigma_y^2) f_{sys}(\boldsymbol{N}|\boldsymbol{\rho}, \boldsymbol{\phi})$ corresponding to the count data, and $f(\boldsymbol{m}|\boldsymbol{\lambda}, \boldsymbol{\phi})$ is the likelihood for the ring-recovery data. The posterior distribution, given all the observed data (count data and ring-recovery data) is,

$$\pi(\boldsymbol{N}, \boldsymbol{\rho}, \boldsymbol{\phi}, \boldsymbol{\lambda}, \sigma_y^2) \propto f(\boldsymbol{y}, \boldsymbol{N}, \boldsymbol{m}|\boldsymbol{\rho}, \boldsymbol{\phi}, \boldsymbol{\lambda}, \sigma_y^2) p(\boldsymbol{\rho}, \boldsymbol{\phi}, \boldsymbol{\lambda}, \sigma_y^2).$$

We specify the same priors on the parameters as those described above for the count data alone, with the analogous priors on the regression coefficients for the recovery probabilities.

*Comment**

Consider the form of the posterior distribution once again for the integrated data. We have that,

$$\begin{aligned}\pi(\boldsymbol{N},\boldsymbol{\rho},\boldsymbol{\phi},\boldsymbol{\lambda},\sigma_y^2|\boldsymbol{y},\boldsymbol{m}) &\propto f(\boldsymbol{y},\boldsymbol{N},\boldsymbol{m}|\boldsymbol{\rho},\boldsymbol{\phi},\boldsymbol{\lambda},\sigma_y^2)p(\boldsymbol{\rho},\boldsymbol{\phi},\boldsymbol{\lambda},\sigma_y^2)\\ &= f(\boldsymbol{y},\boldsymbol{N}|\boldsymbol{\rho},\boldsymbol{\phi},\sigma_y^2)f(\boldsymbol{m}|\boldsymbol{\phi},\boldsymbol{\lambda})p(\boldsymbol{\rho},\boldsymbol{\phi},\boldsymbol{\lambda},\sigma_y^2)\\ &\propto f(\boldsymbol{y},\boldsymbol{N}|\boldsymbol{\rho},\boldsymbol{\phi},\sigma_y^2)\pi(\boldsymbol{\rho},\boldsymbol{\phi},\boldsymbol{\lambda},\sigma_y^2|\boldsymbol{m}),\end{aligned}$$

where $\pi(\boldsymbol{\lambda},\boldsymbol{\phi},\boldsymbol{\rho},\sigma_y^2|\boldsymbol{m})$ denotes the posterior distribution of all the parameters, given *only* the ring-recovery data. Compare this with the expression for the posterior distribution for all the parameters, given all the data, shown below

$$\pi(\boldsymbol{\rho},\boldsymbol{\phi},\boldsymbol{\lambda},\sigma_y^2,\boldsymbol{N}|\boldsymbol{y},\boldsymbol{m}) \propto f(\boldsymbol{y},\boldsymbol{N}|\boldsymbol{\rho},\boldsymbol{\phi},\boldsymbol{\lambda},\sigma_y^2)p(\boldsymbol{\rho},\boldsymbol{\phi},\boldsymbol{\lambda},\sigma_y^2).$$

Thus, we can see that the posterior distribution for the parameters, given both the count data and ring-recovery data, can be simply interpreted as the posterior distribution of the parameters, given the count data, where the prior on the parameters is simply the posterior distribution of the parameters, given the ring-recovery data (i.e. $p(\boldsymbol{\rho},\boldsymbol{\phi},\boldsymbol{\lambda},\sigma_y^2) = \pi(\boldsymbol{\rho},\boldsymbol{\phi},\boldsymbol{\lambda},\sigma_y^2|\boldsymbol{m})$). Note that this simplification holds as a result of the independence assumption between the two data sets, so that the likelihood can be decomposed into the product of the two separate likelihood terms, corresponding to each set of data. This can be compared to the prior specification issue discussed in Section 4.2.5 in the presence of independent data.

Results

Table 10.2 provides the posterior means and corresponding standard deviations for the model parameters from the analysis of the combined data, together with the corresponding estimates under the analyses of the two data sets individually. The comparatively large posterior standard deviations for the majority of parameters under the count data alone confirms our earlier assertion about the lack of information in the count data concerning the survival probabilities. For example, for the count data, the posterior mean for the regression coefficient β_a, corresponding to the influence on the covariate $fdays$, is very close to zero (which represents no relationship). Using the integrated data and incorporating the ring-recovery data we see that the biological interpretation is much clearer, with a negative relationship, as we would expect. We note also the similarity in the parameter estimates under the ring-recovery model and the combined analysis as these data are much more informative concerning the demographic rates relative to the count data. Finally, we observe that by combining the two data sources, the posterior standard deviations for the productivity parameters decrease dramatically, even though there is no direct information in the ring-recovery data relating to these parameters. This is a direct result of the increased precision of the other parameters.

Figure 10.3(a) provides the posterior mean and 95% HPDI for the true

Table 10.2 The Posterior Mean and Standard Deviation (in Brackets) for the Regression Coefficients for Each of the Estimable Parameters for the Lapwing Data for Model $\phi_1(fdays)/\phi_a(fdays)/\rho(t)/\lambda(t)$

Parameter	Integrated Data	Count Data	Ring-Recovery Data
α_1	0.537 (0.068)	1.260 (1.512)	0.535 (0.068)
β_1	−0.197 (0.059)	−1.784 (0.966)	−0.208 (0.062)
α_a	1.536 (0.068)	2.261 (0.382)	1.531 (0.070)
β_a	−0.241 (0.039)	0.064 (0.171)	−0.311 (0.062)
α_ρ	−1.130 (0.087)	−2.053 (0.501)	−
γ_ρ	−0.264 (0.048)	−0.613 (0.254)	−
α_λ	−4.565 (0.035)	−	−4.567 (0.035)
γ_λ	−0.351 (0.039)	−	−0.345 (0.039)
σ_y^2	28427 (8653)	21395 (6297)	

Note: We consider using only count or ring-recovery data and the integrated analysis (combined count data and ring-recovery data). A dash (-) denotes that the parameter is not estimable for the given data.

underlying population levels for the juveniles and adults; and (b) the posterior mean and 95% HPDI for the growth rate r_t. We can compare these with Figures 10.1(a) and (b) for the count data alone. The posterior credible intervals are smaller for the integrated data analysis compared to the analysis of the count data alone, demonstrating the increased amount of information contained within the ring-recovery data on the survival probabilities which permeates throughout all the parameter estimates. The largest difference in the estimates for population size lies in the posterior precision of the number of the juvenile birds aged 1. This is unsurprising, since these are not directly observed within the data and all inference is made from the adult population size for the count data. However, for the integrated data there is external information relating to first-year survival probabilities from the ring-recovery data, as well as to adult survival probabilities. In addition, we note that the pattern of the growth rates appears to differ between analyses, although once more, the posterior credible intervals for the count data appear to be significantly larger than for the integrated data. The four particularly low growth rates for the integrated data (in years 1978, 1981, 1984 and 1985) correspond to the harsh winters. This is not observed for the count data alone. This is a direct result of the uncertainty relating to the parameter β_a, corresponding to the impact of *fdays* on the adult survival probability.

LESLIE MATRIX-BASED MODELS 325

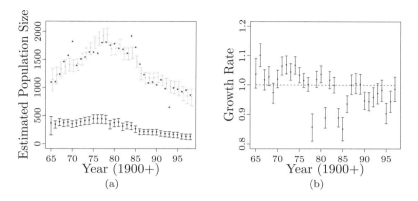

Figure 10.3 (a) The posterior mean (*) and 95% HPDI (vertical lines) of the number of juvenile birds aged 1 (in black) and adults aged at least 2 (in grey) for the integrated data analysis. The *'s give the data y over time. (b) The posterior mean (*) and 95% HPDI (vertical lines) of the growth rate r_t in year $t = 1965, \ldots, 1997$ for the integrated data analysis. The horizontal dashed line corresponds to no change in growth rate.

Ignoring Observation Error

To demonstrate the importance of incorporating the observation error within the parameter estimation process, we rerun the analysis, assuming that there is no observation error (i.e. we essentially set $\sigma_y^2 = 0$). Table 10.3 provides the corresponding results obtained for the integrated analysis with and without observation error for comparison.

Comparing the posterior estimates of the regression parameters, between the two analyses, we can clearly see that some of them are reasonably comparable, with significant overlap between the posterior distributions. However, the posterior estimates, particularly for the survival probabilities and productivity rates, are very different. The differences between the two models (with and without observation error) in terms of the posterior parameter estimates are more clearly illustrated by comparing the demographic rates. Figure 10.4 provides the corresponding posterior estimates for the survival probabilities, recovery probabilities and productivity rates.

From Figures 10.4(a) and (c) we note that there appears to be only relatively minimal difference between the estimates of the first-year survival probabilities and recovery probabilities obtained between the models with and without observation error, with significant overlapping of the 95% HPDIs. This is essentially because the majority of information relating to these parameters is obtained from the ring-recovery data, whose model has remain unchanged. However, from Figures 10.4(b) and (d) we can clearly see that the posterior estimates of the adult survival probabilities and productivity rates are vastly

Table 10.3 The Posterior Mean and Standard Deviation (in Brackets) for the Regression Coefficients for the Lapwing Data for Model $\phi_1(fdays)/\phi_a(fdays)/\rho(t)/\lambda(t)$ (With and Without Observation Error)

Parameter	Integrated Data (With Observation Error)	Integrated Data (No Observation Error)
α_1	0.537 (0.068)	0.388 (0.066)
β_1	-0.197 (0.059)	-0.290 (0.024)
α_a	1.536 (0.068)	0.711 (0.037)
β_a	-0.241 (0.039)	-0.053 (0.010)
α_ρ	-1.130 (0.087)	-0.196 (0.047)
γ_ρ	-0.264 (0.048)	-0.156 (0.014)
α_λ	-4.565 (0.035)	-4.671 (0.033)
γ_λ	-0.351 (0.039)	-0.302 (0.039)
σ_y^2	28427 (8653)	—

different between the two models (with and without observation error). In particular, the survival probabilities are significantly reduced when we ignore the observation error, so that they are generally more comparable with the first-year survival probabilities (which is biologically unrealistic). In addition, the productivity rates are significantly higher when we remove the observation error. Note that the estimated number of first years are also significantly (and unrealistically) higher when we ignore the observation error (typically a magnitude of 2 to 4 times higher). We can explain these results as follows. The higher numbers of first years are needed in order to compensate for the high fluctuations in the estimated number of adults (for example, between years 1992 and 1994 the number of birds are recorded as 981, 648 and 993), in order to allow for the large changes without explaining this via observation error. This in turn means a high productivity rate, to produce the relatively large number of first years, and a reduction in the adult survival probabilities to stabilise the number of adult birds in the population. Thus, the incorporation of an observation error can be vital in the analysis of count data. The larger the amount of observational error, the larger the amount of potential bias that may be obtained, if we ignore this additional level of uncertainty.

Model Discrimination and Averaging

Within the previous integrated analysis of the lapwing data we assume the underlying model for the demographic parameters in terms of the covariate

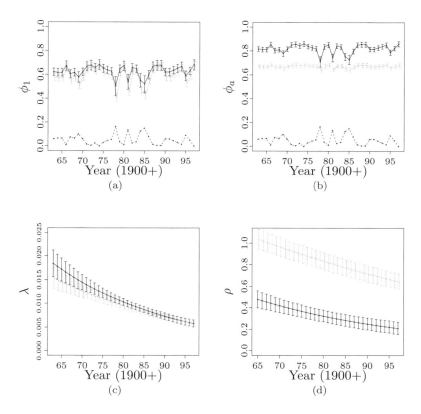

Figure 10.4 The posterior estimates of (a) first-year survival probability, (b) adult survival probability, (c) recovery probability and (d) productivity for the state-space model (in black) and the model assuming no observation error (in grey). The dashed line in plots (a) and (b) provides a scaled plot of the frost days covariate, *fdays*.

dependence. However, in practice this may not be known a priori. In addition, the dependence structure itself may be of interest. For example, for at-risk species, not only is the change in the underlying population size over time of interest, but also the underlying system process that is driving the population. Identifying potential factors that influence the survival probabilities or fecundity rates may allow better conservation policies to be introduced to manage the population.

For the UK lapwing population, we consider two possible covariates: *fdays* (again used as a surrogate for the harshness for the winter) and time (to reflect any trends in the demographic rates). Incorporating the linear trend and the frost days covariate on the logistic scale into the model for first-year survival probability we have,

$$\text{logit } \phi_{1,t} = \alpha_1 + \beta_1 f_t, +\gamma_1 u_t.$$

Here α_1, β_1 and γ_1 are parameters to be estimated and f_t and u_t are the (normalised) covariate values for $fdays$ and time, respectively, at time t. Similar parameterisations for adult survival and recoveries are adopted with suitable subscript notation. For the productivity rate, we assume a logarithmic relationship, and specify,

$$\log \rho_t = \alpha_\rho + \beta_\rho f_{t-1} + \gamma_\rho u_t.$$

The productivity in the spring of year t is regressed on the severity of the winter prior to the spring, (i.e. in year $t-1$). This reflects the fact that the productivity of the lapwings in the spring may be affected by the severity of the winter immediately prior to the spring. In this circumstance we may be particularly interested in:

1. identifying the factors that influence each of the demographic parameters (i.e. performing a model selection procedure); and
2. identifying the nature of the dependence of the covariate on the demographic parameters (i.e. positive or negative), conditional on a dependence being present.

Priors

There are a total of $2^2 = 4$ possible marginal models for each of the demographic parameters corresponding to each possible combination of covariate dependence. Thus, there are a total of $4^4 = 256$ models overall. We assume that each of these models are equally likely a priori. For each demographic parameter, conditional on the covariate being present in the model, we specify a $N(0, 10)$ prior on the regression coefficient (as specified for the fixed model in the previous analyses). Note that we shall also implement a prior sensitivity analysis.

RJMCMC Algorithm

We implement an RJMCMC algorithm similar to that given in Example 6.2. First, we update each of the parameters in the model using the MH algorithm. We then update the model for each of the demographic parameters. In particular, we cycle through each demographic parameter and consider each covariate in turn. If the given demographic parameter is dependent on $fdays$, we propose to remove the dependence; otherwise, if the covariate is absent, we propose to add the covariate to the model. Following this RJMCMC step we repeat the algorithm for *time*. In each RJMCMC step, the dimension of the model is only changed by one (i.e. adding/removing a single covariate dependence). To increase the acceptance probability of the reversible jump step, and hence improve mixing within the Markov chain, we use the updating algorithm.

We initially perform a pilot tuning step. We fix the model to the saturated model with full covariate dependence for each of the demographic parameters and run a standard MCMC algorithm, updating each parameter using a random walk Metropolis step. From this initial MCMC simulation, we record the corresponding posterior mean and variance of each regression coefficient. We then consider the reversible jump algorithm, in the presence of model uncertainty. Without loss of generality, suppose that we propose to add $fdays$ dependence to the first-year survival probability. We propose the new value,

$$\beta_1' \sim N(\mu, \sigma^2),$$

where μ and σ^2 are the corresponding posterior mean and variance of β_1 obtained from the initial pilot run for the saturated model. All other parameters remain the same. The corresponding Jacobian is simply equal to unity, and the probability of moving between the current and proposed model is also equal to one. The corresponding acceptance probability reduces to,

$$\min\left(1, \frac{f(\boldsymbol{x}|\boldsymbol{\theta}', m')p(\beta_1')}{f(\boldsymbol{x}|\boldsymbol{\theta}, m)q(\beta_1')}\right),$$

where the current and proposed models are denoted by m and m', with corresponding model parameters $\boldsymbol{\theta}$ and $\boldsymbol{\theta}'$.

Results

The RJMCMC simulations are run for 1 million iterations with the first 10% discarded as burn-in for the integrated data. Starting the chain in overdispersed starting points obtained essentially identical results, and we conclude that the posterior distribution has converged (see Section 5.4.2). The models identified with the largest posterior support are presented in Table 10.4 and the corresponding marginal posterior model probabilities for each demographic parameter are provided in Table 10.5.

We can see from Table 10.4 that the covariate dependence on the adult survival probability and productivity rates appear to be very highly correlated. When the adult survival probability is time dependent (irrespective of whether it is also dependent on $fdays$), the productivity rate is not time dependent; whereas when the adult survival probability is independent of time (but possibly dependent on $fdays$), the productivity rate is time dependent. In particular, we can calculate the marginal model probabilities for time dependent (and $fdays$ dependent) adult survival probability and constant productivity rate ($\phi_a(fdays, t), \rho$), and an $fdays$ dependent adult survival probability (but not time dependent) and time dependent productivity rate ($\phi_a(fdays), \rho(t)$). These are calculated in the standard manner, by simply summing over the posterior probabilities for the models that have these particular time (and $fdays$) dependence on the adult survival probability and productivity rate. We find that the marginal probability for ($\phi_a(fdays, t), \rho$) is equal to 0.396 and for ($\phi_a(fdays), \rho(t)$) is 0.216. Considering only the marginal models for the individual demographic parameters hides this intricate detail (we also discuss

Table 10.4 The Posterior Probabilities of the Models with Largest Posterior Support for the Lapwing Data

Model				Posterior Probability
ϕ_1	$\phi_a(fdays,t)$	$\lambda(fdays,t)$	ρ	0.211
ϕ_1	$\phi_a(fdays)$	$\lambda(fdays,t)$	$\rho(t)$	0.138
$\phi_1(fdays)$	$\phi_a(fdays,t)$	$\lambda(t)$	ρ	0.125
ϕ_1	ϕ_a	$\lambda(fdays,t)$	$\rho(t)$	0.099
ϕ_1	$\phi_a(t)$	$\lambda(fdays,t)$	ρ	0.076
ϕ_1	$\phi_a(t)$	$\lambda(fdays,t)$	$\rho(fdays)$	0.056
$\phi_1(fdays)$	$\phi_a(fdays)$	$\lambda(t)$	$\rho(t)$	0.052

Note: The terms in brackets correspond to the dependence of the parameter – $fdays$ (frost days) and t (time).

Table 10.5 The Marginal Posterior Model Probabilities for Each of the Demographic Parameters for the Lapwing Data

(a) First-Year Survival – ϕ_1

Model	Posterior Probability
ϕ_1	0.739
$\phi_1(fdays)$	0.225
$\phi_1(fdays,t)$	0.027
$\phi_1(t)$	0.009

(b) Adult Survival – ϕ_a

Model	Posterior Probability
$\phi_a(fdays,t)$	0.465
$\phi_a(fdays)$	0.241
$\phi_a(t)$	0.153
ϕ_a	0.141

(c) Recovery – λ

Model	Posterior Probability
$\lambda(fdays,t)$	0.755
$\lambda(t)$	0.245
$\lambda(fdays)$	0.000
λ	0.000

(d) Productivity – ρ

Model	Posterior Probability
ρ	0.477
$\rho(t)$	0.350
$\rho(fdays)$	0.105
$\rho(fdays,t)$	0.068

this issue in Section 6.6 for a different analysis of the same data). For example, suppose that the marginal models were (approximately) independent. Then,

LESLIE MATRIX-BASED MODELS

we would expect,

$$\pi(\phi_a(fdays,t), \rho|\boldsymbol{y}, \boldsymbol{m}) \approx \pi(\phi_a(fdays,t)|\boldsymbol{y}, \boldsymbol{m})\pi(\rho|\boldsymbol{y}, \boldsymbol{m}).$$

The posterior marginal probabilities for the covariate dependence for each of the demographic parameters are given in Table 10.5. We have that,

$$\pi(\phi_a(fdays,t)|\boldsymbol{y}, \boldsymbol{m})\pi(\rho|\boldsymbol{y}, \boldsymbol{m}) = 0.465 \times 0.477 = 0.222,$$

whereas we have $\pi(\phi_a(fdays,t), \rho|\boldsymbol{y}, \boldsymbol{m}) = 0.396$. Clearly these are very different, as a result of the non-independence of the posterior marginal probabilities for the adult survival probability and productivity rate. A similar result holds for the marginal probability for $(\phi_a(fdays), \rho(t))$.

There does appear to be some uncertainty in relation to the underlying model for the lapwings. We are able to account for this model uncertainty in estimating the true population sizes by considering a model-averaged estimate. This is provided in Figure 10.5(a) and for the growth rate in Figure 10.5(b). The posterior uncertainty of the population size and growth rates are greater in the presence of model uncertainty compared to the fixed model considered in Section 10.2.1, reflecting both parameter and model uncertainty.

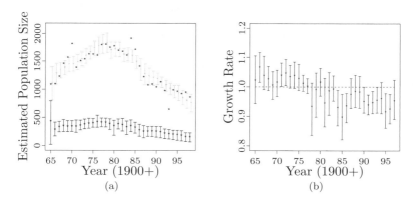

Figure 10.5 (a) Lapwing data: the model-averaged posterior mean (*) and 95% HPDI (vertical lines) of the number of juvenile birds aged 1 (in black) and adults aged at least 2 (in grey) for the integrated data analysis. The *'s give the data $\boldsymbol{y}$ over time. (b) The posterior mean (*) and 95% HPDI (vertical lines) of the model-averaged growth rate r_t in year $t = 1965, \ldots, 1997$ for the integrated data analysis. The horizontal dashed line corresponds to no change in growth rate.

We consider a prior sensitivity analysis of the results by considering different priors on the regression coefficients. In particular we consider the prior specification $N(0, 2.344)$ for each coefficient, conditional on the parameter being present in the model (see Section 8.2.1). The models with largest posterior support are provided in Table 10.6 and the corresponding marginal posterior

Table 10.6 The Posterior Probabilities of the Models with Largest Posterior Support for the Lapwing Data

	Model			Posterior Probability
ϕ_1	$\phi_a(fdays, t)$	$\lambda(fdays, t)$	ρ	0.198
ϕ_1	$\phi_a(fdays)$	$\lambda(fdays, t)$	$\rho(t)$	0.158
$\phi_1(fdays)$	$\phi_a(fdays, t)$	$\lambda(t)$	ρ	0.111
ϕ_1	$\phi_a(fdays, t)$	$\lambda(fdays, t)$	$\rho(fdays)$	0.059
ϕ_1	ϕ_a	$\lambda(fdays, t)$	$\rho(t)$	0.058
$\phi_1(fdays)$	$\phi_a(fdays)$	$\lambda(t)$	$\rho(t)$	0.051

Note: The terms in brackets correspond to the dependence of the parameter – *fdays* (frost days) and t (time), assuming independent $N(0, 2.334)$ priors on the regression coefficients.

probabilities for the covariate dependence for each demographic parameter are provided in Table 10.7. The model-averaged estimates of the parameters, including the total population size, are essentially unchanged. There are some minor changes in the posterior probabilities of the different models identified (compare Tables 10.4 and 10.6), however the general interpretation of the results remains the same. Thus, the posterior distribution seems fairly insensitive to the priors considered within this analysis.

□

Comment

The Leslie-based matrix approach can only be used where there is a linear relationship between the size of the population between years. Thus, not all system processes can be described using a Leslie-based matrix, for example, density dependent demographic rates. The Bayesian approach is easily extended to allow for more complex situations, using the standard approach, as we discuss next.

10.3 Non-Leslie-Based Models

Many biological systems can be modelled using a Leslie matrix; however, these cannot describe all system processes of interest. For example, density dependence and other non-linear processes cannot be expressed in the form of a Leslie matrix. In such circumstances it is possible simply to define the underlying transition equations via a set of equations to describe the changing population size over time. The fitting of Leslie-based system processes and non-Leslie-based system processes within the Bayesian approach are performed in

Table 10.7 The Marginal Posterior Model Probabilities for Each of the Demographic Parameters for the Lapwing Data

(a) First-Year Survival – ϕ_1

Model	Posterior Probability
ϕ_1	0.673
$\phi_1(fdays)$	0.253
$\phi_1(t)$	0.050
$\phi_1(fdays, t)$	0.023

(b) Adult Survival – ϕ_a

Model	Posterior Probability
$\phi_a(fdays, t)$	0.524
$\phi_a(fdays)$	0.284
$\phi_a(t)$	0.103
ϕ_a	0.089

(c) Recovery – λ

Model	Posterior Probability
$\lambda(fdays, t)$	0.772
$\lambda(t)$	0.228
$\lambda(fdays)$	0.000
λ	0.000

(d) Productivity – ρ

Model	Posterior Probability
ρ	0.427
$\rho(t)$	0.368
$\rho(fdays)$	0.127
$\rho(fdays, t)$	0.079

Note: We assume independent $N(0, 2.334)$ priors on the regression coefficients.

exactly the same way. Given any set of transition equations and observation processes, the true (unknown) population sizes are treated as missing data (or auxiliary variables) and the joint posterior distribution formed over both these auxiliary variables and model parameters. The MCMC algorithm is constructed with this given stationary distribution and the posterior (marginal) distributions of the parameters (and true population sizes) estimated.

10.3.1 System Process

When the system process cannot be modelled using a Leslie matrix, we can simply specify the transition equations, in a manner analogous to the system process specified in Section 10.2. For example, the density dependence model of Dennis and Taper (1994) can be expressed in the form,

$$N_t = N_{t-1} \exp(\beta_0 + \beta_1 N_{t-1}) \exp(\epsilon_t),$$

where N_t denotes the population size at time t (assuming only a single state), β_0 and β_1 are parameters to be estimated and ϵ_t denotes the stochastic element of the system process, such that,

$$\epsilon_t \sim N(0, \sigma^2).$$

Note that this model is often reparameterised for simplicity as an additive model, by considering the log-transformation, so that,

$$p_t = p_{t-1} + \beta_0 + \beta_1 \exp(p_{t-1}) + \epsilon_t, \qquad (10.2)$$

where $p_t = \log N_t$ and the associated error (ϵ_t) has a normal distribution. Clearly, setting $\beta_1 = 0$ removes the first-order density dependence, reducing the model to a simple linear trend with normal errors (which could be modelled using a Leslie matrix).

There are many other possible models that may be used to describe density dependence. For example, Reddingius (1971) suggests the model,

$$p_t = p_{t-1} + \beta_0 + \beta_1 p_{t-1} + \epsilon_t,$$

where $\epsilon_t \sim N(0, \sigma^2)$. This is very similar to the previous model, where essentially the density dependence is on a different scale. The form of the transition equations to be used should be motivated and developed from biological knowledge of the system.

10.3.2 Observation Process

The observation process is defined in the standard way, describing how the observed count data relate to the true underlying population size. For example, for the above Dennis and Taper model (or Reddingius model), we might specify that the observed count data, y_t are of the form,

$$y_t \sim N(N_t, \sigma^2),$$

where σ^2 is to be estimated. In some instances, the observation process standard error may be known (or can be estimated), and typically in such circumstances will be known over time (see, for example, King et al. 2008b). This can easily be incorporated into the observation process, by simply removing the estimation of the observation process error parameter.

10.3.3 Comments

Clearly, the density dependence model defined should represent the underlying biological knowledge of the system. It can be easily extended to allow for higher-order (or longer-range) density dependence of the same form. For example, for the above case, suppose that the density dependence extends to the kth previous population size. The model can be extended to,

$$p_t = p_{t-1} + \beta_0 + \sum_{\tau=1}^{k} \beta_\tau \exp(p_{t-\tau}) + e_t. \qquad (10.3)$$

Jamieson and Brooks (2004a) extend this idea to where the value of k is unknown a priori and calculate posterior model probabilities to discriminate between competing models for 10 species of duck from the Waterfowl Breeding Population and Habitat Survey (WBPHS, US Fish and Wildlife Service 2003).

NON-LESLIE-BASED MODELS 335

Three species (Northern Pintail, *Anas acuta*; Redhead, *Aythya americana*; and Canvasback, *Aythya valisineria*) appeared to show some form of density dependence. This approach once more allows for model uncertainty to be incorporated into the estimates of population size in addition to parameter uncertainty using model-averaged estimates (see Section 6.5 and Example 10.4 for the Canvasback ducks). However, note that the calculation of the likelihood for the system process needs to be defined over the same time period, irrespective of the model fitted. For example, suppose that we wish to consider two possible models: $k = 0$ (corresponding to no density dependence) and $k = 1$. For $k = 1$, we either need to introduce the additional parameter p_0 (and specify a corresponding prior) or specify a prior on p_1 and calculate the likelihood from $t = 2, \ldots, T$ (for $k = 0$ and $k = 1$).

Model discrimination techniques can also be used when there is additional uncertainty, for example, in the actual form of the density dependence (as well as the order of the process). In such circumstances, it is possible simply to treat each possible density dependence structure defined by the different transition equations as different models and once more use the standard Bayesian model discrimination techniques (Jamieson and Brooks 2004a). Clearly, the set of possible models that are initially defined should reflect the underlying density dependence structures that are deemed to be biologically plausible a priori.

Example 10.4. Canvasback Ducks

The data are of the form of annual estimates of the population size of North American duck species on their breeding grounds from 1955 to 2002. We fit a state-space model to the population estimates of canvasbacks. In particular, we perform the analysis within WinBUGS to fit population models of density dependence that simultaneously account for both process and observation error. We consider the Dennis and Taper model, given in Equation (10.2), where the value of k is unknown, a priori. We implement an RJMCMC algorithm to compute the posterior probability for each of a set of possible values of k, and thereby estimate the probability of the presence of density dependence (i.e. the probability that $k > 0$) in a population. We initially describe the priors specified on the parameters, before we present the results.

Priors

We need to specify priors on the parameters in the model: for the system process, the regression parameters ($\boldsymbol{\beta} = \{\beta_0, \ldots, \beta_k\}$); and for the observation process, the error variance, σ^2. In addition, we need to specify a prior on the models, i.e. specify a prior on k. Finally, we need to place a prior on the initial population sizes, N_t, such that $t = 1, \ldots, k_{max}$, where k_{max} is the maximum value that k can take.

In particular, we specify the same independent priors on each parameter in the model (i.e. conditional on the value of k). Thus, we specify the general priors in the form:

$$\begin{aligned}
k &\sim U[1, k_{max}], \\
\beta_\tau | k &\sim N(\mu_{\beta_\tau}, \sigma^2_{\beta_\tau}) & \tau = 1, \ldots, k; \\
\sigma^2 &\sim \Gamma^{-1}(\alpha, \beta); \\
N_t &\sim N(0.540, 0.130) & t = 1, \ldots, k_{max};
\end{aligned}$$

where $U[,]$ denotes the discrete uniform distribution. Here we set $\mu_{\beta_\tau} = 0$, $\sigma^2_{\beta_\tau} = 1$ for $\tau = 1, \ldots, k_{max}$; $\alpha = 0.001 = \beta$; $\mu_{N_t} = 0.540$, $\sigma^2_{N_t} = 0.130$ for $t = 1, \ldots, k_{max}$; and $k_{max} = 5$. Note that the numbers of ducks are expressed $\times 10^6$ and that the distribution is truncated so that $N_t > 0$. Further, priors are not required on N_t, $t = k+1, \ldots, T$ (where T denotes the final time point of the period) due to the Markovian structure of the state process model. Priors for these quantities are implicitly specified when priors are set for N_t, $t = 1, \ldots, k$ (see, for example, Jamieson and Brooks 2004a). The corresponding DAG is provided in Figure 10.6, for general prior parameter values.

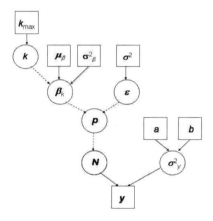

Figure 10.6 The corresponding DAG for the canvasback duck data, using a state-space model and Dennis and Taper system process.

We use WinBUGS and the Jump package to implement the RJMCMC algorithm (see Chapter 7, in particular Section 7.5, and Appendix C). The constructed Markov chain simultaneously traverses the parameter and model space. However, due to the specification of the set of models in the form of a variable selection problem on the parameters $\boldsymbol{\beta}$ (necessary to use the package Jump), all possible combinations of the parameters β_1 to β_5 are permitted to be present in the model. Essentially, the problem is expressed as a covariate variable selection problem on the parameters $\beta_1, \ldots, \beta_5$, where the corresponding covariate values are $\exp(p_t)$ for $t = 1, \ldots, 5$. An indicator function (id) is used to indicate which parameters are in the model. For example, an id of 11000 would indicate that the parameters β_1 and β_2 are in the model (corresponding to a density dependence up to lag 2). However, an id of 10100 corresponds to the model with parameters β_1 and β_3 present (but not β_2). Such a model is not permitted in the set of models considered in Equation (10.3), but can be explored within the covariate selection approach of WinBUGS. Thus, given the output in WinBUGS, we select only those models which are of the correct form. This means that to calculate the posterior probabilities of the models of interest, we simply select out from the list of id's the models we are interested in, and renormalise so that probabilities sum to one (see Section 6.8.2 for further discussion). Note that model-averaged estimates of other unknown quantities (such as the n_t for $t = k+1, \ldots, T$) can also be plotted in WinBUGS. However, these model-averaged estimates will contain both models we are interested in and those that we are not. Thus, it is generally necessary to save the value of the variables of interest generated at each iteration (using the CODA button; see Appendix C) as well as the corresponding id values. This allows the extraction of the sample values for the models of interest, and hence the corresponding posterior model-averaged (or even within-model) estimates to be calculated. Finally, we note that the length of simulations needed will typically be larger when implementing such an approach, since the chain will be exploring models that are simply discarded, and hence the effective sample size will be reduced (potentially significantly).

Results

Summaries of the posterior parameter estimates for the canvasback data for $k = 1, \ldots, 5$ are given in Table 10.8. These posterior estimates are based on an initial burn-in of 50,000 iterations, followed by a further 1 million iterations. The simulations took approximately 25 minutes on an Intel Pentium 1.8 GHz with 512 Mb RAM running Windows XP. We discard 76% of the posterior samples, occurring in models that are not deemed to be plausible, as described above. This means that posterior summary statistics are based on approximately 240,000 iterations. Repeating the simulations obtains essentially identical results, so we assume that the posterior estimates have converged. The model with largest posterior support has only a first-order density dependence (posterior model probability of 0.515; or equivalently a Bayes factor of 2.2 compared to the second-ranked model corresponding to a second-order density dependence). The posterior probability of no density dependence (i.e. $k = 0$) is only 0.030. This means that the posterior probability that there is density dependence at some level (i.e. $k \geq 1$) is 0.970. We note that the prior probability of no density dependence is equal to $\frac{1}{6}$, and that the prior probability of some density dependence (at any level) is $\frac{5}{6}$. Thus, the corresponding Bayes factor of density dependence compared to no density dependence is 6.47, and we conclude that for the given prior specification, there is some evidence for density dependence.

Table 10.8 Posterior Model Probabilities for $k = 0, \ldots, 5$, for the Logistic Model Applied to Canvasback Data

k	Posterior Probability
0	0.030
1	0.515
2	0.233
3	0.129
4	0.059
5	0.034

Table 10.9 Posterior Model Probabilities for $k = 0, \ldots, 5$ for the Logistic Model Applied to Canvasback Data, Specifying Independent $N(0, 100)$ Priors on the β_τ Parameters

k	Posterior Probability
0	0.279
1	0.685
2	0.034
3	0.002
4	0.000
5	0.000

Source: Reproduced with permission from Gimenez et al. (2009a, p. 907) published by Springer.

Prior Sensitivity Analysis

We perform a prior sensitivity analysis by considering a different prior specification on $\boldsymbol{\beta}$, the parameters on which we perform model selection. In particular, we specify,

$$\beta_\tau | k \sim N(0, 100), \quad \tau = 1, \ldots, k,$$

i.e. $\sigma^2_{\beta_\tau} = 100$. This is a significantly larger prior variance on the β terms than previously considered. We repeat the RJMCMC simulations, once more using 50,000 burn-in, followed by a further 1 million iterations. The simulations took approximately 17 minutes (this is shorter than the previous simulations; we will see the reason for this when we examine the posterior distribution). The corresponding posterior model probabilities for each possible model is given in Table 10.9.

We once more discard all posterior observations for models that are not

CAPTURE-RECAPTURE DATA

biologically plausible. This results in discarding 45% of the posterior samples. We note that simpler models (i.e. models with fewer parameters) are identified within this analysis than for the initial prior specified above, so that the simulations are faster to perform (since fewer parameters are typically updated within each iteration). The model with the largest posterior support has only a first-order density dependence as for the previous prior, however, the posterior model probability has now increased to 0.685. Alternatively, the model ranked second (with posterior probability 0.279) corresponds to the model with no density dependence. This is in contrast to the previous prior, where this model had the smallest posterior support. This is another example of the prior sensitivity issue discussed in Section 6.4 (and Example 6.8 in particular) in the presence of model uncertainty. The marginal posterior distributions of the parameters $\beta_1, \ldots, \beta_k$ for the model with density dependence of order k are each concentrated within the interval $[0.3, 0.8]$ for $k = 1, \ldots, 5$. Under the prior specification of $\beta_\tau \sim N(0, 100)$, there is relatively little prior mass specified on the range of values supported by the data, compared to the initial prior (i.e. $\beta_\tau \sim N(0, 1)$). This results in smaller posterior support for models containing β_τ for $\tau = 1, \ldots, 5$. In the terminology of Section 6.4, the large contrast between the (diffuse) $N(0, 100)$ prior and corresponding (relatively tight) posterior distributions for the parameters increases our level of "surprise" regarding the presence of these parameters in the model, so that more information is needed in the data to justify the presence of these parameters (overcoming our level of "surprise").

□

10.4 Capture-Recapture Data

In this section, we show how to analyse capture-recapture data in a state-space modelling framework to estimate survival. We assume that n individuals are involved in the study with T encounter occasions. Let f_i denote the occasion where individual i is encountered for the first time. A general state-space formulation of the CJS model is therefore given by the coupling of a system and an observation process, described at the start of the chapter.

10.4.1 System process

The system process describes the demographic phenomenon of interest here, namely survival. Let $X_{i,t}$ be the binary random variable taking values 1 if individual i is alive at time t and 0 if it is dead at time t. Let $\phi_{i,t}$ be the probability that an animal i survives to time $t+1$ given that it is alive at time t ($t = 1, \ldots, T-1$). Then the state equation is given by,

$$X_{i,t+1}|X_{i,t} \sim Bernoulli(X_{i,t}\phi_{i,t})$$

for $t \geq f_i$. In the CJS model, all calculations are made conditional on the first capture f_i of each individual i, which means that $X_{i,f_i} = 1$ with probability 1. If individual i is alive at time t, then it has probability $\phi_{i,t}$ of being alive at time $t+1$ and probability $1 - \phi_{i,t}$ otherwise, which translates into $X_{i,t+1}$ is distributed as $Bernoulli(\phi_{i,t})$ given $X_{i,t} = 1$. Now if individual i is dead at time t, it remains dead, which translates into $X_{i,t+1}$ is distributed as $Bernoulli(0)$ given $X_{i,t} = 0$.

10.4.2 Observation process

The observation process relates the current state to its observation through detectability. Let $Y_{i,t}$ be the binary random variable taking values 1 if individual i is encountered at time t and 0 otherwise. Let $p_{i,t}$ be the probability of detecting individual i at time t ($t = 2, \ldots, T$). Then the observation equation is given by,

$$Y_{i,t} | X_{i,t} \sim Bernoulli(X_{i,t} p_{i,t})$$

for $t \geq f_i$, with $p_{i,f_i} = 1$ since the reasoning is conditional on the first capture f_i of each individual i. If individual i is alive at time t, then it has probability $p_{i,t}$ of being encountered and probability $1 - p_{i,t}$ otherwise, which translates into $Y_{i,t}$ is distributed as $Bernoulli(p_{i,t})$ given $X_{i,t} = 1$. Now if individual i is dead at time t, then it cannot be encountered, which translates into $Y_{i,t}$ is distributed as $Bernoulli(0)$ given $X_{i,t} = 0$.

10.4.3 Implementation Using MCMC

Statistical inference requires the likelihood of the state-space model specified above. We have essentially defined the true (partially unknown) states to be denoted by the $\boldsymbol{X} = \{X_{i,t} : t = f_i, \ldots, T; i = 1, \ldots, n\}$; and the observed process the corresponding $\boldsymbol{Y} = \{Y_{i,t} : t = f_i, \ldots, T; i = 1, \ldots, n\}$. We form the joint posterior distribution over the model parameters, $\boldsymbol{\theta}$, (the recapture and survival probabilities) and unknown states, $\boldsymbol{Y}$, given by,

$$\pi(\boldsymbol{\theta}, \boldsymbol{Y} | \boldsymbol{X}) \propto f_{obs}(\boldsymbol{X} | \boldsymbol{\theta}, \boldsymbol{Y}) f_{sys}(\boldsymbol{Y} | \boldsymbol{\theta}) p(\boldsymbol{\theta}).$$

The terms f_{obs} and f_{sys} are the likelihood terms for the observation and system process, respectively. In our case, these likelihood terms are simply a product of Bernoulli density function. See Gimenez et al. (2007) for further discussion and the corresponding WinBUGS code.

Example 10.5. Dippers
Here we fit the CJS model (see Section 2.3.2) to the dipper data (see Example 1.1). Uniform priors are used for both the survival and recapture probabilities. The simulations are run for 10,000 iterations following a burn-in of 5,000 iterations in order to obtain posterior summary statistics of interest. Independent replications from over-dispersed starting points suggested that the posterior estimates had converged. Posterior summaries for the recapture and survival probabilities are provided in Table 10.10 (and can be compared with the pos-

CAPTURE-RECAPTURE DATA

terior density plots provided in Figure 4.5 of Example 4.6 for this particular example).

Table 10.10 Estimated Survival and Recapture Probabilities for the Dipper Data Using the CJS Model Fitted via the State-Space Model

Parameter	Posterior Mean (SD)
ϕ_1	0.725 (0.136)
ϕ_2	0.449 (0.071)
ϕ_3	0.484 (0.062)
ϕ_4	0.627 (0.060)
ϕ_5	0.602 (0.058)
ϕ_6	0.706 (0.139)
p_2	0.667 (0.133)
p_3	0.868 (0.080)
p_4	0.879 (0.064)
p_5	0.876 (0.056)
p_6	0.904 (0.050)
p_7	0.756 (0.145)

There is a clear decrease in survival from 1982–1983 to 1983–1984, corresponding to a major flood during the breeding season in 1983 (Lebreton et al. 1992). Note that the final survival probability, ϕ_6, as well as the last recapture probability, p_7, are estimated with low precision. The fact that these two parameters cannot be separately estimated is not surprising since the CJS model is known to be parameter redundant; see Section 2.8 and Example 2.7. Also, the first survival probability and the first detection probability are poorly estimated, due to the fact that very few individuals were marked at the first sampling occasion (approximately 7% of the full data set). See, for comparison, Figure 4.5.

□

10.4.4 Comments

The state-space formulation naturally separates the nuisance parameters (the encounter probabilities) from the parameters of actual interest, i.e. the survival probabilities, the latter being involved exclusively in the system process. Such a clear distinction between a demographic process and its observation makes the description of a biological dynamic system much simpler and allows complex models to be fitted (Clark et al. 2005; Gimenez et al. 2007; Zheng et al. 2007). For example, Gimenez et al. (2007) describes how state-space models can be fitted to multi-state capture-recapture data and ring-recovery data, in addition to the capture-recapture data described above. Alternatively, and of particular importance, Royle (2008) used the state-space modelling approach to incorporate individual heterogeneity in both survival and recapture probabilities in the CJS model.

10.5 Summary

We considered state-space models within this chapter. These models separately estimate the error associated with the observation process from the system process (describing the changing population over time). Typically, within the classical framework, normal assumptions are made relating to the processes and the Kalman filter is applied for inference. In the Bayesian framework, irrespective of the distributional form of the process and system errors (for example, normal, binomial, Poisson, log-normal etc.), the analysis follows the same procedure. We appeal to the ideas associated with Chapter 8 and impute the missing states within the system in a missing data framework. This allows us to calculate explicitly the likelihoods associated with both the observation and system processes. This approach also allows non-linear processes to be fitted to the data in exactly the same manner.

We also show how we can integrate state-space models with additional data; in our case ring-recovery data was used, but the approach is perfectly general. The use of additional (independent) data can appreciably improve the parameter estimates in terms of posterior precision. This appears to be particularly the case for parameters that are not directly observable. Once more model discrimination techniques can be used, in the presence of model uncertainty, permitting model-averaged estimates of the states to be calculated, incorporating both parameter and model uncertainty.

Finally, we demonstrate how we can reformulate capture-recapture data into a state-space framework. This allows the natural separation of the nuisance parameters (recapture probabilities) from the parameters of interest (survival probabilities).

10.6 Further Reading

Integrated analyses allow the joint analysis of different data sets within a single united and consistent framework. Additional integrated analyses include joint capture-recapture and ring-recovery data (Catchpole et al. 1998; King and Brooks 2003a), multi-site capture-recapture and ring-recovery data (King and Brooks 2003b; King and Brooks 2004a); multi-site recruitment, capture-recapture, ring-recovery and multi-site census data (Borysiewicz et al. 2009); joint capture-recapture, ring-recovery, abundance data and count data (Reynolds et al. 2009) and capture-recapture data and constant effort site data (Cave et al. 2009b). Within the Bayesian framework integrated data analyses are typically easily implemented (assuming independence of data sources) as a result of the modularisation aspect of the likelihood.

FURTHER READING

State-space models are becoming increasingly used within a wide range of ecological application areas (Borchers et al. 2010) from modelling animal movement (Anderson-Sprechler and Ledolte 1991; Jonsen et al. 2005; Newman 1998) to occupancy models (Royle and Kéry 2007) and fisheries (Millar and Meyer 2000; Newman 2000; Rivot et al. 2004). These models provide a natural separation of the different processes and have been extended to the more traditional CJS models (Gimenez et al. 2007; Royle 2008): separating the recapture (or recovery) process from the underlying survival of the individual animals. This framework is easily extended, for example, to the case of multi-state/multi-event models (Chilvers et al. 2009) or individual heterogeneity (Royle 2008). WinBUGS code is available for many of these applications (see, for example, Gimenez et al. 2007; Royle 2008; Chilvers et al. 2009; Brooks et al. 2004), and can be easily adapted for different problems. However, fitting such models within WinBUGS where there is a state-space component can be very time consuming, particularly if the parameters are highly correlated.

CHAPTER 11

Closed Populations

11.1 Introduction

Estimating population size has a long history, and is applicable in many different fields ranging from the size of "hidden" populations, such as drug users or disease carriers, to the estimation of trends in wildlife population size over a period of time. The topic was introduced in Section 2.2.4. We are here primarily concerned with the estimation of the size of a closed population, which is where there are no births, deaths, or migration beyond the study area in the population throughout the study period. We shall consider capture-recapture data in which animals are generally observed over a short time period, in order that the closure assumptions are satisfied. As usual, the capture-recapture data collected are generally presented in the form of the capture histories for each individual animal observed within the study (see Williams *et al*, 2002, Chapter 14, for example). The capture history of an individual is simply a sequence of 1's and 0's, corresponding to whether the individual was, or was not, observed at each capture event. Clearly, to estimate the population size we need to estimate the number of individuals with capture histories represented by a sequence of only 0's, i.e. we wish to estimate the number of individuals that were not observed within the study period. We combine this with the number of individuals we do observe to obtain an estimate of the total population size.

The assumptions that we make concerning the catchability of the animals throughout the study period generally have a direct impact on the corresponding estimate of the population size. These assumptions essentially define the underlying model that we fit to the data. Otis et al. (1978) discuss three different types of influence on the catchability of the animals: time effects, behavioural effects and heterogeneity effects. Time effects allow for the possibility that the capture probability of an animal may depend on time; behavioural effects allow the initial capture of the animal to influence its future capture probability (often referred to as trap response), and heterogeneity allows for capture probabilities to differ between individuals. The presence or absence of each of these effects results in eight basic models that can be fitted to the data. For the four possible models without individual heterogeneity, a closed form expression for the likelihood can be derived and the corresponding MLEs of the parameters, including the total population size, can be calculated. However, difficulties arise when considering models that include individual heterogeneity. This has resulted in a variety of different methods to provide estimates of the total population size. Alternatively, a numerical approximation can be derived for the likelihood, (see Coull and Agresti 1999, and the logistic-normal-binomial (LNB) model of Table 3.1) or a number of mixture models may be used (see Pledger 2000, a variety of models presented in Table 3.1 and Morgan and Ridout 2008).

11.2 Models and Notation

The different possible models each represent competing biological hypotheses. These relate to the dependence of the capture probabilities of the individuals on different factors: time effects (t), behavioural effects (b), and individual heterogeneity effects (h). The different models correspond to the different combinations of factors that are present. The models are indexed by subscripts, detailing which of the effects (if any) are present, so that we have models, M_0, M_t, M_b, M_{tb}, M_h, M_{th}, M_{bh} and M_{tbh}.

We let $p_{i\tau}$ denote the capture probability for individual $i = 1, \ldots, N$ at time $\tau = 1, \ldots, T$. Additionally, let $\mathcal{F}(i)$ denote the first time that individual i is observed. Thus, $p_{i\tau}$ can be interpreted as the initial capture probability for times $\tau = 1, \ldots, \mathcal{F}(i)$, and the recapture probability for all future times $\tau = \mathcal{F}(i) + 1, \ldots, T$ (i.e. following the initial capture of the individual, assuming that $\mathcal{F}(i) < T$). The saturated model M_{tbh} has corresponding capture probabilities of the form,

$$\text{logit } p_{i\tau} = \mu + \alpha_\tau + \beta Y_{i\tau} + \gamma_i, \tag{11.1}$$

where,

$$Y_{i\tau} = \begin{cases} 0 & \text{if } \tau \leq \mathcal{F}(i), \text{ and} \\ 1 & \text{if } \tau > \mathcal{F}(i), \end{cases}$$

and γ_i denotes individual random effects, such that,

$$\gamma_i \sim N(0, \sigma_\gamma^2).$$

Thus, $Y_{i\tau}$ is the indicator function for whether individual i is observed for the first time before or at time τ, and hence corresponds to the behavioural aspect of the model.

The model parameters to be estimated are: μ, the mean underlying capture probability; α_τ, the year effects for the capture; β, the behavioural response (*trap response*) for recaptures; and σ_γ^2, the individual random effects variance. Submodels of this saturated model are obtained by setting some parameter values equal to zero. In particular, if there are no heterogeneity effects, then we set $\gamma_i = 0$ for all individuals, $i = 1, \ldots, N$. If there are no behavioural effects, we set $\beta = 0$. Finally, when there are no year effects, we set $\alpha_\tau = 0$, for all $\tau = 1, \ldots, T$. The overall model is constructed by adding combinations of each of these possible restrictions to the saturated model. Note that it may be biologically plausible to consider additional models, for example, where the capture probabilities are time independent and the recapture probabilities time dependent. In this case, we could consider a time dependent β term.

MODEL FITTING

The set of plausible models is clearly data dependent, and we restrict our consideration to the standard eight possible models, though extending the methodology to allow for additional models is relatively straightforward.

We assume that the animals behave independently of one another, so that the capture probabilities of the animals are conditionally independent given the parameter values. For notational convenience, we let $\boldsymbol{\theta} = \{\mu, \boldsymbol{\alpha}, \beta, \sigma_\gamma^2, N\}$, where $\boldsymbol{\alpha} = \{\alpha_1, \ldots, \alpha_T\}$ and set $\boldsymbol{\gamma} = \{\gamma_1, \ldots, \gamma_N\}$. We have that the joint probability distribution of the capture histories of all individuals in the population, (denoted by $\boldsymbol{x}$) given the parameters, $\boldsymbol{\theta}$, and individual effects, $\boldsymbol{\gamma}$, can be specified as,

$$f(\boldsymbol{x}|\boldsymbol{\theta},\boldsymbol{\gamma}) \propto \frac{N!}{(N-n)!} \prod_{\tau=1}^{T} \prod_{i=1}^{N} p_{i\tau}^{x_{i\tau}}(1-p_{i\tau})^{(1-x_{i\tau})} \qquad (11.2)$$

where n denotes the total number of observed individuals within the study, and,

$$x_{i\tau} = \begin{cases} 1 & \text{if individual } i \text{ is observed at time } \tau; \\ 0 & \text{if individual } i \text{ is not observed at time } \tau. \end{cases}$$

If individual i is not observed within the study, then $x_{i\tau} = 0$ for all $\tau = 1, \ldots, T$.

11.3 Model Fitting

For the models without individual heterogeneity present, (i.e. models M_0, M_t, M_b and M_{tb}), $\boldsymbol{\gamma}$ is known, since we have $\gamma_i = 0$ for all individuals and $f(\boldsymbol{x}|\boldsymbol{\theta},\boldsymbol{\gamma})$ $(= g(\boldsymbol{x}|\boldsymbol{\theta})$, dropping the dependence on the known $\boldsymbol{\gamma}$). Given the explicit expression for the likelihood, we can specify the priors on the parameters, form the posterior distribution and use MCMC to summarise and explore the posterior distribution. However, before we consider the priors in more detail, we consider the more complex models, where individual heterogeneity is present.

We consider the individual effects as random effects (see Section 8.5), and form the joint posterior distribution over both the model parameters, $\boldsymbol{\theta} = \{N, \mu, \boldsymbol{\alpha}, \beta\}$, and the individual heterogeneity terms, $\boldsymbol{\gamma}$. We have that,

$$\pi(\boldsymbol{\theta},\boldsymbol{\gamma}|\boldsymbol{x}) \propto f(\boldsymbol{x}|\boldsymbol{\theta},\boldsymbol{\gamma})p(\boldsymbol{\theta},\boldsymbol{\gamma}).$$

The likelihood $f(\boldsymbol{x}|\boldsymbol{\theta},\boldsymbol{\gamma})$ is given in Equation (11.2).

11.3.1 Prior Specification

Priors need to be specified on all model parameters. We discuss each of these in turn.

Prior for the Total Population: N

This is typically the parameter of most interest within the analysis. In some instances, there may be information relating to the total population size from which we may be able to derive a prior (King et al. 2005). Alternatively, if there is no prior information, typically choices for the prior on total population size include Jeffreys prior (see Section 4.2.3) so that,

$$p(N) \propto N^{-1};$$

or a uniform prior,

$$p(N) \propto 1,$$

which is often truncated, i.e. has an upper limit N_{max}, say. An alternative prior parameterisation, particularly in the presence of prior information, may be to assume,

$$N|\lambda \sim Po(\lambda); \qquad \lambda \sim \Gamma(a_1, a_2).$$

Marginalising, by integrating out the λ nuisance parameter, gives,

$$N \sim Neg - Bin(a_1, a_2).$$

This prior specification allows a large variety of possible prior distributions, and can be vague or informative, depending on the choice of a_1 and a_2. For the examples that we shall consider, we shall assume Jeffreys prior for the total population size.

Prior for the Underlying Capture Probability: μ

The parameter μ can be regarded as the underlying capture probability, in the absence of any time, behavioural and/or individual effects. A common prior would be to specify,

$$\mu \sim N(0, 10).$$

However, in the model M_0 this does not present a uniform prior over the interval $[0, 1]$ for the capture probabilities (see Section 8.2.1). Thus, we consider an alternative prior, such that the corresponding induced prior on the capture probability, under model M_0, is uniform on the interval $[0,1]$. Thus, we specify the prior on μ to have probability density function,

$$p(\mu) = \frac{\exp(\mu)}{(1 + \exp(\mu))^2},$$

which induces a $U[0, 1]$ prior on the capture probability under model M_0. Typically, for this model, we would expect a large amount of information to be contained in the data, since there would be $n \times T$ data points (i.e. each capture time for each individual) all containing information about μ. Thus, we would expect the posterior to be data-driven, irrespective of any sensible prior that we would place on μ.

Prior for Time Effects: $\boldsymbol{\alpha}$

It is common for the capture probabilities of animals to be dependent on the capture time. This may be linked to the movement of the animals under study over time, possibly as a result of changing weather conditions, time of day for

MODEL FITTING

collecting the data, etc. Alternatively, the effort in collecting the data over time may vary. For example, the number of traps used, or the length of time that is spent looking for animals may change. If the effort in collecting the data is known explicitly, then this can also be incorporated into the analysis (see, for example, Example 9.3 and King and Brooks 2004a). For each time $t = 1, \ldots, T$ we specify independent priors of the form,

$$\alpha_t | \sigma_\alpha^2 \sim N(0, \sigma_\alpha^2); \qquad \sigma_\alpha^2 \sim \Gamma^{-1}(b_1, b_2).$$

The commonly used vague parameters for the inverse gamma distribution, i.e. $\Gamma^{-1}(0.001, 0.001)$, in this instance produces an unusual, and undesirable, prior distribution over the capture probabilities. Assume that the model for the recapture probabilities is given by the logistic regression, logit $p_{i\tau} = \alpha_\tau$ (so ignoring the other terms in the logistic regression). If we specify the above prior on α_τ, where $b_1 = b_2 = 0.001$, the corresponding prior induced on the recapture probabilities has a large mass on values very close to 0, 0.5 and 1, with little support elsewhere, as shown in Figure 11.1. Thus, we consider the prior where $b_1 = 4$; and $b_2 = 3$, so that the prior mean and variance of σ_α^2 are equal to 1 and 0.25, respectively. The corresponding prior induced on the capture probability is symmetrical around 0.5 and gradually decreases toward values of 0 and 1. Note that the prior specification should be considered for any given problem, and ideally a prior sensitivity analysis performed. Also see King and Brooks (2008) for discussion of the prior specification on the model parameters within this context.

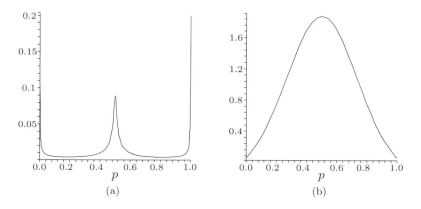

Figure 11.1 The induced prior on the recapture probabilities given the prior specification (a) $\sigma_\alpha^2 \sim \Gamma^{-1}(0.001, 0.001)$ and (b) $\sigma_\alpha^2 \sim \Gamma^{-1}(4, 3)$.

This hierarchical prior for the time effects can be interpreted as a random effect over time, i.e. the parameters are not arbitrarily fixed effects, but instead come from some underlying distribution, and it is parameters in the underlying distribution that are to be estimated (see Section 8.5).

Prior for Behavioural Effects: β

The behavioural effects correspond to any trap response. Essentially, if there is a benefit to the animal in being captured, for example, when there is food in a trap, animals may be more likely to be recaptured following the initial capture. This is said to be a *trap happy* response, and in this case $\beta > 0$. Conversely, there may be a *trap shy* response, with $\beta < 0$, when animals recognize and avoid traps following their initial capture. We have no prior information relating to this parameter when present and so we use a vague prior. Namely, we set,

$$\beta | \sigma_\beta^2 \sim N(0, \sigma_\beta^2); \qquad \sigma_\beta^2 \sim \Gamma^{-1}(4,3).$$

Thus, the prior has mean zero for β, corresponding to there being no preference for a trap happy or trap shy response. Clearly, if we had prior information, possibly due to the nature of the traps, or known behaviour concerning the individuals, this could be incorporated into the prior. For example, if there is prior knowledge that there will be a trap happy/shy response, a positive/negative prior mean could be set for β. An alternative approach could be to use a mixture of two half-normals – one positive and one negative – with prior probabilities on the weights relating to the strength of prior belief of a trap happy/shy response (see, for example, the discussion concerning informative priors in Section 8.2.1 and King et al. 2005).

Prior for Individual Effects: γ

Finally, we consider the individual effects. We assume that the individual effects are random effects, so that the individual effects γ_i come from some distribution, where the parameters of this distribution are of interest. With no prior information relating to these parameters, one possible prior is to set,

$$\gamma | \sigma_\gamma^2 \sim N(0, \sigma_\gamma^2); \qquad \sigma_\gamma^2 \sim \Gamma^{-1}(4,3).$$

Note that this prior appears to be too restrictive for the St. Andrews golf tee data set that we consider in Section 11.4, so that the sensitivity of the posterior distribution on the prior distributions should be checked.

The form of the prior described above can simply be regarded as a hyperprior on the individual effects, where σ_γ^2 is the parameter of interest. An alternative interpretation of the time and individual level heterogeneity defined above is that of a multilevel model, where the random effects relate to different causal factors – time and individual effects.

The corresponding posterior distribution of the parameters is complex, and generally multidimensional. Thus, in order to gain inference concerning the parameters, we use an MCMC algorithm in order to sample from the posterior distribution, and hence obtain estimates of the summary statistics of interest.

11.3.2 MCMC Algorithm

We consider the MCMC algorithm in more detail, due to the additional complexity that arises within this context. We initially present the corresponding DAG in Figure 11.2 for model M_{tbh} (omitting the uppermost level relating to the prior parameters for each parameter to avoid the figure becoming too cluttered). The corresponding DAG for all submodels is obtained by simply omitting the nodes corresponding to the parameters that are not in the given model. For example, for model M_{tb} this means omitting the nodes corresponding to γ and σ_γ^2.

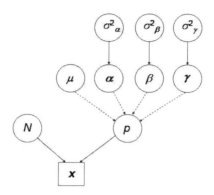

Figure 11.2 The DAG for general model M_{tbh}.

For the MCMC simulations, we use a random walk MH algorithm to update the parameter values (excluding N); see Section 5.3.3. However, the update for the total population size depends on whether there are individual effects present in the model. Suppose that the individual effects are present, so that the random effects are imputed for models containing a heterogeneity effect. Updating the population size also involves updating the number of individual effects that are imputed, and thus involves a change in dimension within the move update, so that we need to use the reversible jump algorithm (Section 6.3.2).

Suppose that the current state of the Markov chain is model M_h with parameter values $(\boldsymbol{\theta}, \boldsymbol{\gamma})$, and that we propose to update N. We propose the new value,

$$N' = N + \epsilon,$$

where ϵ is an integer simulated randomly in the interval $[N-5, N+5]$. If $\epsilon > 0$, we set $\gamma_i' = \gamma_i$ for $i = 1, \ldots, N$ and need to simulate new individual effects for the individual $N+1, \ldots, N+\epsilon$. We simply use the prior distribution, and set,

$$\gamma_i' \sim N(0, \sigma_\gamma^2), \quad \text{for } i = N+1, \ldots, N+\epsilon.$$

Alternatively, if $\epsilon < 0$, we set $\gamma_i' = \gamma_i$ for $i = 1, \ldots, N+\epsilon$. We let $\boldsymbol{\gamma}' = \{\gamma_i' : i = 1, \ldots, N+\epsilon\}$ and $\boldsymbol{\theta}'$ denote the set of proposed parameter values. We accept the move with probability $\min(1, A)$, where,

$$\begin{aligned} A &= \frac{\pi(\boldsymbol{\theta}', \boldsymbol{\gamma}'|\boldsymbol{x})}{\pi(\boldsymbol{\theta}, \boldsymbol{\gamma}|\boldsymbol{x})q(\boldsymbol{\gamma}')} \\ &= \frac{f(\boldsymbol{x}|\boldsymbol{\theta}', \boldsymbol{\gamma}')}{f(\boldsymbol{x}|\boldsymbol{\theta}, \boldsymbol{\gamma})}. \end{aligned}$$

The prior densities cancel in the acceptance probability for all the parameters that remain unchanged. The proposal distribution for the individual effects, q, is simply equal to the corresponding prior disrtibution for the parameters and so also cancel in the acceptance probability. Conversely, if $\epsilon < 0$, then we accept the move with probability equal to $\min(1, A^{-1})$.

In the alternative case, where there are no individual effects, we simply propose the new value N' using the above proposal distribution, and use the standard MH acceptance probability. For further details of the general implementation of the MCMC algorithm see Section 5.3.1.

Example 11.1. Snowshoe Hares

To illustrate the methodology, we fit the set of 8 possible models to a data set of snowshoe hares (Otis et al. 1978; Coull and Agresti 1999). Trappings occurred on 6 consecutive days, with a total of 68 distinct individuals observed throughout the study (see Section 2.2.4). The posterior estimates of the total population size are provided in Table 11.1 for each of the different possible models, along with HPDIs.

Table 11.1 The Posterior Mean and 95% HPDI for the Total Population Size of the Snowshoe Hare Data for the Different Possible Models

Model	Posterior Mean	95% HPDI
M_0	75.68	(70, 83)
M_t	75.12	(70, 82)
M_b	81.73	(70, 98)
M_{tb}	75.07	(68, 87)
M_h	96.22	(76, 123)
M_{th}	93.39	(76, 114)
M_{bh}	106.65	(76, 155)
M_{tbh}	86.38	(69, 115)

These results may be compared with the MLEs of Table 3.1. The binomial result of that table corresponds to M_0, with an MLE of $\hat{N} = 74.4$ (3.4). The other models of that table were different ways of accounting for heterogeneity in recapture probability, corresponding to model M_h. The MLE for the mixture of two binomial distributions, which had the smallest AIC value of Table

MODEL FITTING   353

3.1 was 76.6 (4.13), which is more precise than the estimates of several of the models in Table 11.1.

□

The model fitted to the data should be chosen for biological reasons. However, in this case, the posterior estimates are reasonably consistent between the different models fitted to the data, with overlapping credible intervals, although larger estimates are obtained for models with individual heterogeneity present (this is typically the case). In addition, the models with individual heterogeneity have larger credible intervals, demonstrating the increased uncertainty in the posterior estimates in the presence of individual heterogeneity, as would be expected. A more extreme example of model sensitivity for the estimates of population size can be seen in the following example, relating to St. Andrews golf tees.

Example 11.2. St. Andrews Golf Tees

We consider the data set described by Borchers et al. (2002) involving groups of golf tees (see Table 2.1). A total of eight individuals recorded the location of the golf tees that they observed within a given area, independently of each other, when walking along predefined transects. Each individual was essentially regarded as a capture event. The groups of tees differed with respect to size, colour and visibility, so that there is some heterogeneity within the population being observed. There were a total of 250 groups of golf tees that could be observed within the study, with 162 unique groups recorded by all of the observers. Morgan and Ridout (2008) found that the beta-binomial model was best to describe these data, amongst the alternative mixture models that they considered to describe heterogeneity, resulting in $\hat{N} = 302$. In this case a 95% profile confidence interval was very wide, at (209.37, 7704.74). There are two issues that arise with this example – the prior specification for the individual heterogeneity and model sensitivity for the estimates of total population size. We discuss each in turn.

Prior Specification

The prior specified on the individual variance component ($\Gamma^{-1}(4,3)$) appears to be somewhat restrictive in this case, with values supported by the data having relatively small prior mass. This is easy to identify within a prior sensitivity analysis of this data set, by considering a range of different priors, and noting that the posterior estimate for the variance component changes with respect to the prior. Thus, we consider an alternative prior for the individual random effect variance components, setting,

$$\sigma_\gamma^2 \sim \Gamma^{-1}(2.01, 1.01).$$

This gives the posterior mean for σ_γ^2 as 1, with variance equal to 100, and hence a greater uncertainty corresponding to the value of σ_γ^2. The prior distributions for the other parameters are as before.

Model Sensitivity

To demonstrate the sensitivity of the posterior distribution for the total population size on the underlying model, we consider the two models M_0 and M_h. Table 11.2 gives the corresponding estimates for the total population size for these two different models. Clearly the estimates of the total population size are very different for the two different models fitted to the data. Thus, (without additional external information providing some information relating to the underlying heterogeneities present in the data), model discrimination tools need to be used; see Chapter 6.

Table 11.2 The Posterior Mean and 95% HPDI for the Total Population Size of the St. Andrews Golf Tee Data Set for Models M_0 and M_h

Model	Posterior Mean	95% HPDI
M_0	164	(162, 167)
M_h	235	(193, 281)

Note: The true number of tees is known to be 250.

For this data set the true population size is known to be 250. Thus, the model without individual heterogeneity significantly underestimates the true population size. The general underestimation of the population size when individual heterogeneity is present but not modelled is noted by Link (2003).

□

11.4 Model Discrimination and Averaging

We can discriminate between the competing models via posterior model probabilities or Bayes factors (Section 6.2). In addition, this allows us to provide posterior model-averaged estimates for the total population size, taking into account both parameter and model uncertainty. We note that models that do not fit the data well should be discarded from the analysis. However, this is often a default mechanism within the model averaging procedure. Typically, assuming that some of the candidate models do fit the data well, the corresponding posterior weights (in terms of posterior model probabilities) for the models that fit the data poorly will be very low and often essentially zero. Thus, these models will not influence the overall estimate for the total population size. Finally, we note that presenting a posterior model-averaged mean for the total population size may not be the most appropriate summary statistic to present, since the posterior distribution may be bimodal (with the posterior modes for the total population size corresponding to competing models). This is easily identified by plotting the posterior density for the total population size.

11.4.1 Bayesian Posterior Model Probabilities

RJMCMC Algorithm

Moving between different models generally involves adding or deleting parameters and so we use the reversible jump algorithm. We present a possible updating algorithm to explore the model and parameter space. Within each iteration of the Markov chain we propose three different model update moves. We cycle through each of the effects (time, behavioural and individual) in turn. If the effect is present we propose to remove the relevant term(s); otherwise, if it is not present, we propose to add them. We essentially use the same reversible jump step to update each effect.

For illustration, we shall consider the time effects. Suppose that the Markov chain is in model M_0, with parameters N and μ (the model move will follow similarly for M_b, M_h and M_{bh}). We propose the new parameters,

$$\alpha_\tau | \sigma_\alpha^2 \sim N(0, \sigma_\alpha^2) \text{ for } \tau = 1, \ldots, T; \qquad \sigma_\alpha^2 \sim \Gamma(1,1).$$

We accept this move with probability $\min(1, A)$, where,

$$A = \frac{\pi(N, \mu, \boldsymbol{\alpha}, \sigma_\alpha^2, M_t | \boldsymbol{x})}{\pi(N, \mu, M_0 | \boldsymbol{x}) q(\boldsymbol{\alpha} | \sigma_\alpha^2) q(\sigma_\alpha^2)},$$

in which $\boldsymbol{\alpha} = \{\alpha_\tau : \tau = 1, \ldots, T\}$, $q(\cdot)$ denotes the proposal distribution for the given parameter, and the Jacobian is simply equal to unity. If we accept the move, the Markov chain moves to model M_t with parameters $\{N, \mu, \boldsymbol{\alpha}, \sigma_\alpha^2\}$. Otherwise, if we reject the move, the chain stays in model M_0, with parameters $\{N, \mu\}$. Conversely in the reverse move, where we propose to move from model M_t to model M_0, removing the time dependence, we accept the move with probability equal to $\min(1, A^{-1})$.

We repeat this process for the behavioural effects and the individual effects (conditional on N). Thus, all model moves that are proposed are between neighbouring models. We use the same proposal distributions for these parameters as for the time effects.

Example 11.3. Snowshoe Hares

We return to Example 11.1. Implementing the above algorithm, we obtain both posterior model probabilities and an associated model-averaged estimate for the total population size. Table 11.3 provides the corresponding posterior probability for each model. We see that the model with recapture heterogeneity has by far the largest posterior model probability. The overall model-averaged mean estimate for the total population size is 98 with 95% HPDI (72, 135). The HPDI is now wider than for model M_h, for example, and takes account of the contributions of the alternative models, notably model M_{bh}.
□

Example 11.4. Golf Tees

For the golf tees data set in Example 11.2 we present the posterior probabilities

Table 11.3 The Posterior Probability of Each Model for the Snowshoe Hare Data

Model	Posterior Probability
M_0	0.000
M_t	0.000
M_b	0.000
M_{tb}	0.000
M_h	0.463
M_{th}	0.146
M_{bh}	0.287
M_{tbh}	0.103

for the different models identified, along with the corresponding estimates for total population size in Table 11.4. Clearly, there is very strong evidence that individual heterogeneity is present. This agrees with the experiment, where the groups of golf tees were different colours, sizes and had different degrees of visibility. The corresponding posterior estimate for the number of groups appears to be fairly consistent between the different models identified, and with the true number of groups. This makes it sensible to average over the different models, to obtain a model-averaged mean estimate.

Table 11.4 The Posterior Probability and Corresponding Estimates of the Total Population Size for Each Model Identified with More than 5% Posterior Support, Coupled with the Corresponding Model-Averaged Estimate for the St. Andrews Golf Tees Data

Model	Posterior Probability	Posterior Mean (SD) of N
M_{th}	0.715	232.8 (22.7)
M_{tbh}	0.241	240.7 (26.6)
Model-averaged		235.2 (24.1)
95% HPDI		(194, 288)

□

11.5 Line Transects

Line transect surveys were introduced in Section 2.2.5 and are typically used to obtain estimates of density and/or abundance of a population. These methods are a special case of *distance sampling* (for a detailed description of these

LINE TRANSECTS

methods see for example Buckland et al. 2001, 2004a). These surveys typically involve randomly selecting a set of transect lines, followed by observers travelling down these transect lines and recording each individual that they see, coupled with the perpendicular distance of the individual from the transect line. Typically, distances are truncated, so that only individuals are recorded up to this distance, w. Similar to the collection of capture-recapture data, the observation process is not perfect (i.e. the capture probability is not equal to 1) so that not all individuals up to the truncated distance w are observed within the survey. In line transect data analyses, the probability density function for the observed perpendicular distance of individuals to the line transect can be modelled, and is denoted by $f(y)$, where y denotes the perpendicular distance of the individual to the line transect. Within a classical analysis, the corresponding density of the individuals, D can then be estimated as,

$$\hat{D} = \frac{n\hat{f}(0)}{2L}, \qquad (11.3)$$

where n is the total number of objects detected, L is the total length of the transect lines, and $\hat{f}(0)$ is the estimated probability density function of observed distances evaluated at zero distance.

To illustrate a Bayesian approach for line transect data, we consider a particular application corresponding to an experiment relating to the observation of wooden stakes, where the true number of wooden stakes (and hence density) is known. See Karunamuni and Quinn (1995) for further discussion of a Bayesian analysis of line transect data.

Example 11.5. Line Transect Data: Wooden Stakes

We consider data collected from a line transect study where the population of interest is wooden stakes. These were placed in a sagebrush meadow east of Logan, Utah and a number of graduate students walked a 1000-metre transect line independently of each other. A total of eleven students participated in the study, each recording the perpendicular distance of each observed wooden stake from the transect line. We consider data collected from a single line transect of a particular student, given in Table 11.5. The true density of stakes was 37.5 per hectare. See Buckland et al. (2001) for further details.

We assume that the observed perpendicular distances, $\boldsymbol{y} = \{y_1, \ldots, y_n\}$, are independently distributed from a half-normal distribution, so that, for $i = 1, \ldots, n$,

$$y_i \sim HN^+(0, \sigma^2),$$

where σ^2 is to be estimated. In addition, we assume that the data are not truncated and are continuous (so not grouped into distance intervals). Note that the maximum-likelihood estimator $\hat{f}(0)$ is given by,

$$\hat{f}(0) = \sqrt{\frac{2n}{\pi \sum y_i^2}},$$

(for example, see Buckland et al. 2001).

Table 11.5 Distance Values for the Stakes Line Transect Example

2.02	0.45	10.40	3.61	0.92	1.00	3.40	2.90	8.16	6.47
5.66	2.95	3.96	0.09	11.82	14.23	2.44	1.61	31.31	6.50
8.27	4.85	1.47	18.60	0.41	0.40	0.20	11.59	3.17	7.10
10.71	3.86	6.05	6.42	3.79	15.24	3.47	3.05	7.93	18.15
10.05	4.41	1.27	13.72	6.25	3.59	9.04	7.68	4.89	9.10
3.25	8.49	6.08	0.40	9.33	0.53	1.23	1.67	4.53	3.12
3.05	6.60	4.40	4.97	3.17	7.67	18.16	4.08		

Source: Reproduced with permission from Gimenez et al. (2009a, p. 900) published by Springer.

Within the Bayesian approach, we form the posterior distribution for σ^2 to be,

$$\pi(\sigma^2|\boldsymbol{y}) \propto f(\boldsymbol{y}|\sigma^2)p(\sigma^2).$$

Specifying the prior, $\sigma^2 \sim \Gamma^{-1}(a,b)$, we obtain,

$$\pi(\sigma^2|\boldsymbol{y}) \propto \prod_{i=1}^n \sqrt{\frac{2}{\pi\sigma^2}} \exp\left(-\frac{y_i^2}{2\sigma^2}\right) \times (\sigma^2)^{-(a-1)} \exp\left(-\frac{b}{\sigma^2}\right)$$

$$\propto (\sigma^2)^{-(\frac{n}{2}+a-1)} \exp\left(-\frac{(\frac{1}{2}\sum_{i=1}^n y_i^2 + b)}{\sigma^2}\right).$$

Thus, we have,

$$\sigma^2|\boldsymbol{y} \sim \Gamma\left(\frac{n}{2}+a, \frac{1}{2}\sum_{i=1}^n y_i^2 + b\right).$$

The posterior distribution for σ^2 can be simulated from directly, as it is a standard distribution. However, more generally, if a more complex distribution was obtained, for example, when using a non-conjugate prior distribution, or alternative models for the detection function used (possibly including spatial and/or temporal components), standard MCMC algorithms can be implemented. However, we are not particularly interested in the posterior distribution for σ^2, rather the posterior distribution for the density. This can be obtained explicitly by considering a transformation of variables argument, given the posterior distribution for σ^2. In particular, this involves obtaining an expression for the density function $f(0)$ and then using a transformation of variable argument to calculate the corresponding distribution for D, given the relationship in Equation (11.3). Alternatively, we can obtain a Monte Carlo estimate of the posterior distribution for D, by initially simulating a set of values from $\pi(\sigma^2|\boldsymbol{y})$, which we denote by $(\sigma^2)^1, \ldots, (\sigma^2)^k$. For each simulated value, $i = 1, \ldots, k$, calculate $f(0|(\sigma^2)^i)$, given by,

$$f(0|\sigma^2) = \sqrt{\frac{2}{\pi\sigma^2}}$$

(i.e. evaluate the half-normal density function at $y = 0$, given the value of $(\sigma^2)^i$). Substitute these values into Equation (11.3) to obtain a sample from the posterior distribution for D. We implement this latter approach here. For the observed data, we have $n = 68$, $\sum_{i=1}^{n} y_i^2 = 4551.937$ and $L = 1000$. Finally, we specify the prior parameters $a = 0.001$ and $b = 0.001$. Thus the posterior distribution for σ^2 is of the form,

$$\sigma^2|\boldsymbol{y} \sim \Gamma^{-1}(34.001, 2275.97).$$

We simulate 100,000 values from the posterior distribution for σ^2. To do this we simulate independent values from the $\Gamma(34.001, 2275.97)$ (using the `rgamma` command in R – see Section B.2.2 in Appendix B) and take the reciprocal of the simulated values to be the simulated values of σ^2. Substituting these values into the functions for $f(0|\sigma^2)$ and D, we obtain the corresponding posterior summary statistics provided in Table 11.6. Note that these estimates are fairly insensitive to the prior specified on σ^2, suggesting that the posterior distribution is data-driven. WinBUGS code for this application to line transect data is provided by Gimenez et al. (2009a).

Table 11.6 Results For the Stakes Line Transect Example. The True Value is Provided with the Corresponding MLE and Posterior Mean and Standard Deviation (in Brackets) for D and $f(0)$

	D	$f(0)$
True	0.00375	0.1103
MLE	0.00332	0.0975
Bayesian	0.00330 (0.00028)	0.0972 (0.008)

□

11.6 Summary

In this chapter we consider closed populations, where there are no births, deaths, or migration. Typically, the parameters of particular interest for these populations are the total population size. We consider two different types of data: capture-recapture data and line transect data.

For closed capture-recapture data, the time period of the study is typically very short (in order to satisfy the underlying closed assumption). There are generally three different factors that are regarded to possibly affect the recapture probability of individuals: temporal heterogeneity, behavioural heterogeneity (trap response), and individual heterogeneity. The dependence of the recapture probabilities on these factors is very important, since the estimation of the recapture probabilities has a direct impact on the estimate of the total population size. We consider a logistic regression for the recapture probability on these factors, where the factors present in the model are unknown a priori. This essentially reduces to a variable selection problem on the heterogeneities present and we implement the model discrimination approach described in Chapter 6. Of particular interest is the total population size, for which we can obtain a model-averaged estimate incorporating the model uncertainty into the estimate. As usual with any model averaging, we should be clear that the models averaged are appropriate for the data, and that if there are major differences in estimates from individual models, then the reasons for this need to be understood; see Morgan and Ridout (2008). We note that care needs to be taken when using this approach as it is possible to obtain a bimodal (or multimodal) posterior distribution for the population size, so taking the posterior mean as a point estimate may not be appropriate. In addition, care needs to be taken in the specification of the priors in the model to ensure that the induced prior on the capture probabilities are sensible.

We also consider line transect data and show how we can use a Bayesian approach to estimate the posterior distribution for the density function of the detection distances, thus allowing estimation of the density of the area of interest.

11.7 Further Reading

Distance sampling is widely used to estimate population sizes or densities of populations for many different species of animals, from cetaceans to birds to land mammals. A general introduction to these ideas and methods is provided by Buckland et al. (2001), while more advanced methods associated with these methods is provided by Buckland et al. (2004a). In addition, the freely available computer package DISTANCE (Thomas et al. 2009) allows classical analyses of the data obtained from distance sampling.

For closed capture-recapture data, a number of different approaches have been used to estimate the population size. For models without an individual heterogeneity component, the likelihood is relatively straightforward to write down explicitly, and obtain the corresponding MLEs of the parameters (Borchers et al. 2002). However, (as for random effect models), the inclusion of an individual heterogeneity term makes the likelihood analytically intractable; and hence the maximisation process difficult. A number of model-based approaches have been proposed, in the presence of individual heterogeneity. These include the Rasch model (Coull and Agresti 1999, Bartolucci and Forcina 2001, Fienberg et al. 1999) (M_{th}), covariate models (Huggins 1991, Stanghellini and van der Heijden 2004) (M_{tbh}/M_{th}), mixture models (Pledger 2000) (M_{tbh}), mixture models using the beta-binomial distribution (Dorazio

FURTHER READING

and Royle 2003) (M_h), and a mixture of binomial and beta-binomial distributions (Morgan and Ridout 2008) (M_h). Some of these approaches for estimating the population size can be fitted to data using the freely available computer package MARK. Additionally, a number of Bayesian approaches have been developed to incorporate individual heterogeneity for different models. These include, for example, Durban and Elston (2005) (M_{th}), Royle et al. (2007) (M_h), Tardella (2002) (M_h) and Ghosh and Norris (2005) (M_{bh}). Arnold et al. (2009) also provide a Bayesian analysis of the mixture models proposed by Pledger (2000) (in particular for model M_{th}) where the number of components of a mixture is unknown a priori, and treated as a parameter in the model. They implement an RJMCMC approach and provide model-averaged estimates of the total population size, using the analogous methods to those presented in Section 11.4.

Appendices

APPENDIX A

Common Distributions

This appendix gives summary information for common distributions. For each, we include the WinBUGS command for the distribution (if available) and the R command for simulating a single random deviate from the distribution.

A.1 Discrete Distributions

A.1.1 Discrete Uniform Distribution: $\theta \sim U[a,b]$

Parameters: Lower and upper bounds a and b.

Probability mass function:
$$f(\theta) = \frac{1}{b-a+1},$$
for $\theta = a, a+1, \ldots, b-1, b$.

Properties:
$$\mathbb{E}(\theta) = \frac{a+b}{2}; \qquad \text{Var}(\theta) = \frac{(b-a)(b-a+2)}{12}.$$

WinBUGS command: See `Tricks:AdvancedUseoftheBUGSLanguage` within the WinBUGS `User manual` (in the `Help` menu of the Toolbar) - uses the `dcat` command.

R command: `sample(a:b,1)`.

A.1.2 Bernoulli Distribution: $\theta \sim Bernoulli(p)$

Parameter: Probability $p \in [0,1]$.

Probability mass function:
$$f(\theta) = p^\theta (1-p)^{1-\theta},$$
for $\theta = 0, 1$.

Properties:
$$\mathbb{E}(\theta) = p; \qquad \text{Var}(\theta) = p(1-p).$$

Comments: The Bernoulli distribution is the special case of the binomial distribution where $n=1$.

A.1.3 Binomial Distribution: $\theta \sim Bin(n, p)$

Parameters: Sample size $n \in \mathbb{N}$ and probability $p \in [0, 1]$.

Probability mass function:
$$f(\theta) = \binom{n}{\theta} p^\theta (1-p)^{n-\theta},$$
for $\theta = 0, 1, \ldots, n$.

Properties:
$$\mathbb{E}(\theta) = np; \qquad \operatorname{Var}(\theta) = np(1-p).$$

WinBUGS command: `dbin(p,n)`.

R command: `rbinom(1,n,p)`.

Comments: The conjugate prior for p is a beta distribution.

A.1.4 Beta-Binomial: $\theta \sim Beta - Bin(n, \alpha, \beta)$

Parameters: Sample size n, $\alpha > 0$ and $\beta > 0$.

Probability mass function:
$$f(\theta) = \frac{\Gamma(n+1)\Gamma(a+\theta)\Gamma(n+b-\theta)\Gamma(a+b)}{\Gamma(\theta+1)\Gamma(n-\theta+1)\Gamma(a+b+n)\Gamma(a)\Gamma(b)}$$
for $\theta = 0, 1, \ldots, n$

Properties:
$$\mathbb{E}(\theta) = \frac{n\alpha}{\alpha+\beta}; \qquad \operatorname{Var}(\theta) = \frac{n\alpha\beta(\alpha+\beta+n)}{(\alpha+\beta)^2(\alpha+\beta+1)}.$$

Comments: This distribution is equivalent to $\theta \sim Bin(\mu)$ where $\mu \sim Beta(\alpha, \beta)$.

A.1.5 Poisson Distribution: $\theta \sim Po(\lambda)$

Parameter: Rate $\lambda > 0$.

Probability mass function:
$$f(\theta) = \frac{\exp(-\lambda)\lambda^\theta}{\theta!},$$
for $\theta = 0, 1, \ldots$

DISCRETE DISTRIBUTIONS

Properties:
$$\mathbb{E}(\theta) = \lambda; \qquad \text{Var}(\theta) = \lambda.$$

WinBUGS command: `dpois(`λ`)`.

R command: `rpois(1,`λ`)`.

Comments: The conjugate prior for λ is a gamma distribution.

A.1.6 Geometric Distribution: $\theta \sim Geom(p)$

Parameter: Probability $p \in [0, 1]$.

Probability mass function:
$$f(\theta) = p(1-p)^{\theta-1},$$
for $\theta = 0, 1, \ldots$

Properties:
$$\mathbb{E}(\theta) = \frac{1}{p}; \qquad \text{Var}(\theta) = \frac{(1-p)}{p^2}$$

R command: `rgeom(1,p)`.

Comments: The conjugate prior for p is the beta distribution.

A.1.7 Negative Binomial: $\theta \sim Neg - Bin(\alpha, \beta)$

Parameters: Shape $\alpha > 0$ and inverse scale $\beta > 0$.

Probability mass function:
$$f(\theta) = \binom{\theta + \alpha - 1}{\alpha - 1} \left(\frac{\beta}{\beta+1}\right)^\alpha \left(\frac{1}{\beta+1}\right)^\theta,$$
for $\theta = 0, 1, \ldots$

Properties:
$$\mathbb{E}(\theta) = \frac{\alpha}{\beta}; \qquad \text{Var}(\theta) = \frac{\alpha}{\beta^2}(\beta+1).$$

WinBUGS command: `dnegbin(p,r)`, where `p` $= \beta/(1+\beta)$ and `r` $= \alpha$.

R command: `rnbinom(1,size,prob)`, where `size` $= \alpha$ and `prob` $= \beta/(1+\beta)$.

Comments: This distribution is equivalent to $\theta \sim Po(\lambda)$ where $\lambda \sim \Gamma(\alpha, \beta)$.

A.1.8 Multinomial Distribution: $\boldsymbol{\theta} \sim MN(n, \boldsymbol{p})$

Parameters: Sample size $n \in \mathbb{N}$ and probabilities $p_i \in [0, 1]$; $\sum_{i=1}^{k} p_i = 1$.

Probability mass function:

$$f(\boldsymbol{\theta}) = \frac{n!}{\prod_{i=1}^{k} \theta_i!} \prod_{i=1}^{k} p_i^{\theta_i},$$

for $\theta_i = 0, 1, \ldots, n$ such that $\sum_{i=1}^{k} \theta_i = n$.

Properties:

$$\mathbb{E}(\theta_i) = np_i; \qquad \mathrm{Var}(\theta_i) = np_i(1 - p_i).$$

WinBUGS command: `dmulti(p[],N)`, where `p` denotes the vector with elements $p_1, \ldots, p_n$.

R command: `rmultinom(1,N,prob)`, where `prob` denotes a vector with elements $p_1, \ldots, p_n$.

Comments: The conjugate prior for $\boldsymbol{p}$ is the Dirichlet distribution. The multinomial distribution is an extension to the binomial distribution, where there are a total of k possible values (instead of simply 2).

A.2 Continuous Distributions

A.2.1 Uniform Distribution: $\theta \sim U[a, b]$

Parameters: Lower and upper bounds, a and b.

Probability density function:

$$f(\theta) = \frac{1}{b - a},$$

for $\theta \in [a, b]$.

Properties:

$$\mathbb{E}(\theta) = \frac{a + b}{2}; \qquad \mathrm{Var}(\theta) = \frac{b - a}{12}.$$

WinBUGS command: `dunif(a,b)`.

R command: `runif(1,a,b)`. Note that the default values of `a` and `b` are 0 and 1 if they are not specified. For example, using the command `runif(1)` simulates a random deviate from $U[0, 1]$.

CONTINUOUS DISTRIBUTIONS

A.2.2 Normal Distribution: $\theta \sim N(\mu, \sigma^2)$

Parameters: mean μ and variance $\sigma^2 > 0$ (precision $= 1/\sigma^2$).

Probability density function:

$$f(\theta) = \frac{1}{\sqrt{2\pi\sigma^2}} \exp\left(-\frac{(\theta - \mu)^2}{2\sigma^2}\right),$$

for $\theta \in (-\infty, \infty)$.

Properties:

$$\mathbb{E}(\theta) = \mu; \qquad \text{Var}(\theta) = \sigma^2.$$

WinBUGS command: `dnorm(`μ`,`τ`)`, where $\tau = 1/\sigma^2$.

R command: `rnorm(1,`μ`,`σ`)`, where $\sigma = \sqrt{\sigma^2}$. Note that the default values of μ and σ are 0 and 1 if they are not specified. For example, using the command `rnorm(1)` simulates a random deviate from $N(0, 1)$.

Comments: The conjugate prior for the mean is a normal distribution; and for the variance, an inverse gamma distribution (equivalently, for the precision a gamma distribution).

The positive half-normal distribution is the corresponding (truncated) normal distribution with mean zero defined only over positive values (i.e. for $\theta > 0$). We write $\theta \sim HN^+(0, \sigma^2)$, with corresponding probability density function $g(\theta) = 2f(\phi)$ for $\theta > 0$, where $f(\phi)$ denotes the probability density function such that $\phi \sim N(0, \sigma^2)$. The negative half-normal distribution is defined similarly, over the negative value for θ and write $\theta \sim HN^-(0, \sigma^2)$.

A.2.3 Log-Normal Distribution: $\theta \sim \log N(\mu, \sigma^2)$

Parameters: μ and $\sigma^2 > 0$.

Probability density function:

$$f(\theta) = \frac{1}{\theta\sqrt{2\pi\sigma^2}} \exp\left(-\frac{(\log \theta - \mu)^2}{2\sigma^2}\right),$$

for $\theta \in [0, \infty)$.

Properties:

$$\mathbb{E}(\theta) = \exp\left(\mu + \frac{1}{2}\sigma^2\right); \qquad \text{Var}(\theta) = \exp(2\mu + \sigma^2)(\exp(\sigma^2) - 1).$$

WinBUGS command: `dlnorm(`μ`,`τ`)`, where $\tau = 1/\sigma^2$.

R *command:* `rlnorm(1,`μ`,`σ`)`, where $\sigma = \sqrt{\sigma^2}$. Note that the default values of μ and σ are 0 and 1 if they are not specified. For example, using the command `rlnorm(1)` simulates a random deviate from $\log N(0,1)$.

Comments: If $\theta \sim \log N(\mu, \sigma^2)$, then $\log(\theta) \sim N(\mu, \sigma^2)$. The conjugate prior for μ is a normal distribution; and for σ^2, an inverse gamma distribution.

A.2.4 Student-t Distribution: $\theta \sim t_\nu(\mu, \sigma^2)$

Parameters: degrees of freedom ν, mean μ, and scale $\sigma^2 > 0$.

Probability density function:

$$f(\theta) = \frac{\Gamma\left(\frac{\nu+1}{2}\right)}{\Gamma\left(\frac{\nu}{2}\right)\sqrt{\nu\pi\sigma^2}} \left(1 + \frac{1}{\nu}\left(\frac{(\theta-\mu)^2}{\sigma^2}\right)\right)^{-(\nu+1)/2},$$

for $\theta \in (-\infty, \infty)$.

Properties:

$$\mathbb{E}(\theta) = \mu \quad (\nu > 1); \qquad \text{Var}(\theta) = \frac{\nu}{\nu-2}\sigma^2 \quad (\nu > 2).$$

WinBUGS command: `dt(`μ`,`τ`,`ν`)`, where $\tau = 1/\sigma^2$.

R command: `rt(1,`ν`)`, for $\mu = 0$ and $\sigma^2 = 1$ (written as t_ν).

A.2.5 Beta Distribution: $\theta \sim Beta(\alpha, \beta)$

Parameters: $\alpha > 0$, $\beta > 0$.

Probability density function:

$$f(\theta) = \frac{\Gamma(\alpha+\beta)}{\Gamma(\alpha)\Gamma(\beta)} \theta^{\alpha-1}(1-\theta)^{\beta-1},$$

for $\theta \in [0,1]$.

Properties:

$$\mathbb{E}(\theta) = \frac{\alpha}{\alpha+\beta}; \qquad \text{Var}(\theta) = \frac{\alpha\beta}{(\alpha+\beta)^2(\alpha+\beta+1)}.$$

WinBUGS command: `dbeta(`α`,`β`)`.

R command: `rbeta(1,`α`,`β`)`.

Comments: The $U[0,1]$ distribution is a special case of the beta distribution, where $\alpha = \beta = 1$.

CONTINUOUS DISTRIBUTIONS

A.2.6 Gamma Distribution: $\theta \sim \Gamma(\alpha, \beta)$

Parameters: shape $\alpha > 0$ and scale $\beta > 0$.

Probability density function:

$$f(\theta) = \frac{\beta^\alpha}{\Gamma(\alpha)} \theta^{\alpha-1} \exp(-\beta\theta),$$

for $\theta > 0$.

Properties:

$$\mathbb{E}(\theta) = \frac{\alpha}{\beta}; \qquad \text{Var}(\theta) = \frac{\alpha}{\beta^2}.$$

WinBUGS command: `dgamma(`α`,`β`)`.

R command: `rgamma(1,`α`,`β`)`.

Comments: The exponential distribution with parameter λ (written $Exp(\lambda)$) is a special case of the gamma distribution with $\alpha = 1$ and $\beta = \lambda$.

A.2.7 Inverse Gamma Distribution: $\theta \sim \Gamma^{-1}(\alpha, \beta)$

Parameters: Shape $\alpha > 0$ and scale $\beta > 0$.

Probability density function:

$$f(\theta) = \frac{\beta^\alpha}{\Gamma(\alpha)} \theta^{-(\alpha+1)} \exp(-\beta/\theta),$$

for $\theta > 0$.

Properties:

$$\mathbb{E}(\theta) = \frac{\beta}{\alpha-1}; \qquad \text{Var}(\theta) = \frac{\beta^2}{(\alpha-1)^2(\alpha-2)}, \alpha > 2.$$

Comments: If $\omega \sim \Gamma(\alpha, \beta)$, then $\theta = 1/\omega \sim \Gamma^{-1}(\alpha, \beta)$.

WinBUGS command: Set $\theta = 1/\omega$ and then use $\omega \sim$ `dgamma(`α`,`β`)`.

R command: `1/rgamma(1,`α`,`β`)`. Equivalently, simulate `w <- rgamma(1,`α`,`β`)` and set $\theta = 1/\omega$.

A.2.8 Dirichlet Distribution: $\boldsymbol{\theta} \sim Dir(\alpha_1, \ldots, \alpha_k)$

Parameters: $\alpha_i > 0$, for $i = 1, \ldots, k$, and set $\alpha_0 = \sum_{i=1}^{k} \alpha_i$.

Probability density function:

$$f(\boldsymbol{\theta}) = \frac{\Gamma(\alpha_0)}{\prod_{i=1}^{k} \Gamma(\alpha_i)} \prod_{i=1}^{k} \theta_i^{\alpha_i - 1},$$

for $\theta_i \geq 0$, $\sum_{i=1}^{k} \theta_i = 1$.

Properties:

$$\mathbb{E}(\theta_i) = \frac{\alpha_i}{\alpha_0}; \qquad \text{Var}(\theta_i) = \frac{\alpha_i(\alpha_0 - \alpha_i)}{\alpha_0^2(\alpha_0 + 1)}.$$

WinBUGS command: `ddirch(alpha[])` where `alpha` is a vector with elements $\alpha_1, \ldots, \alpha_k$.

A.2.9 Multivariate Normal Distribution: $\boldsymbol{\theta} \sim \mathcal{N}_k(\boldsymbol{\mu}, \Sigma)$

Parameters: mean vector $\boldsymbol{\mu}$ of length k and covariance matrix Σ, where Σ is a positive definite symmetric $k \times k$ matrix.

Probability density function:

$$f(\boldsymbol{\theta}) = \frac{1}{(2\pi)^{\frac{k}{2}} |\Sigma|^{\frac{1}{2}}} \exp\left(-\frac{1}{2}(\boldsymbol{\theta} - \boldsymbol{\mu})^T \Sigma^{-1} (\boldsymbol{\theta} - \boldsymbol{\mu})\right),$$

for $\theta_i \in (-\infty, \infty)$ for $i = 1, \ldots, k$.

Properties:

$$\mathbb{E}(\boldsymbol{\theta}) = \boldsymbol{\mu}; \qquad \text{Var}(\boldsymbol{\theta}) = \Sigma.$$

WinBUGS command: `dmnorm(mu[],T[,])` where `mu` is the vector containing elements $\mu_1, \ldots, \mu_k$ and $T = \Sigma^{-1}$.

Comments: The multivariate normal distribution is the conjugate prior for the mean vector $\boldsymbol{\mu}$, and the inverse Wishart distribution is the conjugate prior for the covariance matrix, Σ (equivalently, the Wishart distribution is the conjugate prior for the inverse of the covariance matrix, Σ^{-1}).

CONTINUOUS DISTRIBUTIONS 373

A.2.10 Wishart Distribution: $\theta \sim W_k(\nu, \Psi)$

Parameters: Degrees of freedom $\nu > 0$ and positive-definite (symmetric) $k \times k$ scale matrix Ψ.

Probability density function:

$$f(\theta) = \frac{|\Psi|^{-\nu/2} |\theta|^{(\nu-k-1)/2} \exp\left(-\frac{1}{2}\text{tr}(\Psi^{-1}\theta)\right)}{2^{\nu k/2} \pi^{k(k-1)/4} \prod_{j=1}^{k} \Gamma\left(\frac{\nu+1-j}{2}\right)},$$

for $k \times k$ positive-definite matrix θ.

Properties:

$$\mathbb{E}(\theta) = \nu \Psi; \qquad \text{Var}(\theta_{ij}) = \nu(\psi_{ij}^2 + \psi_{ii}\psi_{jj}).$$

WinBUGS command: `dwish(R[,], `ν`)` where `R` $= \Psi^{-1}$.

Comments: The Wishart distribution can be seen as a (multivariate) generalisation of the gamma distribution.

A.2.11 Inverse Wishart Distribution: $\theta \sim W_k^{-1}(\nu, \Psi)$

Parameters: Degrees of freedom $\nu > 0$ and positive-definite (symmetric) $k \times k$ scale matrix Ψ.

Probability density function:

$$f(\theta) = \frac{|\Psi|^{-\nu/2} |\theta|^{-(\nu+k+1)/2} \exp\left(-\frac{1}{2}\text{tr}(\Psi^{-1}\theta^{-1})\right)}{2^{\nu k/2} \pi^{k(k-1)/4} \prod_{j=1}^{k} \Gamma\left(\frac{\nu+1-j}{2}\right)},$$

for $k \times k$ positive-definite matrix θ.

Properties: Let $\Sigma = \Psi^{-1}$. Then,

$$\mathbb{E}(\theta) = \frac{\Sigma}{\nu-k-1}; \qquad \text{Var}(\theta_{ij}) = \frac{(\nu-k+1)\sigma_{ij}^2 + (\nu-p-1)\sigma_{ii}\sigma_{jj}}{(\nu-p)(\nu-p-1)^2(\nu-p-3)}.$$

WinBUGS command: Set $\theta = \omega^{-1}$ (i.e. use the command `inverse(`ω`)`) and then use $\omega \sim$ `dwish(R[,], `ν`)` where `R` $= \Psi^{-1}$.

Comments: The inverse Wishart distribution can be seen as a (multivariate) generalisation of the inverse gamma distribution.

If $\omega \sim W_k(\nu, \Psi)$, then $\theta = \omega^{-1} \sim W_k^{-1}(\nu, \Psi)$.

APPENDIX B

Programming in R

We give a brief introduction to the computer package R, with particular focus on the coding of (RJ)MCMC algorithms. The example codes provided are intended to act as a building block, providing the basic structure of general (RJ)MCMC code, an understanding of the MCMC updating steps and their corresponding implementation. Unlike WinBUGS and MARK, all aspects of the MCMC algorithm need to be explicitly coded, using the R language. However, this does provide complete control over the MCMC simulations, with no constraints on the models fitted to the data. To demonstrate how to write an (RJ)MCMC program, we provide a general structure for the algorithms, which complements the detailed description of R code for the examples in Chapter 7. All R codes can be downloaded from the book's Web site.

B.1 Getting Started in R

The statistical computer package R is freely available and is downloadable from the CRAN (Comprehensive R Archive Network) Web site:

http://cran.r-project.org/

The Web page also provides links to useful R manuals and updates to the computer package are frequently posted on the general Web site for R:

http://www.r-project.org/

The built-in help file in R is very useful and can be accessed via the Help menu in the Toolbar. In the pull-down menu, R functions(text)... can be used to find extra information on a particular function, or Search help..., to search the help files for a particular topic. Alternatively, within the main R window, typing help(topic) will open up the help file on the requested topic (if one exists). A useful shorthand, equivalent, method is to use the command ?topic. If the R command is unknown, the following will be returned: No documentation for 'topic' in specified packages and libraries: you could try 'help.search("topic")'. Typing help.search("topic") searches all the commands in R and returns a list of functions linked to the command topic, which can be used to find the relevant function of interest.

When opening an R session it is advisable to change the working directory. This means that the default directory, for reading in data or saving output from an R session will be to the set working directory. To change the working directory, click on File in the toolbar menu, followed by Change dir....

Click on Browse and find the directory that you wish to use for the session. Click on OK in the browser window and then again in the original window.

There are a number of general R commands that are repeatedly used within (RJ)MCMC codes and thus are worth a particular mention. For example, the R commands associated with the different distributions given in Appendix A. Given the basic syntax, we consider specific functions written in R to perform (RJ)MCMC algorithms and describe the general structure of such codes.

B.2 Useful R Commands

We provide a brief overview of some of the most useful commands in R. These commands are used repeatedly through the R codes provided. In particular we consider basic arithmetical operations, R commands associated with distributions and finally some generic commands and mathematical functions.

B.2.1 Arithmetical Operations

Command	Brief Description
+	Addition.
-	Subtraction.
*	Multiplication.
/	Division.
^	Power operator (note ** can also be used).

B.2.2 Distributional Commands

Programming (RJ)MCMC algorithms involves the generation of random variables from different distributions. R can generate observations from a large number of common statistical distributions. Additional distributional commands involve the evaluation of the probability density (or mass) function (very useful in the calculation of acceptance probabilities), cumulative probability distribution function and quantile (or inverse-cumulative) function. Each of these distributional functions has a name beginning with one of the following four one-letter codes (r, p, d and q), indicating the type of function.

r Random number generator. Requires the number of random variables to be simulated, plus the parameters associated with the distribution required.

p Cumulative probability distribution function. Requires a vector of quantiles, plus the parameters associated with the distribution required.

d Probability density (or mass) function. Requires a vector of quantiles, plus the parameters associated with the distribution required.

q Quantile function. Requires a vector of probabilities, plus the parameters associated with the distribution required.

USEFUL R COMMANDS

B.2.3 Useful Commands

Command	Brief Description
#	Comments the remainder of the line.
a:b	To generate a sequence of integer values from a to b (inclusive).
acf	Calculates an estimate of the autocorrelation function (acf). By default, an acf plot is presented; the acf values can be outputted using plot = F as an argument of the function.
array	Creates an array of given dimension.
c	Concatenates given arguments into a vector.
cat	Prints the arguments to the screen. To indicate an end of line in output, use "\n".
cor	Calculates the correlation between two vectors (or matrices).
density	Calculates a kernel density estimate of given data.
dim	Define (or return) the dimension of the given object.
exp	Calculates the exponential of the given object.
for	Defines a loop over which commands are iterated for a given sequence of values.
hist	Plots a histogram of the given data values. By default the frequency is plotted on the y-axis. This can be changed to density using probability=T.
if else	Specify a condition; if satisfied perform the given command; else (optional extra command) perform another command. Common conditional operators include: == (equal to); != (not equal to); < (less than); > (more than); <= (less than or equal to); >= (greater than or equal to).
lines	Adds lines to a given plot.
log	Calculates the (natural) log of the given object.
matrix	Creates a matrix.
mean	Calculates the sample mean of the given data values.
min	Returns the minimum of the given values.
plot	Plots the given data values. By default a scatterplot is drawn. For a line plot use type="l".
points	Adds points to a given plot.
prod	Calculates the product of the given values.
read.table	Read in a table of data values.
rm	Remove the objects given.
scan	Read in a vector of data values.
sd	Calculates the sample standard deviation of given data values.
source	Read in external file into R.
sqrt	Calculates the square root of the given object.
sum	Calculates the sum of the given values.
quantile	Output the given quantiles of a vector. (Very useful for calculating symmetric credible intervals).
var	Calculates the sample variance of the given data values.
write	Writes the output to a given file (or the session window).
write.table	Writes the output in the form of a table to a given file.

B.2.4 Coda Package

The coda add-on package in R can be used to post-process MCMC output; for example, provide trace and density plots of the parameters of interest; estimate summary statistics of interest; and produce some convergence diagnostic estimates (for example, the Gelman-Rubin statistic, Gelman and Rubin 1992, and Geweke's convergence diagnostic, Geweke 1992, but not the BGR statistic Brooks and Gelman 1998). This package needs to be downloaded (and unzipped) from the CRAN Web site (see Section B.1), placing the package in the directory with other R libraries, for example, C:\Program Files\R-2.9.0\library\.

To use this package within R, in a session window, type the command:

> library(coda)

A manual detailing the list of commands within the package coda can be downloaded from the CRAN Web site for reference. For new users of the package, the command codamenu() may be useful, as this provides a series of step-by-step commands to reply to for producing summary statistics, plots, etc. for the given MCMC iterations. However, this command does not provide all possible tools available, and additional commands can be used to obtain, for example, highest posterior density intervals, autocorrelation plots, etc. Both WinBUGS and MARK produce coda output that can be used for further post-processing. Finally, we note that sample MCMC output obtained, from a bespoke R code can be converted into the format necessary for coda using the command mcmc.

B.3 Writing (RJ)MCMC Functions

Writing bespoke code for an (RJ)MCMC algorithm will typically follow a common structure, irrespective of computing language or the particular problem being addressed. The following is a general structure for (RJ)MCMC algorithms:

Part I

- Read in the data
- Read in the prior parameter values
- Read in parameter values for proposal distributions

Part II

- Perform the (RJ)MCMC iterations, looping over the number of iterations. At each iteration this will typically involve:
  - Updating the parameters in turn (assuming a single-update algorithm) using the MH algorithm or Gibbs update;
  - Updating the model using a reversible jump step (if there is model uncertainty), and

R CODE FOR MODEL C/C 379

- Recording the parameter values and current model (if there is model uncertainty).

Part III

- Output the sampled parameter values from the MCMC iterations, calculating posterior summary statistics of interest and/or plotting the corresponding posterior densities (excluding the initial determined burn-in).

The example R codes that we present follow this general structure. We describe in detail a number of examples relating to capture-recapture data. We begin with the dipper data set and model C/C, given in Section 7.1.

B.4 R Code for Model C/C

The R code is provided in Section 7.1, and in particular in Figures 7.1 through 7.5, with both inline comments using the symbol # as a line comment command, and also additional, brief description of the code. Here we provide a detailed description of the main code for the general structure of the MCMC algorithm, before presenting two subroutines relating to the MH update and the calculation of the log-likelihood function. Note that it is easiest to write the code in a separate file, such as "R-dippersMHnorm.R". Changes to the code can then be made by simply editing this file using a suitable editing package. This file then needs to be read into R using the command source. For example, for the above code, we would type source("R-dippersMHnorm.R").

B.4.1 Main Code

We consider the three separate parts of the code as described above as the general structure of the code.

Part I: Naming the Function and Reading Input Values

Figure 7.1 provides the naming of the function as Rdippercode with parameters nt (total number of iterations) and nburn (length of burn-in). We then read in the data values including the size of the data set in terms of the number of years of release (ni) and number of years of recapture (nj). Note that this is written generally, so that we do not assume the same number of years and release. We also (for ease of coding in other places) set the total number of parameters in the model being fitted (nparam). This is simply equal to two in this case, corresponding to p and ϕ.

The next commands essentially read in the data in the form of the standard capture-recapture m-array and put this into the array data. The number of rows (nrow) is ni (number of years of release) and the data are read into the array row-wise (byrow=T). We define a vector for alpha and beta each of length two, with each element equal to one. These will be the priors for the recapture and survival probabilities – $p \sim Beta(\texttt{alpha[1]},\texttt{beta[1]})$ and $\phi \sim Beta(\texttt{alpha[2]},\texttt{beta[2]})$, where `alpha[i]` and `beta[i]` correspond to the ith element of `alpha` and `beta`, respectively. Set the vector `delta` to be of length two, with both elements equal to 0.05. These are the parameters corresponding to the standard deviation of the normal proposal distribution of the random walk MH updating step for the recapture and survival probabilities, respectively. For example, for the recapture probability, p, we propose the candidate value,

$$p' = p + \epsilon,$$

where $\epsilon \sim N(0, \texttt{delta[1]}^2)$, and `delta[1]` has been set to 0.05. Similarly for updating the survival probability, ϕ, the proposal standard deviation is `delta[2]` (again equal to 0.05). In general we may not wish to use the same proposal distribution for all parameters, particularly when they are defined over different ranges (see Example 5.7).

Part II: MCMC Simulations

Given the initial setting up of the data and necessary prior and proposal specifications, we move on to consider the MCMC simulations as given in Figure 7.2. To simplify coding, it is often easy to write additional subroutines (i.e. additional functions) that are called within the code. For example, a function that calculates the log-likelihood value for given parameter values. This can save repeating the same function a number of times in the code and makes the structure of the code easier to follow.

We begin by specifying the initial values for the parameters. The value of the model parameters (p and ϕ) are represented by the elements of the vector `param`. We initially define this vector to have two elements, the first is equal to 0.9, the second to 0.5. The first element, `param[1]`, corresponds to the recapture probability, p, and the second element, `param[2]`, the survival probability, ϕ. These are the initial values that the Markov chain will start in (i.e. these can be thought of as the parameter values at iteration 0). The log-likelihood of the initial state of the chain is then calculated using the separate (subroutine) function `calclikhood`, which uses the input parameters `ni`, `nj`, `data` and `param` (corresponding to the number of years of release and recapture, capture-recapture m-array data and the parameter values). The calculated log-likelihood value is placed into `likhood`. This log-likelihood function is described later with the corresponding R code given in Figure 7.4.

We set `itns` to be an array (or matrix) of dimension `nt` × `nparam` (number of iterations by number of parameters) with each element equal to zero. The array `itns` will be used to store the realisations from the MCMC algorithm. The ith row, `itns[i,]`, corresponds to the ith iteration number; column 1, `itns[,1]`, the values for p and column 2, `itns[,2]`, the values for ϕ. We then define a vector `output` of dimension `nparam+1`. This vector will be used to store the current state of the parameter values for p and ϕ and corresponding log-likelihood value within the Markov chain.

R CODE FOR MODEL C/C

The final set of R commands performs the iterations of the MCMC updating algorithm. This involves cycling through each iteration in turn for t=1,...,nt and updating the parameters using the subroutine function updateparam (with given input values), placing the output of the function into the vector output. The corresponding subroutine is provided in Figure 7.5. We set the parameter values to be the first two elements of the vector output and the log-likelihood value to be the third element of the vector output. We record the parameter values in the tth row of the array itns.

Part III: Outputting the Results

Following completion of the MCMC simulations, we need to output the parameter values. One simple approach is simply to output the stored parameter values for each iteration of the Markov chain. We do this, but also calculate the mean and standard deviation of the simulations and output these values to the screen to save some post-processing of the output. The R commands for doing the post-processing and outputting of the simulations are given in Figure 7.3.

The parameter values p and ϕ are given in the array itns for each iteration of the Markov chain, i.e. from t=1,...,nt. However, this includes the initial burn-in period, defined for iterations t=1,...,nburn. Thus, we initially place the post burn-in simulations into the array subitns (i.e. iterations t=nburn+1,...,nt).

To summarise the (marginal) posterior distribution for the parameters we calculate their corresponding posterior mean and standard deviation. We define mn and std to be a vector of length 2 (both elements equal to zero). We then set the first (second) element of each vector to be the sample mean and standard deviation, post-burn-in, of the parameter p (ϕ) in the MCMC simulations (using functions mean and sd), to be a Monte Carlo estimate of the posterior mean and standard deviation of the parameter. Note that it is possible to do this calculation without first defining the array subitns by simply using, for example, mn[i] <- mean(itns[(nburn+1):nt,i]). However we retain the additional array subitns to be explicit in this example.

We output the posterior mean and standard deviation of the parameters following burn-in, using the command cat. Note that the command "\n" within cat simply means to finish the line, so the next output will begin on the following line. Finally, we output the full set of realisations in the array itns from the MCMC algorithm from the function. The final '}' term relates to the bracket '{' at the end of the very first line of the R code, giving the function name Rdippercode.

This concludes the main part of the code, however we still need to define the two functions calclikhood and updateparam for calculating the log-likelihood value and the MH step for updating the parameter values, respectively. Within the R file of the code, these functions (or subroutines) can simply be added below the main function. We consider each of these functions next.

B.4.2 Subroutine – Calculating Log-Likelihood Function – calclikhood

We initially consider the function calclikhood, which calculates the log-likelihood of capture-recapture data with constant recapture and recovery probabilities. This is given in Figure 7.4. However, the code is written in a general form, so that it is easily modified to calculate the log-likelihood for alternative models, such as C/T, T/C, etc. The code is comparable to the log-likelihood function calculated in the WinBUGS example in Section 7.2, and in particular, Figure 7.7.

The subroutine begins by defining the function and the corresponding input parameters. For ease of identifiability within the main code we name the function calclikhood with input parameters ni (number of years of release), nj (number of years of recapture), data (the capture-recapture m-array) and param (parameter values for p and ϕ). We then set up a vector for the survival and recapture probabilities (denoted by phi and p) and an array for the multinomial cell probabilities (q). Initially we set all the elements to be equal to zero. For each year, i=1,...,nj, we set the corresponding recapture probability (p[i]) to be equal to param[1] and the survival probability (phi[i]) equal to param[2], since we are considering model C/C. It is here that we would change the specification of the parameters, if we considered alternative models. Note that for convenience we use p[i] $\equiv p_{i+1}$ (and phi[i] $\equiv \phi_i$).

We can now calculate the log-likelihood for any specified model, (such as C/T etc.) since we use the general time dependent terms for the recapture and recovery probabilities. For the model C/C, we are considering, the recapture and survival probabilities are simply the same at each time (as set above). We cycle through each year of release i=1,...,ni and calculate the corresponding cell probabilities, q[i,j] for j=i,...,nj. We consider the diagonal elements and above-diagonal elements separately. Note that the cell probabilities (and cell entries) in the lower triangular matrix are simply equal to zero, so that the contribution to the log-likelihood is zero. In other words we calculate,

$$q[i,j] = \begin{cases} 0 & i > j; \quad \text{(below-diagonal)} \\ \text{phi[i]p[i]} & i = j; \quad \text{(diagonal)} \\ \prod_{k=i}^{j} \text{phi[k]} \prod_{k=i}^{j-1} (1-\text{p[k]})\text{p[j]} & i < j \quad \text{(above-diagonal)}. \end{cases}$$

(Recall that phi[i] $\equiv \phi_i$ and p[i] $\equiv p_{i+1}$). The likelihood is of multinomial form, so that the log-likelihood, denoted by likhood in the R code, is calculated as,

$$\text{likhood} = \sum_{i=1}^{\text{ni}} \sum_{j=i}^{(\text{nj}+1)} \text{data}[i,j] \log(q[i,j]),$$

where q[i,nj+1] denotes the probability that an individual released in year i is not observed again within the study; and data[i,nj+1] the corresponding number of individuals. Within the code, the diagonal cell entries are first calculated and then the upper-diagonal elements. Finally, we output the log-

B.4.3 Subroutine – MH Updating Algorithm – updateparam

Figure 7.5 provides the subroutine updateparam, which performs the random walk single-update MH step using a normal proposal distribution. The function updateparam has nine inputs: the number of parameters (nparam), the parameter values (param), the numbers of years of release and recapture (ni and nj), the capture-recapture m-array (data), current log-likelihood value (likhood), beta prior parameter values for p and ϕ (alpha and beta) and the MH proposal parameters corresponding to the standard deviations of the normal distributions in the random walk proposal (delta). This function performs a single iteration of the MCMC algorithm by cycling through each parameter in turn (i = 1,...,nparam), using a random walk single-update MH step, with a normal proposal distribution (see Sections 5.3.2 and 5.3.3).

We begin by keeping a record of the current parameter value, which is being proposed to be updated, in oldparam. We simulate a new candidate value for the parameter from a normal distribution with mean equal to the current value, param[i], and standard deviation, delta[i]. We set this new proposed value to be param[i] (so now param[i] is the proposed value, rather than current value, and is the reason we need to keep a record of the current value in oldparam). Both parameters are probabilities and so lie in the interval [0,1]. Thus, we only need to calculate the acceptance probability if the parameter lies within this range (else it is rejected and we set the acceptance probability to be equal to zero). We calculate the log-likelihood using the new candidate parameter value, and place the value in newlikhood. Note that we already have the log-likelihood value of the current parameter values stored in likhood. To calculate the acceptance probability, we calculate the corresponding logs of the numerator (num) and denominator (den) separately. We set num to be equal to the sum of the log-likelihood of the proposed parameter value and the log of the prior density (beta) evaluated at the proposed value; and den to be the sum of the log-likelihood of the current parameter value and the log of the prior density evaluated at the current parameter value. We omit the normal proposal density function, since these cancel in the acceptance probability, reducing to a Metropolis acceptance probability (see Section 5.3.3). Finally, the acceptance probability, denoted by A, is set equal to the minimum of 1 and exp(num−den), where exp denotes the exponential function (recall that the terms num and den are calculated on the log scale).

To perform the accept/reject step, we simulate a random variable, u, from a $U[0,1]$ distribution. If this random deviate is less than the acceptance probability (i.e. u ≤ A) we accept the proposed parameter value and update the log-likelihood value (the parameter value has already been set to the proposed value). Otherwise we reject the proposed move and the parameter is reset to the original value stored in oldparam.

Finally, we place the parameter values and log-likelihood into the vector `output`: the first element is the recapture probability; the second the survival probability; and the third the corresponding log-likelihood value. The vector `output` is outputted from the function.

B.4.4 Running the R Dipper Code

To run the dipper code in R for 10,000 iterations, with a burn-in of 1000 iterations, we write:

```
> dip <- dippercodeMHnorm(10000,1000)
```

This command outputs the parameter values at each iteration of the Markov chain in `dip`, we can also obtain any other summary statistics of interest, or produce trace plots for checking the length of burn-in. For example, suppose that we wish to look at the trace plot of the parameter p, we would do this with the R command,

```
> plot(dip[,1],type="l")
```

Similarly, if we wished to produce a trace plot of ϕ, we would use `dip[,2]` in the above command (to plot the values in the second column).

To plot the posterior distribution of the parameters we can plot a histogram of the sampled values from the Markov chain. To do this we can use the command:

```
> hist(dip[1001:10000,1],probability=T,xlab="p")
```

This produces a histogram of the post-burn-in values (from iterations 1001 to 10,000) of the first column of the array `dip`, corresponding to parameter p and given in Figure 7.6(a). The additional command `probability=T` means that the y-axis is the density (with the default being simply frequency if this is omitted). Finally the command `xlab` gives the label of the x-axis to be p. The analogous command `hist(dip[1001:10000,1],probability=T,xlab="phi")` produces the histogram in Figure 7.6(b) for ϕ.

If we wished to calculate the posterior correlation between the two parameters we could type,

```
> cor(dip[1001:10000,1],dip[1001:10000,2])
```

Typing this for the above simulations, R gives:

```
[1] 0.01638828
```

Thus, there appears a posterior correlation between p and ϕ of 0.02 (to two decimal places), corresponding to essentially no correlation between the parameters. Note that when calculating the sample correlation, in order to provide a Monte Carlo estimate of the posterior correlation, we again have to remember to omit the initial iterations within the burn-in. Thus, we only consider the values of the parameters between iterations 1001 and 10,000 (taking the burn-in to be 1000). The `coda` package can also be used to obtain posterior summary statistics, etc.

R CODE FOR MODEL C/C

Calculation of Monte Carlo Error for Parameter Estimates*

We demonstrate how we can obtain other output statistics that can be calculated within WinBUGS immediately (via the `Sample Monitor Tool` – see Appendix C). We consider in particular the associated Monte Carlo error of the parameters (see Section 5.4.3). In order to do this we simply need to post-process the output parameter values. However, since we have all the simulated parameter values stored in an array, performing any post-processing is straightforward. In addition, we can easily calculate the associated Monte Carlo error as the number of iterations increases, and plot the change in Monte Carlo error. For example, the function `MCerror` given in Figure B.1 calculates the Monte Carlo error, where we input the (post-burn-in) sampled parameter values, `data`, the number of parameters `nparam` (corresponding to the number of columns in `data`), the total number of batches (`nbatch`), and the number of simulations within each batch (`nits`).

```
###############
# Define function for calculating Monte Carlo error:

MCerror <- function(data, nparam, nbatch, nits)  {

    mcerror <- array(0,dim=c(nbatch-1,nparam))

# Calculate Monte Carlo error for each parameter in turn:

    for (i in 1:nparam) {
        data1 <- array(data[,i],dim=c(nits, nbatch))
        mn <- array(0,nbatch)

# Calculate the mean of each batch of iterations:

        for (j in 1:nbatch) {
            mn[j] <- mean(data1[,j])
        }

# Calculate the MC error as number of batches increases:

        for (j in 2:nbatch) {
            samplevar <- var(mn[1:j])
            mcerror[j-1,i] <- sqrt(samplevar/j)
        }
    }
mcerror
}
```

Figure B.1 R code for calculating the Monte Carlo error of the estimated posterior mean using post-burn-in simulated parameter values.

Within the code we cycle through each parameter in turn (sampled values in the column of the array `data`), placing the first `nits` into the first column of the array `data1`, the next `nits` iterations into the second column, and so on, to the last `nits` iterations into the (last) `nbatch`'th column. Each column corresponds to a batch of parameter values (with `nits` values in each batch). We calculate the mean of each column in the vector `mn`, corresponding to the mean of sampled parameter values in each batch. Finally, we calculate the corresponding Monte Carlo error of the posterior mean of the parameters at iteration value, $j \times \text{nits}$, for $j = 2, \ldots, \text{nbatch}$ and output these values.[†]

We use the above code to calculate the Monte Carlo error for the dipper data (stored in the array `dip`). Note that the input data for `MCerror` is for the post-burn-in sampled parameter values. Thus, we use the command:

`MCerrordip <- MCerror(dip[1001:10000,],2,90,1000)`

In other words, we only consider the post-burn-in parameter values from iterations 1001 to 10,000. There are two parameters (`nparam = 2`), and we consider a total of 90 batches, with 1000 iterations in each batch. The Monte Carlo error for the posterior means of the parameters are then simply the elements in the final row of the array `MCerrordip`. Note that there are (`nbatch`-1) rows, since we cannot calculate the Monte Carlo error when there is only one batch of sample values. We output these to the screen, using the command:

```
> MCerrordip[89,]
[1] 0.0002859932 0.0001869274
```

We interpret the output as the Monte Carlo error associated with the posterior mean of p and ϕ, respectively. In addition, from the way we have defined the function, we can also tabulate or plot the corresponding Monte Carlo errors of p and ϕ as the number of iterations increases (or equivalently, the number of batches used, assuming a fixed number of iterations per batch). Figure B.2 provides the corresponding MC error for p and ϕ as the number of simulations increase. We have plotted the Monte Carlo error against the number of batches used (so that the number of iterations corresponds to the number of batches times 1000).

Clearly, we can see that as the number of iterations increase (and hence the number of batches), the Monte Carlo error decreases. As always, the Monte Carlo error should be compared with the magnitude of the corresponding

---

[†] Recall that the Monte Carlo error of parameter, θ, say is given by,

$$\sqrt{\frac{1}{n}s^2(\theta)},$$

where $n = \text{nbatch}$, and $s^2(\theta)$ denotes the sample variance of θ given by,

$$s^2(\theta) = \frac{1}{(n-1)} \sum_{i=1}^{n} (\overline{\theta}_i - \overline{\theta})^2,$$

such that $\overline{\theta}_i$ denotes the sampled mean of the parameter θ of the ith batch and $\overline{\theta}$ denotes the overall mean of θ from all n batches. Note that the R function `var` calculates the sample variance, $s^2(\theta)$ directly.

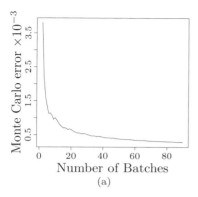 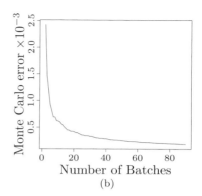

Figure B.2 The Monte Carlo error calculated over cumulative batches, each containing 1000 simulated parameter values for (a) p and (b) ϕ for the dipper data under model C/C.

posterior summary statistics. The Monte Carlo error is clearly very small compared to the estimated posterior mean and standard deviation for the recapture and recovery probabilities, suggesting that the posterior summary statistics have converged (i.e. repeating the simulations would produce essentially identical results with little variation between the posterior estimates).

B.5 R Code for White Stork Covariate Analysis

We now consider the white stork data, initially presented in Exercise 3.5 and discussed further in Example 5.7. This consists of 16 years of capture-recapture data where there is additional covariate information, relating to the amount of rainfall at a given number of weather stations in the population's wintering grounds. We initially consider a single rainfall covariate. We set the survival probability to be of the form,

$$\text{logit } \phi_t = \beta_0 + \beta_1 \text{cov}_t,$$

where cov_t denotes the normalised covariate value at time t. Finally, the model is defined by assuming a constant recapture probability, p.

We specify vague independent priors on all of the parameters,

$$\beta_i \sim N(0, 10) \quad \text{for } i = 0, 1;$$
$$p \sim Beta(1, 1) \equiv U[0, 1].$$

We use a single-update random walk MH algorithm, using a uniform proposal distribution for each parameter p, β_0 and β_1. The MH algorithm that we implement is described in detail in Example 5.7. We once again present the main code for this algorithm, before the separate subroutines for calculating the log-likelihood and MH updating steps. Note that the corresponding WinBUGS algorithm is given in Example C.2.

B.5.1 Main Code

The main structure of the code follows the same format as the R code described in Section B.4 for the dippers capture-recapture data. We begin with reading in the data and prior parameter values, before performing the MCMC iterations and outputting the results from the simulations. We consider each of these sections of the code in turn.

Part I: Reading in the Data and Input Parameters

We begin by defining the function storkcodeMHcov1 and reading in the data in Figure B.3. The size of the capture-recapture m-array is defined by the parameters ni (number of years of release) and nj (number of years of recaptures). We also define the total number of parameters in the model, nparam, which is equal to three in this case for p, β_0 and β_1. The m-array is read in for the stork data and placed in the matrix data. Finally, we read in the set of covariate rainfall values, placing them in the vector, cov.

Given the specification of the data values, we move on to set the prior parameter values and MH proposal values. The corresponding R code is given in Figure B.4. We begin with the priors for the recapture probability, p. In general, we set,

$$p \sim Beta(\texttt{alphap}, \texttt{betap}).$$

In this case, we set alphap = betap = 1, corresponding to the $U[0,1]$ distribution. We specify independent normal priors on the regression coefficients, β_0 and β_1. The prior means are placed into the vector mu (both set equal to zero in this case), and the prior variances placed in sig2 (both equal to 10). The first element of mu and sig2 correspond to the prior parameters for β_0, and the second elements to β_1. Note that for simplicity we set sig to be the square root of sig2, so that sig corresponds to the prior standard deviation on the parameters. Recall that the normal distribution in R uses the mean and standard deviation as the parameters. Finally we set the uniform MH proposal parameters for each parameter, by defining the vector delta. The first element (0.05) corresponds to the proposal parameter for p, the second element (0.1) for β_0, and the third element (0.1) for β_1.

Part II: MCMC Simulations

The corresponding R code of the MCMC algorithm is given in Figure B.5. Note the similarity with the coding in Figure 7.2 for the dipper data – the structure is the same, only the details differ slightly. This is typically the case for MCMC algorithms; the general structure of the code is very similar, but

R CODE FOR WHITE STORK COVARIATE ANALYSIS

```
###############
# Define the function with input parameters:
# nt = number of iterations; nburn  = burn-in

storkcodeMHcov1 <- function(nt,nburn){

# Define the parameter values: ni = number of release years;
# nj = number of recapture years; nparam = number of parameters

ni = 16
nj = 16
nparam = 3

# Read in the data:

data <- matrix(c(
19,2,0,0,0,0,0,0,0,0,0,0,0,0,0,0,5,
0,33,3,0,0,0,0,0,0,0,0,0,0,0,0,0,14,
0,0,35,4,0,0,0,0,0,0,0,0,0,0,0,0,14,
0,0,0,42,1,0,0,0,0,0,0,0,0,0,0,0,26,
0,0,0,0,42,1,0,0,0,0,0,0,0,0,0,0,30,
0,0,0,0,0,32,2,1,0,0,0,0,0,0,0,0,36,
0,0,0,0,0,0,46,2,0,0,0,0,0,0,0,0,16,
0,0,0,0,0,0,0,33,3,0,0,0,0,0,0,0,28,
0,0,0,0,0,0,0,0,44,2,0,0,0,0,0,0,20,
0,0,0,0,0,0,0,0,0,43,1,0,0,1,0,0,10,
0,0,0,0,0,0,0,0,0,0,34,1,0,0,0,0,25,
0,0,0,0,0,0,0,0,0,0,0,36,1,0,0,0,16,
0,0,0,0,0,0,0,0,0,0,0,0,27,2,0,0,22,
0,0,0,0,0,0,0,0,0,0,0,0,0,22,0,0,16,
0,0,0,0,0,0,0,0,0,0,0,0,0,0,15,1,17,
0,0,0,0,0,0,0,0,0,0,0,0,0,0,0,15,8),nrow=ni,byrow=T)

# Read in the covariate values:

cov <- c(0.79,2.04,1.04,-0.15,-0.01,-0.48,1.72,-0.83,
         -0.02,0.14,-0.71,0.50,-0.62,-0.88,-1.00,-1.52)
```

Figure B.3 The R code defining the function and input data for the white stork.

we simply need to input the correct likelihood function for the given example and the MH step used.

We begin by setting the initial parameter values. The values for the three parameters in the model, p, β_0 and β_1, are stored in the vector param (param[1] $\equiv p$; param[2] $\equiv \beta_0$; param[3] $\equiv \beta_1$). Thus, we set the initial state of the chain to be,

$$p = 0.9; \qquad \beta_0 = 0.7; \qquad \beta_1 = 0.3,$$

```
###############
# Read in the priors:
# Beta prior on recapture probability

    alphap <- 1
    betap <- 1

# Normal priors (mean and variance) on regression coefficients

    mu <- c(0,0)
    sig2 <- c(10,10)
    sig <- sqrt(sig2)

# Parameters for MH updates (uniform random walk):

    delta <- c(0.05,0.1,0.1)
```

Figure B.4 The R code specifying the priors on the parameters and proposal parameters for the white stork application.

and calculate the corresponding log-likelihood using the function `calclikhood` (described in Section B.5.2). We define the array `itns`, which will store the values of the parameters at each iteration of the Markov chain, and the vector `output`, which is used to store the current values of the parameters and associated log-likelihood value.

We perform the MCMC updates for the given number of iterations (`nt`). Within each iteration, the function `updateparam` is called to perform the MH step (this function is described in Section B.5.3). The function outputs the vector `output`, with the first three elements corresponding to the parameter values, and the fourth element the log-likelihood value. Finally, we record the parameter values at each iteration in the array `itns`, which are then used to provide Monte Carlo estimates of parameters of interest or to check for convergence, etc.

Part III: Outputting the Results

We output the MCMC simulated parameter values and also calculate the posterior mean and standard deviation of the parameters, which are printed to the screen (to reduce the amount of post-processing of the simulated values). The R code for outputting these results is provided in Figure B.6.

As usual, we begin by removing the burn-in from the sample, before calculating the posterior mean (`mn`) and standard deviation (`std`) and print them to the screen. Finally, we output the full set of simulated values within the MCMC algorithm, for any further post-processing analyses. Once more we note the similarity of the R code with that for the outputting of the results for the dipper data in Figure 7.3. Essentially, the only difference relates to the name of the parameters.

R CODE FOR WHITE STORK COVARIATE ANALYSIS

```
###############
# Set initial parameter values: param[1] = recapture probability;
# param[2:3] = regression coefficients for survival probability

param <- c(0.9,0.7,0.3)

# Calculate log-likelihood of initial state using function
# "calclikhood":

likhood <- calclikhood(ni, nj, data, param, cov)

# Define itns - array to put the sample from posterior distribution;
# output - vector for parameter values and associated log-likelihood:

itns <- array(0, dim=c(nt, nparam))
output <- dim(nparam+1)

# MCMC updates - for each iteration, update parameters using function
# "updateparam". Set parameter values, log-likelihood value of
# current state and record parameter values

for (t in 1:nt){
    output <- updateparam(nparam,param,cov,ni,nj,data,likhood,
                          alphap,betap,mu,sig,delta)

    param <- output[1:nparam]
    likhood <- output[nparam+1]

    for (i in 1:nparam) {
        itns[t,i] <- param[i]
    }
}
```

Figure B.5 The R code that gives the initial values of the parameter values, calculates the corresponding likelihood function and performs the MCMC iterations, for the white stork example.

The main code relies on two subroutines: one for calculating the log-likelihood and the other for performing the MH algorithm within each iteration of the MCMC algorithm. We begin by considering the log-likelihood function.

B.5.2 Subroutine – Calculating Log-Likelihood Function – `calclikhood`

The log-likelihood function is called at the beginning of the MCMC code, to calculate the likelihood of the initial parameter values, and then within the MH updating step of each parameter. The subroutine for calculating the log-likelihood function is presented in Figure B.7. This function is directly comparable with the log-likelihood calculated for the dippers example in Figure

```
###############
# Remove the burn-in from the simulations and calculate
# the mean and standard deviation of the parameters

subitns <- itns[(nburn+1):nt,]

mn <- array(0,nparam)
std <- array(0,nparam)

for (i in 1:nparam) {
    mn[i] <- mean(subitns[,i])
    std[i] <- sd(subitns[,i])
}

# Output the posterior mean and standard deviation of the parameters
# following burn-in to the screen:

cat("Posterior summary estimates for each parameter:   ", "\n")
cat("\n")
cat("mean   (SD)", "\n")
cat("p: ", "\n")
cat(mn[1], "(", std[1], ")", "\n")
for (i in 2:nparam) {
    cat("\n")
    cat("beta_",(i-2), "\n")
    cat(mn[i], "   (", std[i], ")", "\n")
}

# Output the sample from the posterior distribution:

itns
}
```

Figure B.6 R code for outputting the posterior means and standard deviation of the model parameters to the screen and the MCMC realisations of the parameter values stored in the array `itns`.

7.4. The only difference lies in the specification of the survival probabilities. Here, the survival probability in year i, denoted by `phi[i]`, is given by,

$$\texttt{phi[i]} = \frac{\exp(\texttt{cov[i]})}{1+\exp(\texttt{cov[i]})} = \frac{1}{1+\exp(-\texttt{cov[i]})}.$$

Given the specification of the recapture and survival probabilities at each time, the corresponding likelihood, in terms of the calculation of the multinomial cell probabilities (q) follows as given in Figure 7.4, so we omit the details for brevity.

R CODE FOR WHITE STORK COVARIATE ANALYSIS

```
###############
# To calculate the log-likelihood of capture-recapture data

calclikhood <- function(ni, nj, data, param, cov){

# Set up the size of the array containing the survival
# and recapture probabilities and cell probabilities (q)
# (all entries initially equal to 0)

    phi <- array(0,nj)
    p <- array(0,nj)
    q <- array(0,dim=c(ni,nj+1))

# Set the recapture and survival probs for each year:

    for (i in 1:nj) {

        exprn <- param[2] + param[3]*cov[i]
        phi[i] <- 1/(1+exp(-exprn))

        p[i] <- param[1]
    }

# Calculate multinomial cell probabilities (q)
# and log-likelihood contribution:

    for (i in 1:ni){

.....

}
```

Figure B.7 R code for calculating the log-likelihood value for capture-recapture data with the survival probability logistically regressed on a single covariate. The remainder of the likelihood is already given in Figure 7.4, and hence is omitted for brevity.

B.5.3 Subroutine – MH Updating Algorithm – updateparam

The final component of the MCMC code relates to the MH subroutine, which we call updateparam. The corresponding function is provided in Figure B.8. This is very similar to the algorithm presented in Figure 7.5, the MH step for the dipper data, however, we now have a different parameterisation for the survival probabilities (regressed on a covariate), and hence a different prior specification for the parameters associated with the survival probability (i.e. the regression coefficients, β_0 and β_1).

As usual, we cycle through each parameter in turn, propose a new candidate value using a random walk MH step with uniform proposal distribution and

```
###############
# Function for updating the parameters values:

updateparam <- function(nparam,param,cov,ni,nj,data,likhood,
                        alphap,betap,mu,sig,delta){

    for (i in 1:nparam) {
        oldparam <- param[i]
        param[i] <- runif(1, param[i]-delta[i], param[i]+delta[i])

# Automatically reject any moves where recapture prob is outside [0,1]

        if (param[1] >= 0 & param[1] <= 1) {

# To calculate acceptance probability

            newlikhood <- calclikhood(ni, nj, data, param, cov)

# For recapture probability add in prior (beta) terms;
# else for regression coefficients add in prior (normal) terms.
# (Proposal densities cancel since the proposal distribution is symmetric).

            if (i == 1) {
                num <- newlikhood + log(dbeta(param[i],alphap,betap))
                den <- likhood + log(dbeta(oldparam,alphap,betap)) }
            else {
                num <- newlikhood + log(dnorm(param[i],mu[i-1],sig[i-1]))
                den <- likhood + log(dnorm(oldparam,mu[i-1],sig[i-1])) }

            A <- min(1,exp(num-den))
        }

        else { A <- 0 }

# Accept/reject step:

        u <- runif(1)
        if (u <= A) { likhood <- newlikhood }
        else { param[i] <- oldparam }
    }

    output <- c(param, likhood)
    output
}
```

Figure B.8 R code for the MH step for updating p, β_0 and β_1 for the stork capture-recapture data.

keep a record of the current value in oldparam. Note that only the recapture probability parameter is constrained to be in the interval [0,1], with the other two regression coefficient parameters defined over the whole real line. Thus, we only need to ensure that $0 \leq p \leq 1$ within the updating move.

In order to calculate the corresponding acceptance probability, we initially calculate the corresponding log-likelihood of the set of parameter values, and consider the log of the numerator (num) and log of the denominator (den) of the acceptance probability. For num, this is the sum of the log-likelihood using the proposed candidate value and the log prior density for this parameter; for den, this is the sum of the log-likelihood of the current parameter value and the log prior density of the parameter. Note that for the recapture probability, (param[1]), a beta prior is specified; for the regression coefficients, (param[2] and param[3]), a normal prior is specified with means mu[1] and mu[2] and standard deviations sig[1] and sig[2], respectively. The accept/reject step follows as standard (for example, see Section B.4.3 for a detailed description). We note that changing the prior distribution on any of these parameters essentially involves only changing the acceptance probability in the MH step (and the inputting of the prior parameter values).

B.5.4 Running the White Stork R Code

We run the R code for the white stork data for 10,000 iterations, discarding the first 1000 iterations as burn-in. These simulations took approximately 3.5 minutes. Trace plots suggest that this is very conservative, with apparent convergence within even 100 iterations (note that this is aided with the initial starting values specified). The commands in R and corresponding output to the screen for this particular simulation are:

```
> stork <- storkcodeMHcov1(10000,1000)

Posterior summary estimates for each parameter:

mean    (SD)
p:
0.9129703     ( 0.01428288 )

beta_ 0
0.6845587     ( 0.07308035 )

beta_ 1 0.3536333    ( 0.09035605 )
```

The output from the function is stored in the array stork: the first column the set of p values, the second column the β_0 values, and the third column the β_1 values. So, in order to obtain the 95% symmetric credible interval for each parameter, we use the following commands to calculate the lower and upper 2.5% and 97.5% quantiles from the sampled parameter values, with corresponding response from R:

```
> quantile(stork[1001:10000,1],probs=c(0.025,0.975))
      2.5%      97.5%
0.8831229 0.9393124

> quantile(stork[1001:10000,2],probs=c(0.025,0.975))
      2.5%      97.5%
0.5432976 0.8308200

> quantile(stork[1001:10000,3],probs=c(0.025,0.975))
      2.5%      97.5%
0.1794538 0.5288888
```

*Calculating Bayesian p-values**

To assess the goodness-of-fit of the model to the data, we calculate the Bayesian p-values, using the deviance (i.e. $-2 \times$ log-likelihood function) as the discrepancy function (see Section 5.4.6). In this case, the Bayesian p-value is calculated conditional on the number of newly tagged individuals each year (in order to simulate new data comparable to the observed data). Calculating the Bayesian p-value within the MCMC simulations is straightforward, as is the necessary R code. Note that, in general, the p-values can either be calculated within the R code itself, or post-process using the outputted values of the parameters at each iteration. For simplicity, we assume that we calculate the Bayesian p-value post-process, for the simulations run above, and stored in the array stork. The corresponding R function is provided in Figure B.9

We begin by reading in the data and covariate values (as in Figure B.3). We need to calculate these from the observed m-array. We do this by a two-step process. First we calculate the number of birds released each year. Then we calculate the number of these that are new birds, denoted by new. To do this we simply subtract the number of birds that are recaptured at that time from the total number of birds released at that time (since there is no removal of individuals upon recapture). These values are fixed (as they are a function of the observed data), and so we simply calculate these once before cycling through the set of simulated parameter values to calculate the Bayesian p-value.

To calculate the Bayesian p-value, we need a sample of parameter values from the posterior distribution. Thus, we only consider the sampled parameter values following a suitable burn-in period. For each iteration, following the defined burn-in, we set the values of the vector param to be the simulated parameter values from the MCMC iterations. We then calculate the multinomial cell probabilities for capture-recapture data, given the set of parameter values, as in Figure B.7, and hence this is omitted for brevity. Given the multinomial

R CODE FOR WHITE STORK COVARIATE ANALYSIS

```
###############
# To calculate Bayesian p-value using deviance

storkpvalue <- function(nt, nburn, stork){

# Read in data and set of covariate values
...
# Calculate R/new - number of birds/new birds released:

    param <- array(0,nparam)
    R <- array(0,ni)
    new <- array(0,ni)
    for (i in 1:ni) {
        R[i] <- sum(data[i,i:(nj+1)])
        new[i] <- R[i] - sum(data[1:(i-1),(i-1)]) }

    for (t in 1:(nt-nburn)) {              # Iterate over simulations
        for (i in 1:nparam) {
            param[i] <- stork[t+nburn,i]  }  # Set parameter values

# Calculate the multinomial cell probabilities, q ...
# Set up arrays, simulate data anc calculate deviances:

        newR <- array(0,ni)
        outp <- array(0,dim=c(nt-nburn,3))
        data1 <- array(0,dim=c(ni,nj+1))

        for (i in 1:ni) {
            newR[i] <- new[i] + sum(data1[1:(i-1),(i-1)])
            data1[i,] <- rmultinom(1,newR[i],q[i,])
            for (j in i:(nj+1)) {
                outp[t,1] <- outp[t,1] - 2*data[i,j]*log(q[i,j])
                outp[t,2] <- outp[t,2] - 2*data1[i,j]*log(q[i,j]) }
        }
    if (outp[t,1] <= outp[t,2]) { outp[t,3] <- 1 }
    else { outp[t,3] <- 0 }
    }
outp
}
```

Figure B.9 R code for simulating capture-recapture data given the model parameter value with the aim of calculating the Bayesian p-value for capture-recapture data.

cell probabilities (sampled from the posterior distribution) we can simulate data. For each year of release, `i=1,...,ni`, we calculate `newR[i]`, corresponding to the sum of the number of newly tagged birds and recaptured birds in year `i`. We then simulate the ith row of the simulated m-array from a multinomial distribution with sample size `newR[i]` and probabilities `q[i,]` (i.e. the ith row of multinomial cell probabilities).

Given the sample data simulated from the posterior distribution of the parameters, we compare this simulated data with the observed data, using the deviance as the discrepancy function. Thus, we calculate the deviance of both the simulated data and observed data, given the sampled parameter values from the posterior distribution. Finally, we output the deviance values of the observed and simulated data along with an indicator function of whether the deviance of the observed data is less than the deviance of the simulated data. The proportion of times that this indicator is equal to one is the Bayesian p-value.

We run the above Bayesian p-value code on the outputted MCMC simulations stored in the vector `stork` obtained in Section B.5.4 (discarding the first 1000 iterations) and placing the outputted values into the array `pstork`. To calculate the corresponding Bayesian p-value we use the command:

```
> sum(pstork[,3])/length(pstork[,3])
[1] 0.5533333
```

Rerunning the Bayesian p-value simulations a number of times always obtained values within a range of ± 0.01. This would suggest the model appears to fit the data well, and does not indicate any lack of fit. The corresponding plot of the simulated deviance and observed deviance are provided in Figure B.10, and again does not indicate any lack of fit for the model.

B.6 Summary

In this Appendix we consider the computer package R and provide sample MCMC code for the dipper and white stork capture-recapture data. The R codes follow the same general structure, which makes it easier initially to write an MCMC algorithm, but also to adapt existing codes to other problems. The built-in functions in R, particularly the random number generators, summary statistics functions (such as `mean` and `sd`), and plotting functions significantly simplify the writing of the MCMC code and outputting of the corresponding results.

Programming bespoke code allows the user to have complete control over the models, updating algorithm and output. However, this comes at the cost of having to program (and debug) the (RJ)MCMC algorithm, which can be time consuming and is of a more specialist nature. In addition, typically, all the pilot tuning needs to be performed by the individual for each given problem, whereas, for example, WinBUGS does this automatically in the initial adaptive step. However, for more advanced problems (for example, using models that cannot be fitted in WinBUGS, or where WinBUGS is very slow in performing the MCMC iterations), writing bespoke code may be the only feasible way forward. In addition, once an initial MCMC simulation has been written in R, it is much easier to adapt this existing code for alternative problems. For example, suppose that we wish to consider ring-recovery data (as opposed to

SUMMARY

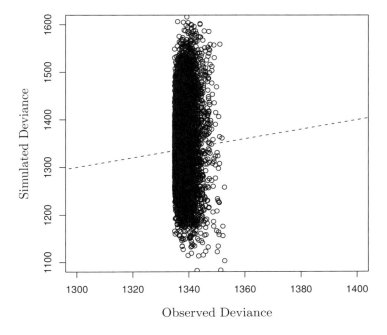

Figure B.10 The plot of the observed log-likelihood value (x-axis) versus the simulated log-likelihood value (y-axis). The dashed line represents "x=y". The proportion of points below the line corresponds to the Bayesian p-value statistic.

capture-recapture data within this chapter). The main changes to the code involve the different likelihood function, but this is in a self-contained subroutine, and can be easily modified. There may also be additional (minor) modifications, such as the number of parameters and prior specifications, but these are quickly and easily performed.

APPENDIX C

Programming in WinBUGS

This appendix gives a brief introduction to WinBUGS (Lunn et al. 2000, 2006). WinBUGS implements up-to-date and powerful MCMC algorithms that are suited to a wide range of target distributions for analysing complex models. This avoids coding the MCMC algorithms by hand, but requires the specification of a likelihood and priors for model parameters. We initially describe the general computer package WinBUGS, including how to download the program, general structure of WinBUGS code and a brief overview of how to run MCMC simulations. To demonstrate how to write a WinBUGS program, we initially provide a general structure for the code, followed by a detailed description of WinBUGS code for a number of examples. We focus on the analysis of capture-recapture data, but the ideas can easily be applied to other problems. Finally, we also provide an example of RJMCMC coding, which is relatively new add-on package to WinBUGS. Note that all timings refer to a 1.8-Ghz computer.

C.1 WinBUGS

Despite the widespread use of MCMC, surprisingly few programs are freely available for its implementation. This is partly because of the fact that algorithms are generally problem specific (particularly in the case of MH, for example) and there is no automatic mechanism for choosing the best implementation procedure for any particular problem. However, one program has overcome many of the problems and is widely used. It is known by the acronym BUGS (Bayesian inference Using Gibbs Sampling), see Gilks et al. (1992), Spiegelhalter et al. (1996) and Lunn et al. (2000). The WinBUGS (BUGS for Windows) package is able to implement MCMC methodology for a wide variety of problems.

The package provides a declarative R-type language for specifying statistical models. The program processes the model and data and sets up the sampling distributions required. Finally, appropriate sampling algorithms are implemented to simulate values of the unknown quantities in the model. WinBUGS is able to handle a wide range of standard distributions, and in the cases where full conditional distributions are non-standard, alternative methods such as adaptive rejection sampling (Gilks and Wild 1992) and slice sampler, (Neal 2003) are used to sample these components. A menu-driven set of R functions (coda) is supplied with WinBUGS in order to calculate convergence diagnostics and both graphical and statistical summaries of the simulated samples.

Overall, the package is extremely easy to use and is capable of implementing the Gibbs sampler algorithm for a wide range of problems. It is also freely available from the WinBUGS Web site at

`http://www.mrc-bsu.cam.ac.uk/work/bugs/`

We provide a brief overview of the WinBUGS package, including the format of a WinBUGS file, running MCMC simulations within WinBUGS and the results that can be obtained.

C.1.1 Getting Started with WinBUGS

A typical WinBUGS session has three steps: setting up the code, running the code and obtaining posterior estimates.

Setting up WinBUGS code

1. For WinBUGS code, one or more files are needed that contain:
   i) The model specification
   ii) Data values
   iii) Initial parameter values (these can also be generated within WinBUGS)

2. Given a WinBUGS program, the first step is to read it into the WinBUGS package:
   i) Click on the `File` button in the tool bar and then `Open`. This will open up a pop-up window, from which one can find a WinBUGS file.
   ii) Find the file, then either double-click on the file, or single-click on it and then `Open`. This will load open the file into a WinBUGS window.

3. The next step is to read the model into WinBUGS:
   i) Click on the `Model` button on the toolbar and then `Specification`. A `Specification Tool` pop-up menu will appear, as shown in Figure C.1.
   ii) Highlight the model specification part of the code (only the beginning of the model needs to be highlighted), and click `check model` in the `Specification Tool` window. The phrase `model is syntactically correct` should appear in the bottom left of the WinBUGS window. If not, there is an error in the specification of the model, and it will need to be corrected.
   iii) If one plans to run multiple chains, then this needs to be specified here, since the number becomes fixed during the following compilation process – note that to run the Brooks-Gelman-Rubin (BGR) convergence diagnostic in WinBUGS, at least 3 replications are needed.

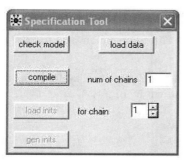

Figure C.1 The Specification Tool pop-up menu produced by Specification in the Model button on the toolbar.

4. Once the model is correct, we need to read in the data:
   i) Highlight the word list corresponding to the data in the file and click on load data in the Specification Tool window. The phrase data loaded should appear in the bottom left corner of the WinBUGS window; otherwise an error message will appear.
   ii) Click compile in the Specification Tool. The phrase model compiled should now appear in the bottom left corner; otherwise an error message will appear stating why there is a problem with the code.
5. The final input step is to read in the starting parameter values for the Markov chain. There are two options:
   i) Highlight the word list in the file that corresponds to the initial parameter values and then click on load inits. The phrase model is initialized should appear in the bottom left corner.
   ii) Or, alternatively, it is also possible to allow WinBUGS to generate starting values for the parameters. To do this, click on gen inits in the Specification Tool. If WinBUGS can generate all the parameter values, the phrase initial values generated will appear in the bottom left corner. However, note that WinBUGS is not always capable of selecting suitable starting values. If this is the case, an error message will appear in the bottom left corner, specifying the parameters which could not be initialised. In some cases we may wish to read in some initial parameter values and allow WinBUGS to generate the rest. This can be done by simply following the above procedure to read in the initial parameter values, where the list in the file contains only some parameter values. Then following reading in these values, we click on gen inits to generate the remaining parameter values.
6. Once the model has been compiled and the data read in, we are ready to start.

Running WinBUGS Code

1. It is useful to pull down a few more pop-up menus from the toolbar at the beginning:

   i) Click on **Update** from the pull-down **Model** menu. The **Update Tool** is used to run the simulations and is shown in Figure C.2.

Figure C.2 The **Update Tool** is used to run the simulations and is obtained by clicking on **Update** from the pull-down **Model** menu

   ii) Click on **Samples** from the **Inference** pull-down menu. The **Sample Monitor Tool** window allows the specification of the parameters to be recorded and is shown in Figure C.3. Note that it is necessary to specify the parameters to be recorded prior to running any simulations.

Figure C.3 The **Sample Monitor Tool** window allows the specification of the parameters to be recorded, and is obtained by clicking on **Samples** from the **Inference** pull-down menu

2. Now, in order to use these to run the simulations and monitor the parameters:

   i) In the **Sample Monitor Tool** enter the names of the parameters individually, pressing **set** after each parameter. Note that for vectors, entering the vector name ensures that all elements of the parameter vector are recorded.

ii) To run the simulations we use the `Update Tool`, inputting the number of iterations to be run (`update`) and how often to refresh the updates to write to the screen (`refresh`). We press `update` to run this many iterations. Clicking on `update` whilst the MCMC algorithm is running will pause the simulation. If `update` is clicked again, the MCMC simulations will continue for as many iterations again. Whilst the `adapting` box in the `Update Tool` window is checked, the simulation is going through an automatic adaptation procedure and no summary statistics are available during this period. This is typically fairly short with a maximum (default) value of 4000 iterations.

iii) Note that initially it is often useful to run very short simulations to check the length of time that they take before running a large number of iterations.

Obtaining Posterior Estimates in WinBUGS

1. Once the simulations are run and the adaptation period is finished, it is possible to monitor the statistics of interest:

    i) Highlight the relevant parameter in the `node` box of the `Sample Monitor Tool` window. Click on the arrow to the right of the `node` box to get a scrollable list of the parameters entered in earlier. Note that using the symbol * in the `node` box represents all specified parameters.

    ii) The values for `beg` and `end` specify the start and end of the sample values to be used to construct summary statistics. The `thin` box allows thinning of the output (see Section 5.4.5). Setting `thin=10`, for example, means that the sample statistics are based on every 10th sample value. The values for `beg` and `end` specify the start and end of the sample values to be used to construct summary statistics.

2. The `Sample Monitor Tool` allows a variety of sample statistics to be calculated:

    - `trace`: Provides a dynamic trace plot of recent parameter values which updates continuously during the simulation.
    - `history`: Provides a static trace plot of parameter values for the entire simulation.
    - `density`: Produces kernel density estimates of the posterior marginal densities of the selected parameters.
    - `stats`: Provides a table of posterior means, variances and other summary statistics.
    - `coda`: Dumps out a list of monitored parameter values for input into alternative packages (including `R`) for calculating posterior summary statistics.

- **quantiles**: Plots out the running mean of the selected parameters, interquartile range and symmetric credible intervals (80%, 90%, 95%).
- **bgr diag**: Calculates the BGR convergence diagnostic if three or more chains have been run. Plots settling to a value close to unity indicates convergence.
- **auto cor**: Plots the autocorrelation function for the selected parameters. This measures the between-iteration dependence sampled values.

We now consider a couple of examples to demonstrate the use of WinBUGS and the corresponding output that can be obtained.

Example C.1. WinBUGS Example – Dippers

To illustrate the WinBUGS package we fit model C/C to the dipper data set of Example 1.1. Example WinBUGS code is provided in Figures 7.7 and 7.8 for the three components needed: model specification, data and initial values. The WinBUGS codes have been annotated for clarity. This is done by simply using the # sign which acts as a line comment, as with R.

The code is read into WinBUGS and compiled as described above, setting the number of chains to be equal to three. Initial values for the survival and recapture parameters are specified as shown in Figure 7.8, using a list with two components. Three different sets of initial values, are specified, one for each of the three chains. Note that the chain could be started at the same parameter values; however, it is better to begin the chains at different values as an additional check for convergence, see Section 5.4.2. For example, it may be possible to identify multimodality with chains getting "stuck" in different modes, with different modes most likely to be identified by starting at different values.

Recall that before we performing any simulations, the parameters that are to be monitored need to be set. This is done via the Sample Monitor Tool, where the parameters in this case are the survival probability, phi, and recapture probability, p. The simulations are initially run for 10,000 iterations, using the Update Tool. The corresponding trace plot of the full MCMC simulation is given using the history button within the Sample Monitor Tool and is given in Figure 7.9.

Several checks can be made regarding convergence. The plots for the BGR diagnostic can be produced in WinBUGS using the bgr diag button within the Sample Monitor Tool. The numerical values can be obtained by selecting the plot followed by a ctrl-left-mouse-click (for further details, see the *"The Inference Menu"* in the *WinBUGS User Manual* which you can obtain by pressing F1).

To summarise the posterior distribution for the parameter, use the button stats in the Sample Monitor Tool, which provides a table of posterior means, standard deviations and other summary statistics; see Figure 7.11

To plot the marginal posterior distribution for each parameter we can use the density button in the Sample Monitor Tool. Figure C.4 provides the kernel density estimates of the marginal densities of p and ϕ. We can compare the corresponding representations with Figure 7.6, of the corresponding kernel density estimates obtained and plotted within R.

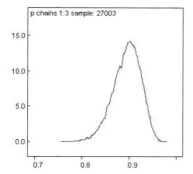 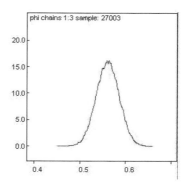

Figure C.4 Kernel density estimates for sampled observations of the posterior distribution of p (left panel) and ϕ (right panel) produced in WinBUGS using the `density` button within the `Sample Monitor Tool`.

To calculate the posterior correlation between two parameters, we click on `Correlations` from the pull-down `Inference` menu, and use `Correlations Tool` to produce a scatter plot of the sampled observations or to print the value of the correlation.

□

Example C.2. WinBUGS Example - White Storks We consider the annual capture-recapture data relating to a population of white storks provided in Exercise 3.5 and discussed further in Example 5.7. In particular, we explore the relationship between the adult survival probabilities and rainfall. Thus, we specify the survival probability to be of the form,

$$\text{logit } \phi_t = \beta_0 + \beta_1 w_t,$$

where w_t denotes the normalised covariate value at time t. As a result of a previous study of the data (Kanyamibwa et al. 1990), we simplify the model, assuming a constant recapture probability over time.

To implement this model in WinBUGS, we first need to extend the likelihood used in the previous example to incorporate time dependent survival parameters, as shown in Figure C.5. This code is written in a general form, so that it is easily modified to calculate the likelihood for alternative models, such as T/C. A prior needs to be specified for this new parameter. Note that, given the priors specified on the regression coefficients, this induces a prior on the survival probabilities for any given set of covariate values. We use independent normal distributions with mean 0 and large variance 10 as non-informative prior distributions for the regression parameters (see Section

###############
Specify the model

model {

# To set constant recapture probability, denoted by pconstant
    for (i in 1 : nj) {
        p[i] <- pconstant
        p1[i] <- 1 - p[i]
    }

Set prior on recapture probability
    pconstant ~ dbeta(1,1)

# Calculate the cell probabilities
    for (i in 1 : ni){

Calculate the diagonal, above diagonal and below diagonal elements
        q[i, i] <- phi[i]*p[i]

        for(j in (i+1) : nj){
            q[i,j] <- prod(phi[i:j])*prod(p1[i:(j-1)])*p[j]
        }

        for(j in 1 : (i - 1)){
            q[i, j] <- 0
        }

Probability of an animal not being seen again
        q[i, nj + 1] <- 1 - sum(q[i, 1:nj])
    }
...
```

**Figure C.5** Example WinBUGS code of the model specification for the white stork data set. The code is similar to that used in the dipper example (see Figure 7.7). Recall that, in general, phi[i] $\equiv \phi_i$ and p[i] $\equiv p_{i+1}$ (which in this case is equal to pconstant $\equiv p$ for constant recapture probability).

8.2.1 for further discussion of the prior specification issue in relation to logistic regression). Note that in WinBUGS for the normal distribution, the mean and precision (1/variance) are the input parameters. Recall that we should normalise (or standardise) the covariate values (i.e. rainfall measurements), see, for instance, Example 4.11. WinBUGS code is provided in Figure C.7 for reading in the data and initial parameter values.

The simulations are run for 10,000 iterations, with a burn-in of 1000 iterations. Posterior estimates for the recapture probability and the regression parameters are obtained using the **stats** button and are given in Table C.1.
```

Table C.1 Posterior Estimates for the Analysis of the White Storks Data

Parameter	Mean	standard Deviation	95% Symmetric Credible Interval	Median
beta[1]	0.670	0.066	(0.547, 0.800)	0.668
beta[2]	0.371	0.091	(0.191, 0.575)	0.369
pconstant	0.912	0.015	(0.881, 0.939)	0.913

Note that we can obtain posterior estimates of ϕ by simply monitoring phi in the Sample Monitoring Tool. The recapture probability is relatively high for capture-recapture studies. We also note that the credible interval of the slope (parameter beta[1]) does not contain the value 0. This suggests that the temporal variations in survival are correlated with rainfall variation. In the next section, we explore a more formal way to carry out covariate selection in order to assess the effect of environmental covariates on demographic parameters.

```
###############
# Provide the logistic regression equation for survival probabilities

for (i in 1:nj) {
    logit(phi[i]) <- beta[1] + beta[2] * w[i]
}

# Specify the prior on the logistic regression coefficients

for (j in 1:2) {
    beta[j] ~ dnorm(0,0.1)
}
```

Figure C.6 WinBUGS code for incorporating a covariate in the survival probabilities where beta[1] $\equiv \beta_0$ and beta[2] $\equiv \beta_1$.

□

C.1.2 RJMCMC in WinBUGS

The additional Jump package (Lunn et al. 2006, 2009) is available from the WinBUGS development site link on the left-hand side of the main WinBUGS Web site. Alternatively, the URL to this development site is:

http://www.winbugs-development.org.uk/

Click on the link to Jump and follow the instructions for downloading and installing the file. See Section 7.5 for a description using the Jump package in WinBUGS.

###############
Read in the white stork data and the covariate values

```
list(ni=16,nj=16,m=structure(.Data=
c(19,2,0,0,0,0,0,0,0,0,0,0,0,0,0,0,5,
0,33,3,0,0,0,0,0,0,0,0,0,0,0,0,0,14,
0,0,35,4,0,0,0,0,0,0,0,0,0,0,0,0,14,
0,0,0,42,1,0,0,0,0,0,0,0,0,0,0,0,26,
0,0,0,0,42,1,0,0,0,0,0,0,0,0,0,0,30,
0,0,0,0,0,32,2,1,0,0,0,0,0,0,0,0,36,
0,0,0,0,0,0,46,2,0,0,0,0,0,0,0,0,16,
0,0,0,0,0,0,0,33,3,0,0,0,0,0,0,0,28,
0,0,0,0,0,0,0,0,44,2,0,0,0,0,0,0,20,
0,0,0,0,0,0,0,0,0,43,1,0,0,1,0,0,10,
0,0,0,0,0,0,0,0,0,0,34,1,0,0,0,0,25,
0,0,0,0,0,0,0,0,0,0,0,36,1,0,0,0,16,
0,0,0,0,0,0,0,0,0,0,0,0,27,2,0,0,22,
0,0,0,0,0,0,0,0,0,0,0,0,0,22,0,0,16,
0,0,0,0,0,0,0,0,0,0,0,0,0,0,15,1,17,
0,0,0,0,0,0,0,0,0,0,0,0,0,0,0,15,8),
.Dim=c(16,17)),
w=c(0.79,2.04,1.04,-0.15,-0.01,-0.48,1.72,-0.83,
-0.02,0.14,-0.71,0.50,-0.62,-0.88,-1.00,-1.52))
```

Read in the initial starting values for the model parameters:

```
list(beta=c(0,0),pconstant=0.5)
```

Figure C.7 WinBUGS code for reading in the white stork data set and initial parameter values.

C.2 Calling WinBUGS from R

Calling WinBUGS from other programs may be useful, for example, in order to read complex sets of data and initial values, and avoid specifying the parameters to be monitored in each run, in addition to post-processing the results in other software for displaying complex graphics or performing Monte Carlo studies from running WinBUGS iteratively in a loop. Here, we give an illustration of the use of the R package R2WinBUGS (Sturtz et al. 2005).

To run R2WinBUGS:

1. Write a WinBUGS model in an ASCII file.
2. Open an R session.
3. Prepare the inputs to the "bugs" function and run it. A WinBUGS window will pop up and R will freeze up. The model will now run in WinBUGS. You will see things happening in the Log window within WinBUGS. When WinBUGS is finished, its window will close and R will work again.
4. If an error message appears, re-run with "debug = TRUE".

CALLING WINBUGS FROM R   411

Example C.3. Running R2WinBUGS code – White Storks
As an illustration of how to call WinBUGS from R, we shall consider again
the analysis of the stork data that was carried out in Example C.2. The
WinBUGS model is stored in a file CJScov.bug. We assume that we model
parameters are the logistic regression parameters (beta[1], beta[2]) and
constant recapture probability (pconstant). The capture-recapture data are
stored in an m-array in storks.dat and the covariate values in rainfall.dat.
Figure C.8 provides the commands for loading the R2WinBUGS package, reading
in the data, specifying the initial parameter values and the parameters to be
monitored (phi, beta and pconstant) and running the MCMC simulations,
for 10,000 iterations with a burn-in of 1000 iterations.

We run the analysis using 10,000 iterations and 1000 burn-in. These simulations took approximatively 10 seconds. The results can be printed using the command print(kita.sim). It is also possible to plot the posterior distribution of the regression parameter corresponding to the rainfall effect (see Figure C.9), or calculate 95% credible interval ($[0.194; 0.537]$) and median (0.364) of this regression parameter. These outputs are easily obtained using the commands in Figure C.10.

□

C.2.1 Calculating Bayesian p-values from WinBUGS Output*

We illustrate how WinBUGS and R can be used in combination to calculate the Bayesian p-values to assess the goodness-of-fit of a model to the data (see Section 5.4.6). The procedure is similar to what we have seen in Appendix B (in Figure B.9), except that here we use the simulated parameter values from the MCMC iterations obtained from WinBUGS. This illustrates an attractive feature of calling WinBUGS from R.

We assume that the simulations were run as in Section C.2 using 1 chain, and stored in a list named stork. The R script to calculate the Bayesian p-value post-process is provided in Figure C.11. The data and covariate values are read in as above. The number of iterations and the length of the burn-in period are obtained directly from the WinBUGS output by a simple extraction of some components of an R list. We calculate the number of birds released each year and the number of these that are new birds as in Figure B.9, and so this is omitted for brevity.

We then need a sample of parameter values from the posterior distribution obtained from WinBUGS. We proceed in the same way as with R, and consider the sampled parameter values following a suitable burn-in period. For each iteration, we set the values of the vector param to be the simulated parameter values from the MCMC iterations that are stored in the component sims.matrix of the list stork. One needs to make sure that the first component of vector param is the recapture probability, and the other two are the regression parameters. To do this we just need a quick look at the general

```r
###############
# Load R2WinBUGS package
library(R2WinBUGS)

# Read in data

# Number of years of releases and recaptures
ni <- 16
nj <- 16

# Storks recapture data (m-array)
m <- as.matrix(read.table("storks.dat"))

# Rainfall covariate
w <- as.vector(matrix(scan("rainfall.dat"),ncol=ni,nrow=1))

# Create list of data
dataz <- list ("m","ni","nj","cov")

# List of initial values

init1 <- list (beta=c(-0.2,-0.2), pconstant=0.1)
init2 <- list (beta=c(0.2,0.2), pconstant=0.9)
init3 <- list (beta=c(0,0), pconstant=0.5)

# Concatenate list of initial values
inits <- list(init1,init2,init3)

# Parameters to be monitored
parameters <- c("phi", "beta", "pconstant")

# Run the MCMC analysis
kita.sim <- bugs(dataz, inits, parameters, "CJScov.bug", n.chains=3,
n.iter=10000, n.burnin = 1000)
```

Figure C.8 The R2WinBUGS code for the white stork data with the survival probability regressed on a single covariate. Three Markov chains are run for 10,000 iterations each with a burn-on of 1000 iterations.

outputs using the summary function on stork. Also, note that when calling WinBUGS, one needs to set the thinning option to 1 by using n.thin=1 in function bugs to ensure that all values following the burn-in period are stored.

The multinomial cell probabilities for the capture-recapture data, given the set of parameter values for a particular data set, are calculated in the standard way (see for example the WinBUGS code in Figure 7.7). Similarly, the discrepancy function (in this case the deviance) is calculated in as a function of these multinomial cell probabilities and data (for either the observed or

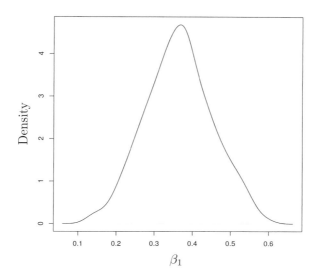

Figure C.9 The posterior distribution of the regression parameter corresponding to the rainfall effect on survival of storks obtained using WinBUGS and R.

```
###############
# Posterior distribution of the rainfall effect

attach.bugs(kita.sim)

plot(density(beta[,2]),main='posterior probability of the rainfall effect')

# Calculate various posterior numerical summaries:
# Credible interval
quantile(beta[,2],probs=c(0.025,0.975))

# Median
median(beta[,2])
```

Figure C.10 The R code for calculating the posterior summary statistics and plotting the marginal posterior distribution for β_1 for the parameters of the model for the white stork data, where the survival probabilities are regressed on a single covariate.

simulated data), which can be used to calculate the corresponding Bayesian p-value (see for example Figure B.9 for the analogous R code). Finally, we output the deviance values of the observed and simulated data along with an indicator function of whether the deviance of the observed data is less than

414   PROGRAMMING IN WINBUGS

```
###############
# Number of iterations and burn-in
nt <- stork$n.iter
nburn <- stork$n.burnin

# Number of parameters in the model
nparam <- 3

# Calculate R/new - number of birds (new birds) released: ...

    for (t in 1:(nt-nburn)) {              # Iterate over simulations

# Set parameter values
            param[1] <- stork$sims.matrix[t,19]   # recapture probability
            param[2] <- stork$sims.matrix[t,17]   # intercept
            param[3] <- stork$sims.matrix[t,18]   # slope

# Set the recapture and survival probs for each year:

...

# Calculate the multinomial cell probabilities, q

...

# Set up arrays, simulate data and calculate deviances:

...

  if (outp[t,1] <= outp[t,2]) { outp[t,3] <- 1 }
     else { outp[t,3] <- 0 }

}

pstork <- outp sum(pstork[,3])/length(pstork[,3])
```

Figure C.11 R code for simulating capture-recapture data given the model parameter values obtained from WinBUGS with the aim of calculating the Bayesian p-value for capture-recapture data.

the deviance of the simulated data. The proportion of times that this indicator is equal to one is the Bayesian p-value.

We run the above Bayesian p-value script on the outputted MCMC simulations following a run of WinBUGS from which outputs were stored in the list stork obtained above and placing the outputted values into the array pstork. To calculate the corresponding Bayesian p-value we use the command:

```
> sum(pstork[,3])/length(pstork[,3])
[1] 0.5376667
```

SUMMARY

which is very close to the Bayesian p-value obtained in R in Section B.5.1 (allowing for Monte Carlo error). This p-value suggests that the model fits the data well.

C.3 Summary

In this appendix we introduce the software WinBUGS and provide examples of analyses of the dipper and white stork capture-recapture data sets. The codes all have the same two-stage structure, for specifying the priors and writing the likelihood. This makes it easier to adapt existing codes to other problems; for example, it is easy to incorporate random effects into a fixed effects model (to result in a mixed effects model). We also provide two illustrations of calling WinBUGS from R.

References

Aebischer, N. J. (1999), Multi-way comparisons and generalised linear models of nest success: Extensions of the Mayfield method. *Bird Study* **46** (suppl), 22–31

Al-Awadhi, F., Hurn, M. and Jennison, C. (2004), Improving the acceptance rate of reversible jump MCMC proposals. *Statistics and Probability Letters* **69**, 189–198

Albert, J. (2007), *Bayesian computation with R*. Springer, New York

Anderson-Sprechler, R. and Ledolte, J. (1991), State-space analysis of wildlife telemetry data. *Journal of the American Statistical Association* **86**, 596–602

Arnold, R., Hayakawa, Y. and Yip, P. (2009), Capture-recapture estimation using finite mixtures of arbitrary dimension. *Biometrics* – in press

Baillie, S. R., Brooks, S. P., King, R. and Thomas, L. (2009), Using a state-space model of the British song thrush *Turdus philomelos* population to diagnose the causes of a population decline. In D. L. Thomson, E. G. Cooch and M. J. Conroy (eds.), *Modeling Demographic Processes in Marked Populations*, Springer – Series: Environmental and Ecological Statistics, Volume 3, pp. 543–564

Bairlein, F. (1991), Population studies of white storks *Ciconia ciconia* in Europe, with reference to the western population. In C. Perrins, J. D. Lebreton and G. Hirons (eds.), *Bird population studies: Relevance to conservation and management*, Oxford University Press, Oxford, pp. 207–229

Barbraud, C., Barbraud, J. C. and Barbraud, M. (1999), Population dynamics of the white stork *Ciconia ciconia* in western France. *Ibis* **141**, 469–479

Barker, R. J. (1997), Joint modelling of live recapture, tag-resight and tag-recovery data. *Biometrics* **53**, 666–677

Barker, R. J. (1999), Joint analysis of mark-recapture, resighting and ring-recovery data with age-dependence and marking-effect. *Bird Study* **46** (suppl), 82–91

Barry, S. C., Brooks, S. P., Catchpole, E. A. and Morgan, B. J. T. (2003), The analysis of ring-recovery data using random effects. *Biometrics* **59**, 54–65

Bartolucci, F. and Forcina, A. (2001), Analysis of capture-recapture data with a Rasch-type model allowing for conditional dependence and multidimensionality. *Biometrics* **57**, 714–719

Bayarri, M. J. and Berger, J. O. (1998), Quantifying surprise in the data and model verification. In J. M. Bernardo, J. O. Berger, A. P. Dawid and A. F. M. Smith (eds.), *Bayesian statistics 6*, Oxford University Press, Oxford, pp. 53–82

Bayes, T. (1763), An essay towards solving a problem in the doctrine of chances. *Philosophical Transactions of the Royal Society of London* **53**, 370–418

Bellhouse, D. R. (2004), The Reverend Thomas Bayes, FRS: A biography to celebrate the tercentenary of his birth. *Statistical Science* **19**, 3–43

Bellman, R. and Åström, K. J. (1970), On structural identifiability. *Mathematical Biosciences* **7**, 329–339

Besbeas, P., Freeman, S. N. and Morgan, B. J. T. (2005), The potential of integrated population modelling. *Australian and New Zealand Journal Statistics* **46**, 35–48

Besbeas, P., Freeman, S. N., Morgan, B. J. T. and Catchpole, E. A. (2001), Stochastic models for animal abundance and demographic data. Technical Report UKC/IMS/01/16, University of Kent, Canterbury, England

Besbeas, P., Freeman, S. N., Morgan, B. J. T. and Catchpole, E. A. (2002), Integrating mark-recapture-recovery and census data to estimate animal abundance and demographic parameters. *Biometrics* **58**, 540–547

Besbeas, P., Lebreton, J. D. and Morgan, B. J. T. (2003), The efficient integration of abundance and demographic data. *Journal of the Royal Statistical Society, Series C* **52**, 95–102

Bolker, B. M. (2008), *Ecological models and data in R*. Princeton University Press, Princeton, NJ

Bonner, S. J., Morgan, B. J. T. and King, R. (2009), Continuous covariates in mark-recapture-recovery analysis: A comparison of methods. Technical report, Simon Fraser University

Bonner, S. J. and Schwarz, C. J. (2006), An extension of the Cormack-Jolly-Seber model for continuous covariates with application to *Microtus Pennsylvanicus*. *Biometrics* **62**, 142–149

Borchers, D. L., Buckland, S. T., King, R., Newman, K. B., Thomas, L., Besbeas, P. T., Gimenez, O. and Morgan, B. J. T. (2010), *Estimating Animal Abundance: Open Populations*. In preparation

Borchers, D. L., Buckland, S. T. and Zucchini, W. (2002), *Estimating Animal Abundance: Closed Populations*. Springer, London

Borysiewicz, R. S., Gimenez, O., Hénaux, V., Bregnballe, T., Lebreton, J.-D. and Morgan, B. J. T. (2009), An integrated analysis of multisite recruitment, mark-recapture-recovery and multisite census data. In D. L. Thomson, E. G. Cooch and M. J. Conroy (eds.), *Modeling Demographic Processes in Marked Populations*, Springer – Series: Environmental and Ecological Statistics, Volume 3, pp. 1055–1067

Brooks, S. P. (2003), Bayesian computation: A statistical revolution. *Transactions of the Royal Society, Series A* **361**, 2681–2697

References

Brooks, S. P., Catchpole, E. A. and Morgan, B. J. T. (2000a), Bayesian animal survival estimation. *Statistical Science* **15**, 357–376

Brooks, S. P., Catchpole, E. A., Morgan, B. J. T. and Barry, S. C. (2000b), On the Bayesian analysis of ring-recovery data. *Biometrics* **56**, 951–956

Brooks, S. P., Catchpole, E. A., Morgan, B. J. T. and Harris, M. P. (2002), Bayesian methods for analysing ringing data. *Journal of Applied Statistics*. **29**, 187–206

Brooks, S. P., Friel, N. and King, R. (2003a), Classical model selection via simulated annealing. *Journal of the Royal Statistical Society, Series B* **65**, 503–520

Brooks, S. P. and Gelman, A. (1998), Alternative methods for monitoring convergence of iterative simulations. *Journal of Computational and Graphical Statistics* **7**, 434–455

Brooks, S. P., Giudici, P. and Roberts, G. O. (2003b), Efficient construction of reversible jump MCMC proposal distributions (with discussion). *Journal of the Royal Statistical Society, Series B* **65**, 3–55

Brooks, S. P. and Guidici, P. (2000), MCMC convergence assessment via two-way ANOVA. *Journal of Computational and Graphical Statistics* **9**, 266–285

Brooks, S. P., King, R. and Morgan, B. J. T. (2004), A Bayesian approach to combining animal abundance and demographic data. *Animal Biodiversity and Conservation* **27**, 515–529

Brooks, S. P. and Morgan, B. J. T. (1995), Optimization using simulated annealing. *Statistician* **44**, 241–257

Brooks, S. P. and Roberts, G. O. (1999), On quantile estimation and MCMC convergence. *Biometrika* **86**, 710–717

Brownie, C., Anderson, D. R., Burnham, K. P. and Robson, D. S. (1985), Statistical inference from band recovery data: A handbook. U.S. Fish and Wildlife Resource Publication 156

Brownie, C., Hines, J. E., Nichols, J. D., Pollock, K. H. and Hestbeck, J. B. (1993), Capture-recapture studies for multiple strata including non-Markovian transitions. *Biometrics* **49**, 1173–1187

Buckland, S. T., Anderson, D. R., Burnham, K. P., Laake, J. L., Borchers, D. L. and Thomas, L. (2001), *Introduction to distance sampling*. Oxford University Press, Oxford

Buckland, S. T., Anderson, D. R., Burnham, K. P., Laake, J. L., Borchers, D. L. and Thomas, L. (eds.) (2004a), *Advanced distance sampling*. Oxford University Press, Oxford

Buckland, S. T., Burnham, K. P. and Augustin, N. H. (1997), Model selection: an integral part of inference. *Biometrics* **53**, 603–618

Buckland, S. T., Newman, K. N., Fernández, C., Thomas, L. and Harwood, J. (2007), Embedding population dynamics models in inference. *Statistical Science* **22**, 44–58

Buckland, S. T., Newman, K. N., Thomas, L. and Koesters, N. B. (2004b),

State-space models for the dynamics of wild animal populations. *Ecological Modelling* **171**, 157–175

Burnham, K. P. (1993), A theory for combined analysis of ring-recovery and recapture data. In J.-D. Lebreton and P. M. North (eds.), *Marked individuals in the study of bird populations*, Birkhäuser Verlag, Basel, pp. 199–213

Cameron, C., Barker, R. J., Fletcher, D., Slooten, E. and Dawson, S. (1999), Modelling survival of Hector's dolphins around Banks peninsula, New Zealand. *Journal of Agricultural, Biological, and Environmental Statistics* **4**, 126–135

Carlin, B. P. and Chib, S. (1995), Bayesian model choice via Markov chain Monte Carlo. *Journal of the Royal Statistical Society, Series B* **57**, 473–484

Carlin, B. P. and Louis, T. A. (1996), *Bayes and empirical Bayes methods for data analysis*. Chapman & Hall, London

Carlin, B. P. and Louis, T. A. (2001), *Bayes and empirical Bayes methods for data analysis*. Chapman & Hall, Boca Raton, second edition

Carothers, A. D. (1973), Capture-recapture methods applied to a population with known parameters. *Journal of Animal Ecology* **42**, 125–46

Casella, G. and George, E. I. (1992), Explaining the Gibbs sampler. *Journal of the American Statistical Association* **46**, 167–174

Casella, G. and Robert, C. P. (1996), Rao-Blackwellization of sampling schemes. *Biometrika* **83**, 81–94

Castledine, B. J. (1981), A Bayesian analysis of multiple-recapture sampling of a closed population. *Biometrika* **68**, 197–210

Caswell, H. (2001), *Matrix population models*. Sinauer Associates, Sunderland, MA, second edition

Catchpole, E. A. (1995), Matlab: An environment for analyzing ring-recovery and recapture data. *Journal of Applied Statistics* **22**, 801–816

Catchpole, E. A., Fan, Y., Morgan, B. J. T., Clutton-Brock, T. H. and Coulson, T. (2004), Sexual dimorphism, survival and dispersal in red deer. *Journal of Agricultural, Biological, and Environmental Statistics* **9**, 1–26

Catchpole, E. A., Freeman, S. N. and Morgan, B. J. T. (1995), Modelling age variation in survival and reporting rates for recovery models. *Journal of Applied Statistics* **22**, 597–609

Catchpole, E. A., Freeman, S. N. and Morgan, B. J. T. (1996), Steps to parameter redundancy in age-dependent recovery models. *Journal of the Royal Statistical Society, Series B* **58**, 763–774

Catchpole, E. A., Freeman, S. N., Morgan, B. J. T. and Harris, M. P. (1998), Integrated recovery/recapture data analysis. *Biometrics* **54**, 33–46

Catchpole, E. A., Freeman, S. N., Morgan, B. J. T. and Nash, W. J. (2001), Abalone I: Analysing mark-recapture-recovery data, incorporating growth and delayed recovery. *Biometrics* **57**, 469–477

Catchpole, E. A. and Morgan, B. J. T. (1996), Model selection in ring-recovery models using score tests. *Biometrics* **52**, 664–672

Catchpole, E. A. and Morgan, B. J. T. (1997), Detecting parameter redundancy. *Biometrika* **84**, 187–196

Catchpole, E. A. and Morgan, B. J. T. (2001), Deficiency of parameter-redundant models. *Biometrika* **88**, 593–598

Catchpole, E. A., Morgan, B. J. T., Coulson, T. N., Freeman, S. N. and Albon, S. D. (2000), Factors influencing Soay sheep survival. *Journal of the Royal Statistical Society, Series C* **49**, 453–472

Catchpole, E. A., Morgan, B. J. T., Freeman, S. N. and Peach, W. J. (1999), Modelling the survival of British lapwings *Vanellus vanellus* using weather covariates. *Bird Study* **46** (suppl), 5–13

Catchpole, E. A., Morgan, B. J. T. and Tavecchia, G. (2008), A new method for analysing discrete life history data with missing covariate values. *Journal of the Royal Statistical Society, Series B* **70**, 445–460

Catchpole, E. A., Morgan, B. J. T. and Viallefont, A. (2002), Solving problems in parameter redundancy using computer algebra. *Journal of Applied Statistics*, **29**, 625–636

Cave, V. M., Freeman, S. N., Brooks, S. P., King, R. and Balmer, D. E. (2009a), On adjusting for missed visits in the indexing of abundance from Constant Effort ringing. In D. L. Thomson, E. G. Cooch and M. J. Conroy (eds.), *Modeling Demographic Processes in Marked Populations*, Springer – Series: Environmental and Ecological Statistics, Volume 3, pp. 951–966

Cave, V. M., King, R. and Freeman, S. N. (2009b), An integrated population model from constant effort bird-ringing data. *Journal of Agricultural, Biological, and Environmental Statistics* – in press

Celeux, G., Forbes, F., Robert, C. P. and Titterington, D. M. (2006), Deviance information criteria for missing data models (with discussion). *Bayesian Analysis* **1**, 651–705

Chao, A., Tsay, P., Lin, S., Shau, W. and Chao, D. (2001), Tutorial in biostatistics: The applications of capture-recapture models to epidemiological data. *Statistics in Medicine* **20**, 2123–3157

Chatfield, C. (1995), Model uncertainty, data mining and statistical inference (with discussion). *Journal of the Royal Statistical Society, Series A* **158**(3), 419–466

Chib, S. and Greenberg, E. (1995), Understanding the Metropolis-Hastings algorithm. *The American Statistician* **49**, 327–335

Chilvers, B. L., Wilkinson, I. S. and Mackenzie, D. I. (2009), Predicting life-history traits for female New Zealand sea lions, *Phocartos hookeri*: Integrating short-term mark-recapture data and population modelling. *Journal of Agricultural, Biological, and Environmental Modelling* – in press

Clark, J. S., Ferraz, G., Oguge, N., Hays, H. and Dicostanzo, J. (2005),

Hierarchical Bayes for structured, variable populations: From recapture data to life-history prediction. *Ecology* **86**, 2232–2244

Clobert, J. and Lebreton, J.-D. (1985), Dependance de facteurs de milieu dans les estimations de taux de survie par capture-recapture. *Biometrics* **41**, 1031–1037

Clutton-Brock, T. H. (ed.) (1988), *Reproductive success: Studies of individual variation in contrasting breeding systems.* Chicago University Press, Chicago

Clutton-Brock, T. H., Guinness, F. E. and Albon, S. D. (1982), *Red deer: Behaviour and ecology of two sexes.* Chicago University Press, Chicago

Congdon, P. (2003), *Applied Bayesian modelling.* Wiley, Chichester

Congdon, P. (2006), *Bayesian statistical modelling.* Wiley, Chichester, second edition

Cormack, R. M. (1970), Statistical appendix to Fordham's paper. *Journal of Animal Ecology* **39**, 24–27

Coull, B. and Agresti, A. (1999), The use of mixed logit models to reflect heterogeneity in capture-recapture studies. *Biometrics* **55**, 294–301

Coulson, T. N., Catchpole, E. A., Albon, S. D., Morgan, B. J. T., Pemberton, J. M., Clutton-Brock, T. H., Crawley, M. J. and Grenfell, B. T. (2001), Age, sex, density, winter weather, and population crashes in Soay sheep. *Science* **292**, 1528–1531

Cowles, M. K. and Carlin, B. P. (1996), Markov chain Monte Carlo convergence diagnostics: A comparative review. *Journal of the American Statistical Association* **91**, 883–904

Cox, D. R. and Oakes, D. (1984), *Analysis of survival data.* Chapman & Hall, London

Crainiceanu, C. M., Ruppert, D. and Wand, M. (2005), Bayesian analysis for penalized spline regression using WinBUGS. *Journal of Statistical Software* **14**, 1–24

Cramp, S. and Simmons, K. E. L. (1983), *The Birds of the Western Palearctic, Volume III.* Oxford University Press, Oxford

Crawley, M. J. (2007), *The R book.* Wiley, Chichester

Dempster, A. P., Laird, N. M. and Rubin, D. B. (1977), Maximum likelihood estimation from incomplete data via the EM algorithm (with discussion). *Journal of the Royal Statistical Society, Series B* **39**, 1–38

Dennis, B. and Taper, M. L. (1994), Density dependence in time series observations of natural populations: Estimation and testing. *Ecological Monographs* **64**, 205–224

Dorazio, R. M. and Royle, J. A. (2003), Mixture models for estimating the size of a closed population when capture rates vary among individuals. *Biometrics* **59**, 351–364

Dorazio, R. M., Royle, J. A., Söderström, B. and Glimskär (2006), Estimating species richness and accumulation by modeling species occurrence and detectability. *Ecology* **87**, 842–854

Dupuis, J. A. (1995), Bayesian estimation of movement and survival probabilities from capture-recapture data. *Biometrika* **82**(4), 761–772

Dupuis, J. A., Badia, J., Maublanc, M. and Bon, R. (2002), Survival and spatial fidelity of mouflons (Ovis gmelini): A Bayesian analysis of an age-dependent capture-cecapture model. *Journal of Agricultural, Biological, and Environmental Statistics* **7**, 277–298

Durban, J. W. and Elston, D. A. (2005), Mark-recapture with occasion and individual effects: Abundance estimation through Bayesian model selection in a fixed dimensional parameter space. *Journal of Agricultural, Biological, and Environmental Statistics* **10**, 291–305

Ellison, A. M. (2004), Bayesian inference in ecology. *Ecology Letters* **7**, 509–520

Fewster, R. M., Buckland, S. T., Siriwardena, G. M., Baillie, S. R. and Wilson, J. D. (2000), Analysis of population trends for farmland birds using generalized additive models. *Ecology* **81**, 1970–1984

Fienberg, S. E., Johnson, M. S. and Junker, B. W. (1999), Classical multilevel and Bayesian approaches to population size estimation using multiple lists. *Journal of the Royal Statistical Society, Series A* **162**, 383–405

Fish and Wildlife Service (2003), Waterfowl Population Status. Technical report, Department of the Interior, Washington D.C.

Fisher, R. A. (1935), *The design of experiments.* Oliver and Boyd, London

Freeman, S. N. and Morgan, B. J. T. (1990), Studies in the analysis of ring-recovery data. *The Ring* **13**, 271–288

Freeman, S. N. and Morgan, B. J. T. (1992), A modelling strategy for recovery data from birds ringed as nestlings. *Biometrics* **48**, 217–236

Freeman, S. N., Morgan, B. J. T. and Catchpole, E. A. (1992), On the augmentation of ring-recovery data with field information. *Journal of Animal Ecology* **61**, 649–657

Freeman, S. N. and North, P. M. (1990), Estimation of survival rates of British, Irish and French grey herons. *The Ring* **13**, 139–165

Furness, R. W. and Greenwood, J. J. D. (1993), *Birds as monitors of environmental change.* Chapman & Hall, London

Gaillard, J.-M., Allaine, D., Pontier, D., Yoccoz, N. and Promislow, D. (1994), Senescence in natural populations of mammals: a reanalysis. *Evolution* **48**(2), 509–516

Gamerman, D. (1997), *Markov chain Monte Carlo: Stochastic Simulation for Bayesian Inference.* Chapman & Hall, London

Geiger, D., Heckerman, D. and Meek, C. (1996), Asymptotic model selection for directed networks with hidden variables. In *Proceedings of the twelfth conference on uncertainty in artificial intelligence*, Morgan Kaufmann, San Mateo, CA, pp. 283–290

Gelfand, A. E. and Ghosh, S. K. (1998), Model choice: A minimum posterior predictive loss approach. *Biometrika* **85**, 1–11

Gelman, A. (1996), Inference and monitoring convergence. In W. R. Gilks,

S. Richardson and D. J. Spiegelhalter (eds.), *Markov chain Monte Carlo in practice*, Chapman & Hall, London, pp. 131–143

Gelman, A. (2006), Prior distributions for variance parameters in hierarchical models. *Bayesian Analysis* **1**, 515–534

Gelman, A., Carlin, J. B., Stern, H. S. and Rubin, D. B. (2003), *Bayesian data analysis*. Chapman & Hall, London, second edition

Gelman, A. and Hill, J. (2007), *Data analysis using regression and multilevel/hierarchical models*. Cambridge University Press, New York

Gelman, A., Jakulin, A., Pittau, M. G. and Su, Y.-S. (2008), A weakly informative default prior distribution for logistic and other regression models. *Annals of Applied Statistics* **2**, 1360–1383

Gelman, A. and Meng, X. (1996), Model checking and model improvement. In W. R. Gilks, S. Richardson and D. J. Spiegelhalter (eds.), *Markov chain Monte Carlo in practice*, Chapman & Hall, London, pp. 189–201

Gelman, A., Roberts, G. O. and Gilks, W. R. (1996), Efficient Metropolis jumping rules. In J. M. Bernardo, J. O. Berger, A. P. Dawid and A. F. M. Smith (eds.), *Bayesian statistics 5*, Oxford University Press, Oxford, pp. 599–608

Gelman, A. and Rubin, D. B. (1992), Inference from iterative simulation using multiple sequences. *Statistical Science* **7**, 457–511

Geweke, J. (1992), Evaluating the accuracy of sampling-based approaches to calculating posterior moments. In J. M. Bernardo, J. O. Berger, A. P. Dawid and A. F. M. Smith (eds.), *Bayesian statistics 4*, Oxford University Press, Oxford

Geyer, C. and Thompson, E. (1995), Annealing Markov chain Monte Carlo with applications to ancenstral inference. *Journal of the American Statistical Association* **90**, 909–920

Ghosh, S. K. and Norris, J. L. (2005), Bayesian capture-recapture analysis and model selection allowing for heterogeneity and behavioural effects. *Journal of Agricultural, Biological, and Environmental Statistics* **10**, 35–49

Gilks, W. R., Richardson, S. and Spiegelhalter, D. J. (1996), *Markov chain Monte Carlo in practice*. Chapman & Hall, London

Gilks, W. R. and Roberts, G. O. (1996), Strategies for improving mcmc. In W. R. Gilks, S. Richardson and D. J. Spiegelhalter (eds.), *Markov Chain Monte Carlo in Practice*, Chapman & Hall, London, pp. 89–114

Gilks, W. R., Thomas, D. and Spiegelhalter, D. J. (1992), Software for the Gibbs sampler. *Computing Science and Statistics* **24**, 439–448

Gilks, W. R. and Wild, P. (1992), Adaptive rejection sampling for Gibbs sampling. *Journal of the Royal Statistical Society, Series C* **41**, 337–348

Gimenez, O., Bonner, S. J., King, R., Parker, R. A., Brooks, S. P., Jamieson, L. E., Grosbois, V., Morgan, B. J. T. and Thomas, L. (2009a), WinBUGS for population ecologists: Bayesian modelling using Markov chain Monte Carlo. In D. L. Thomson, E. G. Cooch and M. J. Conroy (eds.),

References

Modeling Demographic Processes in Marked Populations, Springer – Series: Environmental and Ecological Statistics, Volume 3, pp. 885–918

Gimenez, O., Choquet, R. and Lebreton, J.-D. (2003), Parameter redundancy in multistate capture-recapture models. *Biometrical J* **45**, 704–722

Gimenez, O., Covas, R., Brown, C. R., Anderson, M. D., Brown, M. B. and Lenormand, T. (2006a), Nonparametric estimation of natural selection on a quantitative trait using mark-recapture data. *Evolution* **60**, 460–466

Gimenez, O., Crainiceanu, C., Barbraud, C., Jenouvrier, S. and Morgan, B. J. T. (2006b), Semiparametric regression in capture-recapture modelling. *Biometrics* **62**, 691–698

Gimenez, O., Morgan, B. J. T. and Brooks, S. P. (2009b), Weak identifiability in models for mark-recapture-recovery data. In D. L. Thomson, E. G. Cooch and M. J. Conroy (eds.), *Modeling Demographic Processes in Marked Populations*, Springer – Series: Environmental and Ecological Statistics, Volume 3, pp. 1055–1067

Gimenez, O., Rossi, V., Choquet, R., Dehais, C., Doris, B., Varella, H., Vila, J.-P. and Pradel, R. (2007), State-space modelling of data on marked individuals. *Ecological Modelling* **206**, 431–438

Gimenez, O., Viallefont, A., Catchpole, E. A., Choquet, R. and Morgan, B. J. T. (2004), Methods for investigating parameter redundancy. *Animal Biodiversity and Conservation* **27**, 561–572

Givens, G. H. and Hoeting, J. A. (2005), *Computational statistics*. Wiley-Interscience, Hoboken, NJ

Goodman, L. A. (1974), Exploratory latent structure analysis using both identifiable and unidentifiable models. *Biometrika* **61**, 215–31

Goudie, I. B. J. and Goudie, M. (2007), Who captures the marks for the Petersen estimator? *Journal of the Royal Statistical Society, Series A* **170**, 825–839

Gramacy, R. B., Samworth, R. J. and King, R. (2009), Importance tempering. *Statistics and Computing* – in press

Green, P. and Silverman, B. (1994), *Nonparametric regression and generalized linear models*. Chapman & Hall, New York

Green, P. J. (1995), Reversible jump Markov chain Monte Carlo computation and Bayesian model determination. *Biometrika* **82**, 711–732

Green, P. J. and Mira, A. (2001), Delayed rejection in reversible jump Metropolis-Hastings. *Biometrika* **88**, 1035–53

Hammond, P. S., Mizroch, S. A. and Donovan, G. P. (1990), Individual recognition of cetaceans: Use of photo-identification and other techniques to estimate population size. Technical report, International Whaling Commission, Special issue 12

Heisey, D. M. and Nordheim, E. V. (1990), Biases in the Pollock and Cornelius method of estimating nest survival. *Biometrics* **46**, 855–862

Heisey, D. M. and Nordheim, E. V. (1995), Modelling age-specific survival in nesting studies, using a general approach for doubly-censored and truncated data. *Biometrics* **51**, 51–60

Hestbeck, J. B., Nichols, J. D. and Malecki, R. A. (1991), Estimates of movement and site fidelity using mark-resight data of wintering Canada geese. *Ecology* **72**, 523–533

Hinde, A. (1998), *Demographic methods*. Arnold, London

Hoeting, J. A., Madigan, D., Raftery, A. E. and Volinsky, C. T. (1999), Bayesian model averaging: A tutorial. *Statistical Science* **14**, 382–401

Huggins, R. H. (1991), Some practical aspects of a conditional likelihood approach to capture experiments. *Biometrics* **47**, 725–732

Jamieson, L. and Brooks, S. (2004a), Density dependence in North American ducks. *Animal Biodiversity and Conservation* **27**, 113–128

Jamieson, L. and Brooks, S. P. (2004b), Assessing density and covariate dependence: a Bayesian assessment of North American ducks. Technical report, University of Cambridge

Jennison, C. (1993), Discussion on the meeting on the Gibbs sampler and other Markov chain Monte Carlo methods. *Journal of the Royal Statistical Society, Series B* **55**, 54–56

Johnson, D. S. and Hoeting, J. A. (2003), Autoregressive models for capture-recapture data: A Bayesian approach. *Biometrics* **59**, 341–350

Jonsen, I. D., Flemming, J. M. and Myers, R. A. (2005), Robust state-space modelling of animal movement data. *Ecology* **86**, 2874–2880

Kalman, R. R. (1960), A new approach to linear filtering and prediction problems. *Transactions of ASME Journal of Basic Engineering* **82**, 35–45

Kanyamibwa, S., Schierer, A., Pradel, R. and Lebreton, J. D. (1990), Changes in adult survival rates in a western European population of the White Stork *ciconia ciconia*. *Ibis* **132**, 127–135

Karunamuni, R. and Quinn, T. (1995), Bayesian estimation of animal abundance for the line transects sampling. *Biometrics* **51**, 1325–1337

Kass, R. E. and Raftery, A. E. (1995), Bayes factors. *Journal of the American Statistical Association* **90**, 773–793

Kennedy, M. (ed.) (1990), *Australia's endangered species: The extinction dilemma*. Simon and Schuster, Brookvale, New South Wales

Kerman, J. and Gelman, A. (2006), Tools for Bayesian data analysis in R. *Statistical Computing and Graphics* **17**, 9–13

Kéry, M. and Royle, J. A. (2007), Hierarchical Bayes estimation of species richness and occupancy in spatially replicated surveys. *Journal of Applied Ecology* **45**, 589–598

King, R. (2010), Statistical ecology. In S. P. Brooks, X. Meng, A. Gelman and G. Jones (eds.), *Handbook of Markov chain Monte Carlo; Methods and Applications*, Chapman & Hall, London

King, R., Bird, S. M., Brooks, S. P., Hutchinson, S. J. and Hay, G. (2005),

References

Prior information in behavioural capture-recapture methods: Demography influences injectors' propensity to be listed on data sources and their drugs-related mortality. *American Journal of Epidemiology* **162**, 1–10

King, R. and Brooks, S. P. (2002a), Bayesian model discrimination for multiple strata capture-recapture data. *Biometrika* **89**, 785–806

King, R. and Brooks, S. P. (2002b), Model selection for integrated recovery/recapture data. *Biometrics* **58**, 841–851

King, R. and Brooks, S. P. (2003a), A note on closed form likelihoods for Arnason-Schwarz models. *Biometrika* **90**(2), 435–444

King, R. and Brooks, S. P. (2003b), Survival and spatial fidelity of mouflon: The effect of location, age and sex. *Journal of Agricultural, Biological, and Environmental Statistics* **8**, 486–531

King, R. and Brooks, S. P. (2004a), Bayesian analysis of the survival of Hector's dolphins. *Animal Biodiversity and Conservation* **27**, 343–354

King, R. and Brooks, S. P. (2004b), A classical study of catch-effort models for Hector's dolphins. *Journal of the American Statistical Society* **99**, 325–333

King, R. and Brooks, S. P. (2008), On the Bayesian estimation of a closed population size in the presence of heterogeneity and model uncertainty. *Biometrics* **64**, 816–824

King, R., Brooks, S. P. and Coulson, T. (2008a), Analysing complex capture-recapture data in the presence of individual and temporal covariates and model uncertainty. *Biometrics* **64**, 1187–1195

King, R., Brooks, S. P., Mazzetta, C., Freeman, S. N. and Morgan, B. J. T. (2008b), Identifying and diagnosing population declines: A Bayesian assessment of lapwings in the UK. *Journal of the Royal Statistical Society, Series C* **57**, 609–632

King, R., Brooks, S. P., Morgan, B. J. T. and Coulson, T. (2006), Bayesian analysis of Soay sheep survival. *Biometrics* **62**, 211–220

Kotz, S. and Johnson, N. L. (eds.) (1992), *Breakthroughs in Statistics*. Springer Verlag, New York

Lang, S. and Brezger, A. (2004), Bayesian P-splines. *Journal of Computational and Graphical Statistics* **13**, 183–212

Laplace, P. S. (1786), Sur les naissances, les mariages et les morts. *Histaire de I'Academie Royale des Sciences,* 1783, Paris, 693–702

Lebreton, J. D., Burnham, K. P., Clobert, J. and Anderson, D. R. (1992), Modelling survival and testing biological hypotheses using marked animals: A unified approach with case studies. *Ecological Monographs* **62**, 67–118

Lee, P. M. (2004), *Bayesian statistics: An introduction*. Arnold, London, third edition

Lefkovitch, L. P. (1965), The study of population growth in organisms grouped by stages. *Biometrics* **21**, 1–18

Leslie, P. H. (1945), On the use of matrices in certain population mathematics. *Biometrika* **33**, 183–212

Leslie, P. H. (1948), Some further notes on the use of matrices in certain population mathematics. *Biometrika* **35**, 213–245

Lindley, D. V. (1965), *Introduction to probability and statistics from a Bayesian viewpoint*. Cambridge University Press, Cambridge

Link, W. A. (2003), Nonidentifiability of population size from capture-recapture data with heterogeneous detection probabilities. *Biometrics* **59**, 1123–1130

Link, W. A. and Barker, R. J. (2005), Modelling association among demographic parameters in analysis of open population capture-recapture data. *Biometrics* **61**, 46–54

Link, W. A. and Barker, R. J. (2008), Efficient implementation of the Metropolis-Hastings algorithm, with application to the Cormack-Jolly-Seber model. *Environmental and Ecological Statistics* **15**, 79–87

Link, W. A. and Barker, R. J. (2009), *Bayesian Inference with ecological applications*. Academic Press

Link, W. A., Cam, E., Nichols, J. D. and Cooch, E. G. (2002), Of BUGS and birds: Markov chain Monte Carlo for hierarchical modelling in wildlife research. *Journal of Wildlife Management* **66**, 277–291

Liu, J. S. (2001), *Monte Carlo strategies in scientific computing*. Springer

Loison, A., Festa-Bianchet, M., Gaillard, J. M., Jorgenson, J. T. and Jullien, J. M. (1999), Age-specific survival in five populations of ungulates: Evidence of senescence. *Ecology* **80**, 2539–2554

Lunn, D. J., Best, N. and Whittaker, J. (2009), Generic reversible jump MCMC using graphical models. *Statistics and Computing* – in press

Lunn, D. J., Thomas, A., Best, N., and Spiegelhalter, D. (2000), WinBUGS – a Bayesian modelling framework: Concepts, structure, and extensibility. *Statistics and Computing* **10**, 325–337

Lunn, D. J., Whittaker, J. C. and Best, N. (2006), A Bayesian toolkit for genetic association studies. *Genetic Epidemiology* **30**, 231–247

Madigan, D. and Raftery, A. E. (1994), Model selection and accounting for model uncertainty in graphical models using Occam's window. *Journal of the American Statistical Association* **89**, 1535–1547

Marin, J. and Robert, C. R. (2007), *Bayesian core: A practical approach to computational Bayesian statistics*. Springer, New York

Marinari, E. and Parisi, G. (1992), Simulated tempering: A new Monte Carlo scheme. *Europhysics Letters* **19**, 451–458

Massot, M., Clobert, J., Pilorge, T., Lecomte, J. and Barbault, R. (1992), Density dependence in the common lizard: Demographic consequences of a density manipulation. *Ecology* **73**, 1742–1756

McAllister, M. K. and Kirkwood, G. P. (1998a), Bayesian stock assessment: A review and example application using the logistic model. *ICES Journal of Marine Science* **55**, 1031–1060

McAllister, M. K. and Kirkwood, G. P. (1998b), Using Bayesian decision analysis to help achieve a precautoinary approach for managing dolphin fisheries. *Canadian Journal of Fisheries and Aquatic Science* **55**, 2642–2661

McCarthy, M. A. (2007), *Bayesian methods for ecology.* Cambridge University Press, Cambridge

McCrea, R. S. and Morgan, B. J. T. (2009), Multi-site mark recapture model selection using score tests – in revision for *Biometrics*

Millar, R. B. (2004), Sensitivity of Bayes estimators to hyper-parameters with an application to maximum yield from fisheries. *Biometrics* **60**, 536–542

Millar, R. B. and Meyer, R. (2000), Non-linear state space modelling of fisheries biomass dynamics by using Metropolis-Hastings within-Gibbs sampling. *Journal of the Royal Statistical Society, Series C* **49**, 327–342

Mingoti, A. S. (1999), Bayesian estimator for the total number of distinct species when quadrat sampling is used. *Journal of Applied Statistics* **26**, 469–483

Morgan, B. J. T. (1984), *Elements of simulation.* Chapman & Hall, London

Morgan, B. J. T. (1992), *Analysis of quantal response data.* Chapman & Hall, London

Morgan, B. J. T. (2008), *Applied stochastic modelling.* Chapman & Hall, London, second edition

Morgan, B. J. T., Lebreton, J. D. and Freeman, S. N. (1997), Ornithology, Statistics in. *Encyclopedia of Statistical Sciences, Update Volume* **1**, 438–447

Morgan, B. J. T., Palmer, K. J. and Ridout, M. S. (2007a), Negative score test statistic. *The American Statistician* **61**, 285–288

Morgan, B. J. T., Revell, D. J. and Freeman, S. N. (2007b), A note on simplifying likelihoods for site occupancy models. *Biometrics* **63**, 618–621

Morgan, B. J. T. and Ridout, M. S. (2008), A new mixture model for recapture heterogeneity. *Journal of the Royal Statistical Society, Series C* **57**, 433–446

Morgan, B. J. T. and Thomson, D. L. (eds.) (2002), *Statistical Analysis of data from marked bird populations: special issue of Journal of Applied Statistics, Vol 29, Nos 1-4.* Taylor and Francis

Morgan, B. J. T. and Viallefont, A. (2002), Ornithological data. *Encyclopedia of Environmetrics* **3**, 1495–1499

Moyes, K. (2007), *Demographic Consequences of Individual Variation.* Ph.D. thesis, University of Kent

Neal, R. (2003), Slice sampling (with discussion). *Annals of Statistics* **31**, 705–767

Neal, R. M. (1996), Sampling from multimodal distributions using tempered transition. *Statistics and Computing* **6**, 353–366

Neal, R. N. (2001), Annealed importance sampling. *Statistics and Computing* **11**, 125–129

Newman, K. B. (1998), State space modelling of animal movement and mortality with application to salmon. *Biometrics* **54**, 1290–1314

Newman, K. B. (2000), Hierarchical modeling of salmon harvest and migration. *Journal of Agricultural, Biological, and Environmental Modelling* **5**, 430–455

Newman, K. B. (2003), Modelling paired release-recovery data in the presence of survival and capture heterogeneity with application to marked juvenile salmon. *Statistical Modelling* **3**, 157–177

Newman, K. N., Buckland, S. T., Lindley, S. T., Thomas, L. and Fernández, C. (2006), Hidden process models for animal population dynamics. *Ecological Applications* **16**, 74–86

Newman, K. N., Fernández, C., Thomas, L. and Buckland, S. T. (2009), Monte carlo inference for state-space models of wild animal populations. *Biometrics* **65**, 572–583

Ngo, L. and Wand, M. (2004), Smoothing with mixed model software. *Journal of Statistical Software* **9**, 1–54

Nichols, J., Hines, J. and Blums, P. (1997), Tests for sensecent decline in annual survival probabilities of common pochards, *Aythya ferina*. *Ecology* **78**(4), 1009–1018

Nichols, J. D., Sauer, J. R., Pollock, K. H. and Hestbeck, J. B. (1992), Estimating transition probabilities for stage-based population projection matrices using capture-recapture data. *Ecology* **73**, 306–312

North, P. M. and Morgan, B. J. T. (1979), Modelling heron survival using weather data. *Biometrics* **35**, 667–681

O'Hagan, A. O. (1998), Eliciting expert beliefs in substantial practical applications. *The Statistician* **47**, 21–36

O'Hara, R. B., Arjas, E., Toivonen, H. and Hanski, I. (2002), Bayesian analysis of metapopulation data. *Ecology* **83**, 2408–2415

O'Hara, R. B., Lampila, S. and Orell, M. (2009), Estimation of rates of births, deaths, and immigration from mark-recapture data. *Biometrics* **65**

O'Hara, R. B. and Silanpaa, M. J. (2009), A review of Bayesian variable selection methods, what, how and which. *Bayesian Analysis* **4**, 85–118

Otis, D. L., Burnham, K. P., White, G. C. and Anderson, D. R. (1978), Statistical inference from capture data on closed animal populations. *Wildlife Monographs* **62**, 1–135

Peach, W. J., Baillie, S. R. and Balmer, D. E. (1998), Long-term changes in the abundance of passerines in Britain and Ireland as measured by constant effort mist-netting. *Bird Study* **45**, 257–275

Peskun, P. H. (1973), Optimum Monte Carlo sampling using Markov chains. *Biometrika* **60**, 607–612

Pledger, S. (2000), Unified maximum likelihood estimates for closed capture-recapture models using mixtures. *Biometrics* **56**, 434–442

Pledger, S., Pollock, K. H. and Norris, J. L. (2003), Open capture-recapture models with heterogeneity: I. Cormack-Jolly-Seber model. *Biometrics* **59**, 786–794

Polacheck, T., Hilborn, R. and Punt, A. E. (1993), Fitting surplus production models: Comparing methods and measuring uncertainty. *Canadian Journal of Fisheries and Aquatic Sciences* **50**, 2597–2607

Pollock, K. (1991), Modeling capture, recapture and removal statistics for estimation of demographic parameters for fish and wildlife populations: Past, present and future. *Journal of the American Statistical Association* **86**, 226–238

Pollock, K. (2000), Capture-recapture models. *Journal of the American Statistical Association* **95**, 293–296

Pollock, K. H., Bunck, C. M., Winterstein, S. R. and Chen, C. L. (1995), A capture-recapture survival analysis model for radio-tagged animals. *Journal of Applied Statistics* **22**, 661–672

Pollock, K. H. and Cornelius, W. L. (1988), A distribution-free nest survival model. *Biometrics* **44**, 397–404

Pollock, K. H., Nichols, J. D., Brownie, C. and Hines, J. E. (1990), Statistical inference for capture-recapture experiments. *Wildlife Monographs* **107**, 1–97

Pradel, R. (1996), Utilization of capture-mark-recapture for the study of recruitment and population growth rate. *Biometrics* **52**, 703–709

Pradel, R. (2005), Multievent: An extension of multistate capture-recapture models to uncertain states. *Biometrics* **61**, 442–447

Propp, J. G. and Wilson, D. B. (1996), Exact sampling with coupled Markov chain and applications to statistical mechanics. *Random Structure and Algorithms* **9**, 223–252

Punt, A. E. and Hilborn, R. (1997), Fisheries stock assessment and decision analysis: the Bayesian approach. *Reviews in Fish Biology and Fisheries* **7**, 35–63

Raftery, A. E. and Lewis, S. M. (1996), Implementing MCMC. In W. R. Gilks, S. Richardson and D. J. Spiegelhalter (eds.), *Markov chain Monte Carlo in practice*, Chapman & Hall, London, pp. 115–130

Reddingius, J. (1971), Gambling for existence: A discussion of some theoretical problems in animal population ecology. *Acta Biotheoretica* **20**, 1–208

Reiersøl, O. (1950), Identifiability of a linear relation between variables which are subject to error. *Econometrica* **18**, 375–389

Reynolds, T., King, R., Harwood, J., Frederiksen, M., Wanless, S. and Harris, M. (2009), Integrated data analyses in the presence of emigration and tag-loss. *Journal of Agricultural, Biological, and Environmental Statistics* – in press

Richardson, S. and Green, P. J. (1997), On Bayesian analysis of mixtures with an unknown number of components. *Journal of the Royal Statistical Society, Series B* **59**, 731–792

Ripley, B. D. (1987), *Stochastic simulation*. Wiley, Chichester

Rivot, E. and Prévost, E. (2002), Hierarchical Bayesian analysis of capture-mark-recapture data. *Canadian Journal of Fisheries and Aquatatic Sciences* **59**, 1768–1784

Rivot, E., Prévost, E. and Parent, E. (2001), How robust are Bayesian posterior inferences based on a Ricker model with regards to measurement errors and prior assumptions about parameters? *Canadian Journal of Fisheries and Aquatatic Sciences* **58**, 2284–2297

Rivot, E., Prévost, E., Parent, E. and Bagliniére, J. L. (2004), A Bayesian state-space modelling framework for fitting a salmon stage-structured population dynamics model to multiple time series of field data. *Ecological Modelling* **179**, 463–485

Rizzo, M. L. (2007), *Statistical computing with R*. Chapman & Hall, London

Robert, C. P. (1994), *The Bayesian choice: A decision-theoretic motivation*. Springer, New York

Robert, C. P. (2007), *The Bayesian choice: From decision-theoretic foundations to computational implementation*. Springer, New York, second edition

Robert, C. R. and Casella, G. (2004), *Monte Carlo statistical methods*. Springer, New York, second edition

Roberts, G. O. (1996), Markov chain concepts related to sampling algorithms. In W. R. Gilks, S. Richardson and D. J. Spiegelhalter (eds.), *Markov Chain Monte Carlo in Practice*, Chapman & Hall, London, pp. 45–58

Roberts, G. O. and Sahu, S. K. (1997), Updating schemes, covariance structure, blocking and parameterisation for the Gibbs sampler. *Journal of the Royal Statistical Society, Series B* **59**, 291–318

Rothenberg, T. J. (1971), Identification in parametric models. *Econometrica* **39**, 577–591

Royle, J. A. (2006), Site occupancy models with heterogeneous detection probabilities. *Biometrics* **62**, 97–102

Royle, J. A. (2008), Modeling individual effects in the Cormack-Jolly-Seber model: A state-space formulation. *Biometrics* **64**, 364–370

Royle, J. A. and Dorazio, R. M. (2008), *Hierarchical modeling and inference in ecology*. Academic Press

Royle, J. A., Dorazio, R. M. and Link, W. A. (2007), Analysis of multinomial models with unknown index using data augmentation. *Journal of Computational and Graphical Statistics* **16**, 67–85

Royle, J. A. and Kéry, M. (2007), A Bayesian state-space formulation of dynamic occupancy models. *Ecology* **88**, 1813–1823

Rue, H. (2001), Fast sampling of Gaussian Markov random fields. *Journal of the Royal Statistical Society, Series B* **63**, 325–338

Ruppert, D. (2002), Selecting the number of knots for penalized splines. *Journal of Computational and Graphical Statistics* **11**, 735–757

References

Ruppert, D., Wand, M. P. and Carroll, R. (2003), *Semiparametric regression*. Cambridge University Press, Cambridge

Savage, L. J. (1954), *The Foundations of Statistics*. Wiley, New York

Schaefer, M. B. (1954), Some aspects of the dynamics of populations important to the management of the commercial marine fisheries. *Bulletin of the Inter-American Tropical Tuna Commission* **1**, 25–26

Schaub, M., Kania, W. and Koppen, U. (2005), Variation of primary production during winter induces synchrony in survival rates in migratory white storks *Ciconi ciconia*. *Journal of Animal Ecology* **74**, 656–666

Schwarz, C. G., Schweigert, J. F. and Arnason, A. N. (1993), Estimating migration rates using tag-recovery data. *Biometrics* **49**, 177–193

Schwarz, C. J. and Seber, G. A. F. (1999), A review of estimating animal abundance III. *Statistical Science* **14**, 427–456

Seber, G. A. F. (1971), Estimating age-specific rates from bird-bands when the reporting rate is constant. *Biometrika* **58**, 491–497

Seber, G. A. F. (1982), *The estimation of animal abundance and related parameters*. MacMillan, New York

Senar, J., Dhondt, A. and Conroy, M. E. (2004), The quantitative study of marked individuals in ecology, evolution and conservation biology: A foreword to the EURING 2003 conference. *Animal Biodiversity and Conservation* **27**, 1–2

Skalski, J. R., Robson, D. S. and Simmons, M. A. (1983), Comparative census procedures using single mark-recapture methods. *Ecology* **665**, 1006–1015

Smith, A. F. M. and Gelfand, A. E. (1992), Bayesian statistics without tears: A sampling-resampling perspective. *The American Statistician* **46**, 84–88

Smith, P. (1991), Bayesian analyses for a multiple capture-recapture model. *Biometrics* **78**, 399–407

Spiegelhalter, D. J., Best, N. G., Carlin, B. P. and van der Linde, A. (2002), Bayesian measures of model complexity and fit (with discussion). *Journal of the Royal Statistical Society, Series B* **64**, 583–616

Spiegelhalter, D. J., Thomas, A. and Best, N. G. (1996), Computation on Bayesian graphical models. In J. M. Bernardo, A. F. M. Smith, A. P. Dawid and J. O. Berger (eds.), *Bayesian statistics 5*, Oxford University Press, Oxford, pp. 407–425

Stanghellini, E. and van der Heijden, P. G. M. (2004), A multiple-record systems estimation methods that takes observed and unobserved heterogeneity into account. *Biometrics* **60**, 510–516

Stephens, M. (2000), Bayesian analysis of mixture models with an unknown number of components – an alternative to reversible jump methods. *Annals of Statistics* **28**, 40–74

Stigler, S. M. (1986), *The History of Statistics: The Measurement of Uncertainty before 1900*. Belknap Press, Cambridge, MA

Sturtz, S., Ligges, U. and Gelman, A. (2005), R2WinBUGS: A package for running WinBUGS from R. *Joural of Statistical Software* **12**, 1–16

Tardella, L. (2002), A new Bayesian method for nonparametric capture-recapture models in presence of heterogeneity. *Biometrika* **89**, 807–817

ter Braak, C. J. F. and Etienne, R. S. (2003), Improved Bayesian analysis of metapopulation data with an application to a tree frog metapopulation. *Ecology* **84**, 231–241

Thomas, L., Buckland, S. T., Newman, K. B. and Harwood, J. (2005), A unified framework for modelling wildlife population dynamics. *Australian and New Zealand Journal of Statistics* **47**, 19–34

Thomas, L., Laake, J. L., Rexstad, E., Strindberg, S., Marques, F. F. C., Buckland, S. T., Borchers, D. L., Anderson, D. R., Burnham, K. P., Burt, M. L., Hedley, S. L., Pollard, J. H., Bishop, J. R. B. and Marques, T. (2009), *Distance 6.0 Release 2*. University of St. Andrews, Research Unit for Wildlife Population Assessment, University of St. Andrews, UK. http://www.ruwpa.st-and.ac.uk/distance

Thurstone, L. L. (1947), *Multiple-factor analysis*. University of Chicago Press, Chicago

Viallefont, A., Lebreton, J.-D., Reboulet, A.-M. and Gory, G. (1998), Parametric identifiability and model selection in capture-recapture models: A numerical approach. *Biometrical Journal* **40**, 313–325

Vounatsou, P. and Smith, A. F. M. (1995), Bayesian analysis of ring-recovery data via Markov chain Monte Carlo simulation. *Biometrics* **51**, 687–708

West, M. and Harrison, J. (1997), *Bayesian forecasting and dynamic models*. Springer-Verlag, New York

White, G. C. and Burnham, K. P. (1999), Program MARK: Survival estimation from populations of marked animals. *Bird Study* **46** suppl, 120–39

Wilby, R. L., O'Hare, G. and Barnsley, N. (1997), The North Atlantic Oscillation and British Isles climate variability, 1865–1996. *Weather* **52**, 266–276

Williams, B. K., Nichols, J. D. and Conroy, M. J. (2002), *Analysis and management of animal populations*. Academic Press, San Diego, CA

Yates, F. (1937), The design and analysis of factorial experiments. Technical Report 35, Imperial Bureau of Soil Science, Harpenden, England

Zheng, C., Ovaskainen, O., Saastamoinen, M. and Hanski, I. (2007), Age-dependent survival analyzed with Bayesian models of mark-recapture data. *Ecology* **88**, 1970–1976

Index

abalone, 40
ACF, 136, 137, 143, 144, 207, 377
age
   as a covariate, 31, 147
   effect on reproductive performance, 13
   effect on survival, 9
age dependent survival, 8, 31, 34, 38, 91, 143, 178, 183, 196, 242, 248, 257
annual demographic rates, 40
annual survival, 5
apparent survival, 28
Arnason-Schwarz model, 28, 277, 286
asymptotic assumptions of classical inference, 7
autocorrelation function, see ACF

batteries
   failure of, 40
Bayes factors, 151, 154, 193, 195, 225, 229, 261, 263, 337, 354
Bayes' Theorem, 7, 70, 71, 150, 244, 252, 313
Bayesian p-values, 138–140, 396–398, 411–415
Bayesian revolution in ecology, 12
best model
   in classical inference, 7
BGR statistic, 126, 127, 189, 207, 211, 235, 237, 402, 406
biological diversity, 4
biological hypotheses, 147, 148, 178, 295, 346
block updates, 132, 133, 233
blue-winged teal, 62, 196
breeding
   skipping, 41
breeding state, 29
British Census, 3
burn-in, 102, 125–127, 141, 157, 161, 199, 209, 211, 212, 235, 236, 314, 379, 381

capture history, see life history data
capture-recapture data, 5, 24, 37, 66, 339, 407
   for dippers, 5, 200, 207
   for white storks, 66, 214, 228, 407
   likelihood, 25, 203, 205
   multi-site, 28, 45
capture-recapture methods
   use in medical studies and epidemiology, 12
catch-effort information, 294
census
   national, 33
Central England temperatures, 15, 43, 91, 177, 248, 318
change-point model, 295, 300
Chapman estimate, 19
cheetah, 18
chi-square distribution
   use in asymptotic tests, 54
classical model selection, 51
closed populations, 18, 28, 345–361
   behavioural effects, 346
   heterogeneity effects, 346
   models, 346
   time effects, 346
clutch sizes, 35
coat colour, 5
coat type as a covariate, 31
coda, 209, 211, 232, 378, 384, 405
combination of likelihoods, 35, 321
Common Bird Census (CBC), 8, 16, 316
common yellow-throat, 58
confidence intervals, 7, 58, 88
confidence regions, 57
conservation policy, 294
conservation science, 12

convergence diagnostic, 126, 127, 189, 210, 211, 401, 402, 406
Cormack-Jolly-Seber (CJS) model, 37, 51, 53, 84, 339, 340
  fitted in MARK, 60
  nested submodels, 55
  parameter redundancy, 54, 84, 341
  probabilities for, 37
Cormack-Seber model
  probabilities for, 37
covariate
  frost days, 30
  lagged, 31
covariate selection, 64, 177, 227, 242, 259, 273
covariates, 30, 91, 147, 241–263, 270, 273
  continuous, 250, 253, 255, 256
  discrete, 250, 253, 254, 277
  environmental, 5, 147, 177, 241, 242, 257, 264, 266, 272
  individual, 5, 147, 242, 250, 256, 257, 266
    time invariant, 250, 251, 253–255
    time varying, 250, 251, 256, 277, 278
  interaction, 243
  missing values, 242, 244, 251, 257, 258, 261, 273, 277, 278
  normalised, 92, 114, 116, 243, 244, 248, 253, 318, 408
  underlying model, 254–256, 274
credible interval, 86, 88
  highest posterior density interval, 86, 87, 212
  symmetric credible interval, 86
culling, 4
curvature of likelihood surface at the maximum, 58

data
  canvasbacks, 72
  capture-recapture, 24
  catch-effort information, 294, 296
  dippers, 4, 24
  ducks, 335
  golf tees, 20
  hares, 20
  Hector's dolphins, 294
  herons, 15
  herring gulls, 55
  herrings, 47
  integrated data, 30, 33, 320
  integrated recovery and recapture, 26
  lapwings, 8, 16, 22
  life history, 26
  lizard, 43
  multi-state, 28
  parsing, 44
  red deer, 31, 45
  ring-recovery, 22
  shags, 26
  Soay sheep, 5, 10
  taxi-cabs, 20
  timed species count, 45
  transect, 21
  voles, 19
  white storks, 65
days below freezing, *see* frost days
dealing with different types of variation, 8
decline
  in lapwing numbers, 8, 176
deficiency
  in parameter redundant models, 38
degrees of freedom
  in chi-square distribution, 54
demographic time-bomb, 3
demography
  human, 3
density dependence, 332–334
  in model for North American ducks, 33
derivative matrix, 39
detection
  heterogeneity of, 46
deterministic optimisation procedures, 56
difference approximations to derivatives, 57
differential equation
  use in determining estimable parameters, 39
dipper data, 4, 53
  CJS model, 55, 60, 96, 154, 340
  environmental covariate, 241
  illustration of likelihood-ratio test, 54
  model averaging, 173

model discrimination, 154, 158
posterior model probabilities, 161, 173
prior sensitivity analysis, 82, 171
R code, 199–203, 379–387
state-space model, 340
three-parameter model, 82, 90, 138, 154, 158, 159, 171, 173
two-parameter model, 74, 88, 90, 110, 127, 130, 133, 138, 154, 158, 159, 171, 173, 199, 379, 406
use of MARK, 60
WinBUGS code, 205–209, 406–407
directed acyclic graph (DAG), 89–93, 98, 122, 234, 321, 336, 351
DISTANCE, 62
distribution
  Bernoulli, 254, 339, 365
  beta, 76, 100, 104, 370
  beta-binomial, 52, 360, 366
  binomial, 20–22, 35, 42, 52, 76, 312, 317, 366
  Dirichlet, 76, 254, 372
  gamma, 76, 371
  geometric, 73, 76, 367
  half-normal, 247, 350, 357, 369
  inverse gamma, 76, 371
  inverse Wishart, 373
  log-normal, 308, 309, 312, 369
  multinomial, 19, 20, 22, 24, 37, 45, 73, 76, 90, 254, 368
  multivariate normal, 41, 58, 76, 132, 134, 245, 256, 372
  negative binomial, 367
  normal, 35, 76, 115, 118, 146, 245, 255, 308, 309, 312, 369
  Poisson, 35, 76, 146, 312, 317, 366
  t, 118, 370
  uniform, 118, 223, 265, 365, 368
  Wishart, 373
divergence of optimisation methods, 57
DNA matching, 18

E-SURGE, 61
eagle, 61
ecology
  population, 3
eigen values
  zero, 61

environmental covariates, 5, 241, 242, 264
equation
  measurement, 34
  transition, 34
estimated standard errors, 34, 58
estimating the population of France, 19
estimation of the size of a closed population, 18, 345–361
EURING, 12
expected information matrix, 58
  use of inverse in classical inference, 58
explicit maximum-likelihood estimates, 45, 56
extinction
  of fish stocks, 4
  of populations, 3

filter
  Kalman, 35, 312
first instance of marking of birds, 17
Fisher's information, 61
fisheries, 4, 11
fitness, 13, 28
flat likelihoods, 36
fledging time, 36
flood
  effect of in dipper data, 55
  effect on survival, 5, 7
fmincon, 57
food availability as a covariate, 30
frost days, 15, 30, 91, 177, 248, 317

generalised linear model, 8
Gibbs sampler, see Markov chain Monte Carlo (MCMC)
global optima, 56
global warming, 3
golf tees, 353, 355
  how many?, 20
goodness-of-fit, 138
Great Auk, 3
Grey heron, 15
grey heron ring-recovery data, 42
grey herons
  survival as function of weather, 30
grey seal, 33
growth curve, 40

growth rate, 309

habitat fragmentation, 28
hares, 352, 355
Hector's dolphins, 28, 194, 294
heronry census, 15
herring
    recovery data for, 47
herring gulls, 55, 64
heterogeneity of detection, 46
heterogeneity of recapture probability
    bias due to, 41
hierarchical models, 95
hierarchical prior, 78, 93, 122, 210, 227, 237, 247, 265, 266, 274, 349
high mortality of first year birds, 23
highest posterior density interval (HPDI), 86, 87, 212
Hirta, 5, 64
horn shape, 5
human demography, 11
hyper-prior, *see* hierarchical prior
hypothesis test, 7, 148

identification of animals through natural markings, 18
increased longevity, 3
incubation time, 36
indicator species for farmland, 8
information criteria, 20, 36, 51, 54, 55, 62
    Akaike's information criterion (AIC), 51, 53, 62, 149
    applied to models for dipper data, 53
    Bayesian information criterion (BIC), 51, 53, 62, 149
    Deviance information criterion (DIC), 148, 149, 195
    penalties, 52
    use of for model selection, 6, 62, 148
integrated analysis, 35
integrated modelling, 33, 41
interaction
    of population density and weather in survival of deer, 32
Isle of May, 26, 179
iterative methods for optimisation, 56

Jacobian, 158
Jeffreys prior, 77, 297, 348

Kalman filter, 35, 312
kernel density estimate, 84

lagged covariate, 31
lapwings, 8
    covariate analysis, 91, 176, 248
    integrated data, 320
    Northern, 41
    ring-recovery data, 22, 91, 248
    state-space model, 316
late recovery of dead animal, 40
Lefkovitch matrix, 315
Leslie matrix, 34, 35, 314, 317
life expectancy
    increase in, 9
life history data, 10, 26, 44, 251, 277, 345
    red deer, 31
    for lizards, 45
    likelihood, 27
life-table analysis, 11, 40
lifetime productivity
    optimisation of, 41
likelihood, 6, 71
    as starting point for inference, 49
    combination, 35
    conditional Schnabel, 20
    flat, 36
    for capture-recapture data, 24
    for Schnabel data, 20
    Schnabel, 21, 49
likelihood construction, 15
likelihood-ratio tests, 54
Lincoln-Petersen estimate, 18
    history of, 19
    maximum likelihood, 42
line transects, 356
link function, 210
    log, 243, 318
    logit, 243, 245, 250, 259, 264, 270, 318, 346
    splines, 270
link functions in modelling survival, 31
Lisbon, 30
lizard data, 43, 284, 286

Index

local optima, 56
logistic regression, 30–32, 64, 114, 116, 144, 162, 195, 213, 243, 248, 250, 264, 265, 318, 321, 328, 349, 409
long-term data sets, 10

m-array, 4, 36, 60
M-SURGE, 61
mallard data, 62
Maple, 39
marginal distribution, 75, 176–178
MARK, 31, 59, 60, 64, 150, 209, 246, 361, 378
mark-recapture-recovery data, 33
Markov chain, 101, 156
Markov chain Monte Carlo (MCMC), 75, 99, 101–124, 280
    acceptance function, 103, 104, 106, 118, 119, 131
    Brooks-Gelman-Rubin (BGR) statistic, 126
    burn-in, 102, 125–127, 141, 157, 199, 209, 211, 212, 235, 236, 314, 379, 381
    candidate value, 103, 106
    convergence, 102, 121, 125, 126, 130, 406
    Gibbs sampler, 119, 121, 130, 280–283, 287
    implementation, 124
    independence sampler, 119
    initial values, 101, 125, 126, 314, 319, 406
    Metropolis algorithm, 119
    Metropolis Hastings, 102, 121, 156, 157, 168, 261, 284
    mixing, 132
    Monte Carlo error, 125, 129, 130, 202, 385
    Monte Carlo estimates, 152
    pilot tuning, 130
    prior sensitivity analysis, 157
    proposal distribution, 103, 118, 119, 130, 131
    random walk, 118, 121
    run lengths, 125
    single-update Metropolis Hastings, 106, 119
    trace plots, 125, 161

Mathematica, 39
Matlab
    use in programming classical model-fitting, 58
maximum
    non-unique, 36
maximum-likelihood estimates, 6
    explicit, 56
    for timed species count data, 45
mean-centering of covariates, 64
measurement equation, 34
Metropolis algorithm, *see* Markov chain Monte Carlo (MCMC)
Metropolis Hastings, *see* Markov chain Monte Carlo (MCMC)
migration, 28
Mike Harris, 26
missing data, 242, 244, 251, 257, 258, 261, 277, 278, 312
    auxiliary variables, 252, 258, 265, 268, 274, 278, 281, 288, 312
mixture of two binomial distributions, 46
MLEs, 6
model averaging, 7, 149, 172–176, 182, 286, 326, 354–356
    disadvantages of, 194
    using AIC, 62
model discrimination, 148–152, 273, 286, 326, 354–356
    Bayes factors, 151
    posterior model probabilities, 149–151
model for adults, 37
model notation, 53, 147, 216, 287
    used in eagle, 61
model selection, 51
model uncertainty, 147, 172, 194
    accounting for by model averaging, 149
models
    state-space, 33
Monte Carlo integration, 99, 101, 124
mortality, 3
mourning dove, 36
movement, 3
multi-site models, *see* multi-state models
multi-state data, 45, 284

440 Index

multi-state models, 29, 277–305
multinomial model for timed species count data, 45
multinomial probabilities, 36
　for CJS model, 37
　for Cormack-Seber model, 38
　omission of final term, 37
multiple sources of information, 33
multiple tests
　need for conservative significance level, 55
multivariate normal approximations
　use in combining likelihoods, 41
multivariate normal distribution
　asymptotic distribution for mles, 58

NAO, 30
national census, 16, 33
nested models, 54, 163
Newton-Raphson, 57
newts
　great crested, 18
non-unique maximum, 36
normal approximation
　robustness of in integrated modelling, 41
　to binomial and Poisson distributions, 35, 146, 312
normal distribution
　for observation error in state-space model, 35
normalising constant, 71
North Atlantic Oscillation, 30
Northern lapwings, 8, 30
　productivity of, 41
notation, 53
numerical optimisation, 56

observation process, *see* state-space model - observation process
Occam's window, 155
optim, 57
optimisation methods
　speed of convergence, 57
over-fitting data, 52
overdispersion
　in state-space model, 35

P-splines, 270
parameter estimation
　in multi-site models, 28
parameter index matrix, 60
parameter redundancy, 36, 41, 84, 317
　convention in MARK, 60
　Cormack-Jolly-Seber (CJS) model, 54, 84, 341
　definition of, 38
　in state-space model, 35
parameters, 6
parsing data, 44
Passenger Pigeon, 3, 36
Patuxent, 12
Pearson goodness-of-fit, 63
pilot tuning, 131, 209
PIM, 60, 62, 211
PIT tags, 18
population density, 5
population dynamics modelling, 11
population ecology, 3
posterior distribution, 7, 71, 150, 156
　conditional distribution, 119, 121, 125
　credible interval, 86, 88
　expectation, 99
　marginal densities, 124
　marginal distribution, 84, 151, 178, 253, 259
　mean, 85
　standard deviation, 86
　summary statistics, 85, 100, 124, 176, 257
　variance, 86
posterior model probabilities, 7, 149–151, 157, 173, 177, 355
　estimation, 152–168
　prior sensitivity analysis, 171
precision, 76
　recapture and recovery components, 33
prediction, 270
prior distribution, 7, 71, 75, 150, 287
　conjugate, 280
　conjugate prior, 76, 121
　eliciting prior beliefs, 75
　hierarchical prior, 78, 93, 122, 210, 227, 237, 247, 265, 266, 274, 349
　informative prior, 76, 78, 247, 284

Index

model space, 150, 179, 185, 260, 287, 297
uninformative prior, 76–78, 245, 284
prior knowledge
  incorporation of, 95
prior sensitivity analysis, 80, 84, 171, 230, 235, 246, 247, 249, 259, 269, 275, 331, 338, 353
  dipper data, 82, 171
  overlap between prior and posterior, 84
  posterior model probabilities, 171
prior specification, 75, 179, 287, 297, 318, 347–350
probability density function, 71, 72
probability model, 4, 6
process matrix, 310
productivity, 3, 33, 35
profile log-likelihoods, 58
pulli, 33

Quasi-Newton, 56

R, 199–203, 375–399
  R2WinBUGS, 410
  simulated annealing, 57
  use in programming classical model-fitting, 59
R2WinBUGS, 410
rabbits
  population size estimation of, 18
radio-marking, 18
  affect on animal behaviour, 40
  in combination with ring-recovery data, 41
radio-tagging data, 72, 79, 81, 87, 89, 104, 107, 120, 131, 137, 139
radio-tracking data
  relationship to human life-time data, 40
random effects, 9, 93, 122, 227, 237, 242, 264–274, 347, 349
  in model for survival of lapwings, 30
recapture probability, 5
recapture probability of unity, 32
red deer, 4, 45
  eestimating probability of movement, 28

survival modelling, 31
red list, 4
relative precision, 33
reporting probability
  decline over time, 30
resighting, 33
reversible jump Markov chain Monte Carlo (RJMCMC), 156–168, 179, 261, 266, 287, 298, 328, 335, 355
  acceptance function, 158, 160, 165–168, 180, 186, 189, 262, 267, 289, 300, 301, 329, 355
  age dependence, 183
  burn-in, 161
  convergence, 189
  covariate selection, 157, 163, 167
  efficient updates, 181, 189–192, 329
  improving convergence, 183
  Jacobian, 158, 160, 165, 168, 180, 186, 262, 289, 290, 292, 299–301, 329, 355
  mixing, 161
  model averaging, 175
  nested model moves, 163
  nested models, 261
  non-nested model moves, 166
  pilot tuning, 190
  proposal distribution, 158, 159, 165, 170, 190, 289–291, 299–301, 355
  ring-recovery data, 168
  temporal dependence, 178
ring-recovery data, 22, 91, 168, 248
  for grey herons, 43
  for herring gulls, 55, 64
  for lapwings, 22
  grey heron, 42
  likelihood, 24, 109
Royal Society for the Protection of Birds, 12
Rum, 45

saturated model, 45
Schnabel census
  examples, 20
  likelihood, 49
  simulated data, 20
score tests, 54
  simpler than likelihood-ratio tests, 62
  used in eagle for model selection, 61

senescence, 9
  modelled as a function of age, 31
  red deer, 31
  Soay sheep, 10
severe winter
  effect on survival, 16, 177, 248
shag data, 26, 33, 34, 178, 183
sheep weight, 5
simplex search, 56
simulating from a posterior
  distribution, 7
site fidelity, 36
skipping breeding, 28, 35, 41
Soay sheep, 5, 40, 64, 253, 259, 265
  use of NAO as covariate, 30
  capture-recapture-recovery data, 253
  covariate analysis, 253
  covariate model, 257
  survival of, 10
splines, 227, 270, 272, 274
  knots, 271
  P-splines, 270
  polynomial basis, 271
  smoothing parameter, 272
St. Kilda, 5, 64
stags, 45
starlings, 30
state-space model, 33, 307–343
  capture-recapture data, 339
  observation process, 308, 313, 316, 334, 340
  projection matrix, 314
  system process, 309–313, 317, 333, 339
  transition equations, 314
  WinBUGS, 314, 335, 340–341
stationary distribution, 101, 102, 156
statistical inference, 4, 6
step up system for model selection, 61
stochastic search procedures, 57
Stykkisholmur, 30
sufficient matrices, 27, 33
sufficient statistics, 27
surveys
  national vs local, 40
SURVIV, 64
survival
  apparent, 28
survival probability, 5

symbolic algebra
  use of in detecting parameter redundancy, 39
system process, see state-space model - system process

taxi-cabs in Edinburgh
  how many?, 20
teal, 62, 196
thinning, 136
time dependent parameters, 37
time varying covariates, 30
timed species count data, 45
trans-dimensional simulated annealing, 57
transect sampling, 21, 357
transition equations, 34, 309, 332
transition kernel, 101–103
trap response, see closed populations - behavioural effects
trends in abundance, 16
twinning, 35

unique optimum, 50
use of age classes in modelling survival, 31
use of marked animals, 17

weight as a covariate, 31
white storks, 268, 272, 407
  R code, 213–226, 387–398
  R2WinBUGS code, 411
  WinBUGS code, 226–231, 407–409
WinBUGS, 274, 340, 378, 401–415
  R2WinBUGS, 410
  Jump, 336, 409

zero eigen values, 61